Applications of Fibonacci Numbers

Applications of
Fibonacci Numbers
Volume 4

Proceedings of 'The Fourth International Conference
on Fibonacci Numbers and Their Applications',
Wake Forest University, N.C., U.S.A.,
July 30–August 3, 1990

edited by

G. E. Bergum

South Dakota State University,
Brookings, South Dakota, U.S.A.

A. N. Philippou

Ministry of Education, Nicosia, Cyprus

and

A. F. Horadam

University of New England,
Armidale, New South Wales, Australia

Springer Science+Business Media, B.V.

Library of Congress Cataloging-in-Publication Data: LC 89-24547

ISBN 978-94-010-5590-1   ISBN 978-94-011-3586-3 (eBook)
DOI 10.1007/978-94-011-3586-3

Cover figure by Dr Sabine Jäger

Printed on acid-free paper

TABLE OF CONTENTS

A REPORT ON THE FOURTH INTERNATIONAL CONFERENCE...   vii
LIST OF CONTRIBUTORS TO THIS PROCEEDINGS   ix
FOREWORD   xv
THE ORGANIZING COMMITTEES   xvii
LIST OF CONTRIBUTORS TO THE CONFERENCE   xix
INTRODUCTION   xxiii

A FIBONACCI-BASED PSEUDO-RANDOM NUMBER GENERATOR
    Peter G. Anderson. 1
ON THE PROOF OF GCD AND LCM EQUALITIES CONCERNING THE GENERALIZED
BINOMIAL AND MULTINOMIAL COEFFICIENTS
    Shiro Ando and Daihachiro Sato 9
SUPERCUBE
    Joseph Arkin, David C. Arney, Lee S. Dewald and Frank R. Giordano. 17
A NOTE ON FUNDAMENTAL PROPERTIES OF RECURRING SERIES
    Joseph Arkin, David C. Arney, Frank R. Giordano and Rickey A. Kolb 33
PERIOD PATTERNS OF CERTAIN SECOND-ORDER LINEAR RECURRENCES MODULO
A PRIME
    David Banks and Lawrence Somer 37
NEARLY ISOSCELES TRIANGLES WHERE THE VERTEX ANGLE IS A MULTIPLE OF
THE BASE ANGLE
    Marjorie Bicknell-Johnson 41
THE RING OF FIBONACCI (FIBONACCI "NUMBERS" WITH MATRIX SUBSCRIPT)
    Odoardo Brugia, Piero Filipponi and Francesco Mazzarella 51
ONE-RELATOR PRODUCTS OF CYCLIC GROUPS AND FIBONACCI-LIKE SEQUENCES
    C. M. Campbell, P. M. Heggie, E. F. Robertson and R. M. Thomas 63
A GENERALIZATION OF THE FIBONACCI SEARCH
    Renato M. Capocelli 69
PASCAL'S TRIANGLE: TOP GUN OR JUST ONE OF THE GANG?
    Daniel C. Fielder and Cecil O. Alford 77
CONVERSION OF FIBONACCI IDENTITIES INTO HYPERBOLIC IDENTITIES VALID
FOR AN ARBITRARY ARGUMENT
    Piero Filipponi and Herta T. Freitag 91
DERIVATIVE SEQUENCES OF FIBONACCI AND LUCAS POLYNOMIALS
    Piero Filipponi and Alwyn F. Horadam 99
A CARRY THEOREM FOR RATIONAL BINOMIAL COEFFICIENTS
    Dan Flath and Rhodes Peele 109
ON CO-RELATED SEQUENCES INVOLVING GENERALIZED FIBONACCI NUMBERS
    Herta T. Freitag and George M. Phillips 121
FIBONACCI AND B-ADIC TREES IN MOSAIC GRAPHS
    Heiko Harborth and Sabine Jäger 127
FIBONACCI REPRESENTATIONS OF GRAPHS
    Heiko Harborth and Arnfried Kemnitz 133
ON THE SIZES OF ELEMENTS IN THE COMPLEMENT OF A SUBMONOID OF
INTEGERS
    Chung-wu Ho, James L. Parish, and Jau-shyong Shiue .139
GENOCCHI POLYNOMIALS
    A. F. Horadam 145
AN APPLICATION OF ZECKENDORF'S THEOREM
    Roger V. Jean 167

A NEW KIND OF GOLDEN TRIANGLE
    Clark Kimberling .... .171
TERMS COMMON TO TWO SEQUENCES SATISFYING THE SAME LINEAR
RECURRENCE
    Clark Kimberling .... .177
RECURRENCE RELATIONS IN EXPONENTIAL FUNCTIONS AND IN DAMPED
SINUSOIDS AND THEIR APPLICATIONS IN ELECTRONICS
    Joseph Lahr .... .189
SOME BASIC PROPERTIES OF THE FIBONACCI LINE-SEQUENCE
    Jack Y. Lee .... .203
DE MOIVRE-TYPE IDENTITIES FOR THE TETRABONACCI NUMBERS
    Pin-Yen Lin .... .215
TWO GENERALIZATIONS OF GOULD'S STAR OF DAVID THEOREM
    Calvin Long and Shiro Ando .... .219
ON TRIANGULAR LUCAS NUMBERS
    Ming Luo 231
A FAST ALGORITHM OF THE CHINESE REMAINDER THEOREM AND ITS
APPLICATION TO FIBONACCI NUMBERS
    Kenji Nagasaka, Jau-Shyong Shiue and Chung-Wu Ho .... .241
GENERATING THE PYTHAGOREAN TRIPLES VIA SIMPLE CONTINUED FRACTIONS
    A. G. Schaake and J. C. Turner 247
ON THE MOEBIUS KNOT TREE AND EUCLID'S ALGORITHM
    A. G. Schaake and J. C. Turner .... .257
GENERALIZED FIBONACCI AND LUCAS FACTORIZATIONS
    A. G. Shannon, R. P. Loh and A. F. Horadam .... .271
ON EVEN FIBONACCI PSEUDOPRIMES
    Lawrence Somer .... .277
POSSIBLE RESTRICTED PERIODS OF CERTAIN LUCAS SEQUENCES MODULO P
    Lawrence Somer .... .289
USING MATRIX TECHNIQUES TO ESTABLISH PROPERTIES OF A GENERALIZED
TRIBONACCI SEQUENCE
    Marcellus E. Waddill .... .299
SUBJECT INDEX 309

A REPORT ON
THE FOURTH INTERNATIONAL CONFERENCE
ON
FIBONACCI NUMBERS AND THEIR APPLICATIONS

Sponsored jointly by the Fibonacci Association and Wake Forest University, The Fourth International Conference on Fibonacci Numbers and Their Applications was held from July 30th to August 3rd, 1990. As the Conference took place at Wake Forest University, our foreign visitors especially gained a most enjoyable insight into one of America's delightful set-ups: a small, highly esteemed, liberal arts University, nestled at the outskirts of a faithfully restored 18th century town - Winston-Salem, N.C.

Immediately upon arrival it became clear to us how carefully and competently -under the leadership of the co-chairmen of the International Committee, A. F. Horadam (Australia) and A. N. Philippou (Cyprus), as well as of the co-chairmen of the Local Committee, F. T. Howard and M. E. Waddill -our Conference had been planned and prepared. Special thanks must also go to G.E. Bergum, editor of our Fibonacci Quarterly Journal, for arranging an outstanding program.

There were about 50 participants, 40 of them presented papers, of these, two were women. From some 13 different lands they came, beside the U.S., the host country, Italy would have won the prize for maximum attendance, then Canada and Scotland, closely followed by Australia and Japan.

Papers related to the Fibonacci numbers and their ramifications, and to recursive sequences and their generalizations, as well as those which analyzed and explained number relationships, were presented. Once again, as had been the case in our previous conference, did the diversity of the papers give testimony to the fertility of Fibonacci-related mathematics, as well as to the fructification of ideas, brought about through our mutual, but at the same time, disparate interests. The interplay between theoretically oriented manuscripts and those which highlighted practical aspects, was again conspicuous and fascinating.

The Conference was held in the new Olin Physical Laboratory which was accessible via overcoming several road hurdles which had become necessitated by construction work across the campus. Although our hosts were most apologetic about this, we saw it as a sign of a vital, dynamic and indeed, growing University.

Once in our medium-sized auditorium, we were intrigued (and assisted) by "the wonders technology had wrought": there were two overhead projectors and blackboards - ugh, whiteboards (!) - came from everywhere; up and down they went, above and below, over and across, sometimes interceded by a screen which appeared from nowhere..., and all of it happened by the touch of a button, skillfully activated by the cognoscenti.

Of course, there was not only food for the mind and the soul, but also for the stomach. Wake Forest University graciously treated us to daily morning and afternoon coffee breaks, and the president, Dr. Thomas K. Hearn, Jr., hosted a wine and cheese reception on campus.

Even though our daily meetings took place from 9:00 a.m. till noon, and from 2:00 p.m. to 5:00 p.m., we did not ALWAYS work. In midweek, the afternoon was freed, and we took off to Doughton Park in the beautiful Blue Ridge Mountains of North Carolina. There the group dispersed to enjoy the magnificent scenery with a choice of several hiking trails that offer spectacular vistas. Those of us who preferred less energetic activities, relaxed at a coffee shop where we did, what we seem to be doing best, or at least, most often, and with pleasure:

exchange mathematical ideas.   All this was followed by a lavish, typically North Carolinian dinner at Shatley Springs.

The next day we celebrated our customary evening banquet.   It was held on campus, and was at once elegant and friendly, somehow reflecting the spirit of our group. We speak with many different foreign accents.   And yet, we all understand each other, professionally, and personally. The magnetism of our beloved discipline has somehow promoted a very special bond of friendship. Many of us had been together in some of the past conferences. Quite a few papers exhibited the resulting kindling of common mathematical interests which culminated in joint authorships.

Maybe, several of you are already gathering your thoughts for our next Conference. *"Auf Wiedersehen"*, then, in 1992 at St. Andrews University, Scotland.

Herta T. Freitag

LIST OF CONTRIBUTORS TO THIS PROCEEDINGS

PROFESSOR CECIL O. ALFORD (pp. 77-90)
SCHOOL OF ELECTRICAL ENGINEERING
GEORGIA INSTITUTE OF TECHNOLOGY
ATLANTA, GEORGIA 30332-0250

PROFESSOR PETER G. ANDERSON (pp. 1-8)
SCHOOL OF COMPUTER SCIENCE AND TECHNOLOGY
ROCHESTER INSTITUTE OF TECHNOLOGY
ONE LOMB MEMORIAL DRIVE
POST OFFICE BOX 9887
ROCHESTER, NY 14623-0887

PROFESSOR SHIRO ANDO (pp. 9-16; 219-230)
COLLEGE OF ENGINEERING
HOSEI UNIVERSITY
3-7-2, KAJINO-CHO
KOGANEI-SHI
TOKYO 184, JAPAN

MR. JOSEPH ARKIN (pp. 17-32; 33-36)
197 OLD NYACK TURNPIKE
SPRING VALLEY, NY 10977

LT. COL. DAVID C. ARNEY (pp. 17-32; 33-36)
DEPARTMENT OF MATHEMATICS
UNITED STATES MILITARY ACADEMY
WEST POINT, NY 10996-1786

MR. DAVID BANKS (pp. 37-40)
213 VALLEY PARK DRIVE
CHAPEL HILL, NC 27514

DR. MARJORIE BICKNELL-JOHNSON (pp. 41-50)
665 FAIRLANE AVENUE
SANTA CLARA, CA 95051

PROFESSOR THOMAS C. BROWN
DEPARTMENT OF MATHEMATICS AND STATISTICS
SIMON FRASER UNIVERSITY
BURNABY, BRITISH COLUMBIA
CANADA V5A 1S6

---

*This list also includes those authors whose papers were published elsewhere or were not accepted.

MR. ODOARDO BRUGIA (pp. 51-62)
FONDAZIONE UGO BORDONI
VIA B. CASTIGLIONE, 59
I-00142 ROMA, ITALY

DR. COLIN M. CAMPBELL (pp. 63-68)
UNIVERSITY OF ST. ANDREWS
DEPARTMENT OF MATHEMATICS AND COMPUTER SCIENCE
MATHEMATICAL INSTITUTE, NORTH HAUGH
ST. ANDREWS KY16 9SS
FIFE, SCOTLAND

PROFESSOR RENATO M. CAPOCELLI (pp. 69-76)
DIPARTIMENTO DI MATEMATICA
UNIVERSITA' DI ROMA "LA SAPIENZA"
00185 ROMA
ITALY

COL. LEE S. DEWALD (pp. 17-32)
DEPARTMENT OF MATHEMATICS
UNITED STATES MILITARY ACADEMY
WEST POINT, NY 10996-1786

PROFESSOR DANIEL C. FIELDER (pp. 77-90)
SCHOOL OF ELECTRICAL ENGINEERING
GEORGIA INSTITUTE OF TECHNOLOGY
ATLANTA, GEORGIA 30332-0250

MR. PIERO FILIPPONI (pp. 51-62; 91-98; 99-108)
FONDAZIONE UGO BORDONI
VIA B. CASTIGLIONE, 59
I-00142 ROMA, ITALY

PROFESSOR DANIEL FLATH (pp. 109-120)
DEPARTMENT OF MATHEMATICS AND STATISTICS
FACULTY COURT SOUTH NUMBER THREE
UNIVERSITY OF SOUTH ALABAMA
MOBILE, ALABAMA 36688

PROFESSOR HERTA T. FREITAG (pp. 91-98; 121-125)
B-40 FRIENDSHIP MANOR
320 HERSHBERGER ROAD, N.W.
ROANOKE, VA 24012

LT. COL. FRANK R. GIORDANO (pp. 17-32; 33-36)
DEPARTMENT OF MATHEMATICS
UNITED STATES MILITARY ACADEMY
WEST POINT, NY 10996-1786

PROFESSOR DR. HEIKO HARBORTH (pp. 127-132; 133-138)
BIENRODER WEG 47
D-3300 BRAUNSCHWEIG
WEST GERMANY

PROF. P. M. HEGGIE (pp. 63-68)
UNIVERSITY OF ST. ANDREWS
DEPARTMENT OF MATHEMATICS AND COMPUTER SCIENCE
MATHEMATICAL INSTITUTE, NORTH HAUGH
ST. ANDREWS KY16 9SS
FIFE, SCOTLAND

PROFESSOR CHUNG-WU HO (pp. 139-144; 241-246)
DEPARTMENT OF MATHEMATICS AND STATISTICS
BOX 1653
SOUTHERN ILLINOIS UNIVERSITY AT EDWARDSVILLE
EDWARDSVILLE, IL 62026-1653

PROFESSOR A. F. HORADAM (pp. 99-108; 145-166; 271-276)
DEPARTMENT OF MATHEMATICS, STATISTICS AND COMPUTER SCIENCE
THE UNIVERSITY OF NEW ENGLAND
ARMIDALE, NEW SOUTH WALES 2351
AUSTRALIA

DR. SABINE JÄGER (pp. 127-132)
BÜLTENWEG 7
D-3300 BRAUNSCHWEIG
WEST GERMANY

PROFESSOR ROGER V. JEAN (pp. 167-170)
DEPARTMENT OF MATHEMATICS AND COMPUTER SCIENCE
UNIVERSITÉ DU QUÉBEC À RIMOUSKI
300, ALLÉE DES URSULINES, REMOUSKI
QUÉBEC, G5L 3A1 CANADA

PROFESSOR ARNFRIED KEMNITZ (pp. 133-138)
WÜMMEWEG 10
3300 BRAUNSCHWEIG
WEST GERMANY

PROFESSOR CLARK KIMBERLING (pp. 171-176; 177-188)
DEPARTMENT OF MATHEMATICS
UNIVERSITY OF EVANSVILLE
1800 LINCOLN AVENUE
EVANSVILLE, INDIANA 47722

LT. COL. RICKEY A. KOLB (pp. 33-36)
DEPARTMENT OF MATHEMATICS
UNITED STATES MILITARY ACADEMY
WEST POINT, NY 10996-1786

DR. JOSEPH LAHR (pp. 189-202)
56, RUE DE L'EGLISE
L-7224 WALFERDANGE
LUXEMBOURG

MR. JACK Y. LEE (pp. 203-214)
FORT HAMILTON HIGH SCHOOL
8301 SHORE ROAD
BROOKLYN, NY 11209

DR. PIN-YEN LIN (pp. 215-218)
TAIWAN POWER COMPANY
16F, 242 ROOSEVELT ROAD, SECTION 3
TAIPEI 10763, TAIWAN R.O.C.

PROFESSOR R. P. LOH (pp. 271-276)
DEPARTMENT OF MATHEMATICS
THE UNIVERSITY OF SYDNEY
P.O. BOX 123
BROADWAY, N.S.W. 2006
AUSTRALIA

PROFESSIR CALVIN T. LONG (pp. 219-230)
DEPARTMENT OF PURE AND APPLIED MATHEMATICS
WASHINGTON STATE UNIVERSITY
PULLMAN, WA 99164-2930

PROFESSOR MING LUO (pp. 231-240)
DEPARTMENT OF MATHEMATICS
CHONGQING TEACHERS' COLLEGE
CHONGQING, SICHUAN PROVINCE
PEOPLE'S REPUBLIC OF CHINA 630047

MR. FRANCESCO MAZZARELLA (pp. 51-62)
FONDAZIONE UGO BORDONI
VIA B. CASTIGLIONE, 59
I-00142 ROMA, ITALY

PROFESSOR KENJI NAGASAKA (pp. 241-246)
COLLEGE OF ENGINEERING
HOSEI UNIVERSITY
3-7-2 KAJINO-CHO, KOGANEI-SHI
184 TOKYO, JAPAN

PROFESSOR JAMES L. PARISH (pp. 139-144)
DEPARTMENT OF MATHEMATICS AND STATISTICS
BOX 1653
SOUTHERN ILLINOIS UNIVERSITY AT EDWARDSVILLE
EDWARDSVILLE, IL 62026-1653

PROFESSOR RHODES PEELE (pp. 109-120)
DEPARTMENT OF MATHEMATICS
AUBURN UNIVERSITY AT MONTGOMERY
7300 UNIVERSITY DRIVE
MONTGOMERY, AL 36117-3596

DR. GEORGE M. PHILLIPS (pp. 121-125)
UNIVERSITY OF ST ANDREWS
THE MATHEMATICAL INSTITUTE
THE NORTH HAUGH
ST. ANDREWS KY16 9SS
FIFE, SCOTLAND

DR. EDMUND F. ROBERTSON (pp. 63-68)
UNIVERSITY OF ST. ANDREWS
DEPARTMENT OF MATHEMATICS AND COMPUTER SCIENCE
MATHEMATICAL INSTITUTE, NORTH HAUGH
ST. ANDREWS KY16 9SS
FIFE, SCOTLAND

PROFESSOR DAIHACHIRO SATO (pp. 9-16)
DEPARTMENT OF MATHEMATICS AND STATISTICS
UNIVERSITY OF REGINA
REGINA SASKATCHEWAN
CANADA S4S 0A2

PROFESSOR A. G. SCHAAKE (pp. 247-256; 257-270)
DEPARTMENT OF MATHEMATICS
WAIKATO POLYTECHNIC
HAMILTON, NEW ZEALAND

PROFESSOR A. G. SHANNON (pp. 271-276)
UNIVERSITY OF TECHNOLOGY, SYDNEY
SCHOOL OF MATHEMATICAL SCIENCES
P.O. BOX 123
BROADWAY, N.S.W. 2007
AUSTRALIA

PROFESSOR JAU-SHYONG SHIUE (pp. 139-144; 241-246)
DEPARTMENT OF MATHEMATICAL SCIENCES
4505 MARYLAND PARKWAY
UNIVERSITY OF NEVADA, LAS VEGAS
LAS VEGAS, NEVADA 89154-4020

DR. LAWRENCE SOMER (pp. 37-40; 277-288; 289-298)
1400 20TH ST., NW #619
WASHINGTON, DC 20036

DR. RICHARD M. THOMAS (pp. 63-68)
DEPARTMENT OF COMPUTING STUDIES
UNIVERSITY OF LEICESTER
UNIVERSITY ROAD
LEICESTER LE1 7RH, ENGLAND

DR. JOHN C. TURNER (pp. 247-256; 257-270)
THE DEPARTMENT OF MATHEMATICS AND STATISTICS
UNIVERSITY OF WAIKATO
PRIVATE BAG
HAMILTON, NEW ZEALAND

PROFESSOR MARCELLUS E. WADDILL (pp. 299-308)
DEPARTMENT OF MATHEMATICS AND COMPUTER SCIENCE
WAKE FOREST UNIVERSITY
BOX 7311 REYNOLDA STATION
WINSTON-SALEM, NC 27109

MR. PAUL WILLIAMS
350 S. OAKLAND AVE #311
PASADENA, CA 91101

FOREWORD

This book contains thirty-three papers from among the thirty-eight papers presented at the Fourth International Conference on Fibonacci Numbers and Their Applications which was held at Wake Forest University, Winston-Salem, North Carolina from July 30 to August 3, 1990. These papers have been selected after a careful review by well known referees in the field, and they range from elementary number theory to probability and statistics. The Fibonacci numbers and recurrence relations are their unifying bond.

It is anticipated that this book, like its three predecessors, will be useful to research workers and graduate students interested in the Fibonacci numbers and their applications.

March 1, 1991

The Editors

Gerald E. Bergum
South Dakota State University
Brookings, South Dakota, U.S.A.

Alwyn F. Horadam
University of New England
Armidale, N.S.W., Australia

Andreas N. Philippou
Minister of Education
Ministry of Education
Nicosia, Cyprus

THE ORGANIZING COMMITTEES

LOCAL COMMITTEE

Howard, Fred T., *Co-Chair*

Waddill, Marcellus E., *Co-Chair*

Hayashi, Elmer K.

Vaughan, Theresa

Harrell, Deborah

INTERNATIONAL COMMITTEE

Horadam, A. F. (Australia), *Co-Chair*

Philippou, A. N. (Cyprus), *Co-Chair*

Ando, S. (Japan)

Bergum, G. E. (U.S.A.)

Bicknell-Johnson, M. B. (U.S.A.)

Campbell, Colin (Scotland)

Filipponi, Piero (Italy)

Kiss, P. (Hungary)

Turner, J. C. (New Zealand)

LIST OF CONTRIBUTORS TO THE CONFERENCE

*ALFORD, CECIL O., (coauthor Daniel C. Fielder) "Pascal's Triangle: Top Gun or Just One of the Gang?"

*ANDERSON, PETER G., "A Fibonacci-Based Pseudo-Random Number Generator."

ANDO, SHIRO, (coauthor Daihachiro Sato) "Mutually Exclusive Sets of Binomial Coefficients Each Pair of Which Gives Equal Product, Equal GCD and Equal LCM Properties Simultaneously."

*ANDO, SHIRO, (coauthor Daihachiro Sato) "On the Proof of GCD and LCM Equalities Concerning the Generalized Binomial and Multinomial Coefficients."

*ANDO, SHIRO, (coauthor Calvin T. Long) "Two Generalizations of Gould's Star of David Theorem."

*ARKIN, JOSEPH, (coauthors David C. Arney, Frank R. Giordano and Rickey A. Kolb) "A Note on Fundamental Properties of Recurring Series."

*ARKIN, JOSEPH, (coauthors David C. Arney, Lee S. Dewald and Frank R. Giordano) "Supercube."

ARKIN, JOSEPH, (coauthors David C. Arney and Frank R. Giordano) "The Original Manuscript of the Generalized Fibonacci Numbers Combined with the Generalized Pascal Triangle."

*ARNEY, DAVID C., (coauthors Joseph Arkin, Frank R. Giordano and Rickey A. Kolb) "A Note on Fundamental Properties of Recurring Series."

*ARNEY, DAVID C., (coauthors Joseph Arkin, Lee S. Dewald and Frank R. Giordano) "Supercube."

ARNEY, DAVID C., (coauthors Joseph Arkin and Frank R. Giordano) "The Original Manuscript of the Generalized Fibonacci Numbers Combined with the Generalized Pascal Triangle."

*BANKS, DAVID, (coauthor Lawrence Somer) "Period Patterns of Certain Second-Order Linear Recurrences Modulo A Prime."

*BICKNELL-JOHNSON, MARJORIE, "Nearly Isosceles Triangles Where the Vertex Angle is a Multiple of the Base Angle."

BROWN, TOM C. "Characterization of the Quadratic Irrationals."

*BRUGIA, ODOARDO, (coauthors Piero Filipponi and Francesco Mazzarella) "The Ring of Fibonacci (Fibonacci "Numbers" with Matrix Subscript)."

*CAMPBELL, COLIN M., (coauthors P. M. Heggie, E. F. Robertson, and R. M. Thomas) "One-Relator Products of Cyclic Groups and Fibonacci-Like Sequences."

*CAPOCELLI, RENATO M., "A Generalization of the Fibonacci Search."

*DEWALD, LEE S., (coauthors Joseph Arkin, David C. Arney and Frank R. Giordano) "Supercube."

*FIELDER, DANIEL C., (coauthor Cecil O. Alford) "Pascal's Triangle: Top Gun or Just One of the Gang?"

*FILIPPONI, PIERO, (coauthor Herta T. Freitag) "Conversion of Fibonacci Identities into Hyperbolic Identities Valid for an Arbitrary Argument."

*FILIPPONI, PIERO, (coauthor A. F. Horadam) "Derivative Sequences of Fibonacci and Lucas Polynomials."

---

*The asterisk indicates that the paper is included in this book.

*FILIPPONI, PIERO, (coauthors Odoardo Brugia and Francesco Mazzarella) "The Ring of Fibonacci (Fibonacci "Numbers" with Matrix Subscript)."

*FLATH, DANIEL, (coauthor Rhodes Peele) "A Carry Theorem for Rational Binomial Coefficients."

*FREITAG, HERTA T., (coauthor Piero Filipponi) "Conversion of Fibonacci Identities into Hyperbolic Identities Valid for an Arbitrary Argument."

*FREITAG, HERTA T., (coauthor George M. Phillips) "On Co-related Sequences Involving Generalized Fibonacci Numbers."

*GIORDANO, FRANK R., (coauthors Joseph Arkin, David C. Arney and Rickey A. Kolb) "A Note on Fundamental Properties of Recurring Series."

*GIORDANO, FRANK R., (coauthors Joseph Arkin, David C. Arney and Lee S. Dewald) "Supercube."

GIORDANO, FRANK R., (coauthors Joseph Arkin and David C. Arney) "The Original Manuscript of the Generalized Fibonacci Numbers Combined with the Generalized Pascal Triangle."

*HARBORTH, HEIKO, (coauthor Sabine Jäger) "Fibonacci and B-adic Trees in Mosaic Graphs."

*HARBORTH, HEIKO, (coauthor Arnfried Kemnitz) "Fibonacci Representations of Graphs."

*HEGGIE, P. M., (coauthors Colin M. Campbell, E. F. Robertson, and R. M. Thomas) "One-Relator Products of Cyclic Groups and Fibonacci-Like Sequences."

*HO, CHUNG-WU, (coauthors Kenji Nagasaka and Jau-Shyong Shiue) "A Fast Algorithm of the Chinese Remainder Theorem and its Application to Fibonacci Numbers."

*HO, CHUNG-WU, (coauthors James L. Parish and Jau-shyong Shiue) "On the Sizes of Elements in the Complement of a Submonoid of Integers."

*HORADAM, A. F., (coauthor Piero Filipponi) "Derivative Sequences of Fibonacci and Lucas Polynomials."

*HORADAM, A. F., (coauthors R. P. Loh and A. G. Shannon) "Generalized Fibonacci and Lucas Factorizations."

*HORADAM, A. F., "Genocchi Polynomials."

*JÄGER, SABINE, (coauthor Heiko Harborth) "Fibonacci and B-adic Trees in Mosaic Graphs."

*JEAN, ROGER V., "An Application of Zeckendorf's Theorem."

*KEMNITZ, ARNFRIED, (coauthor Heiko Harborth) "Fibonacci Representations of Graphs."

*KIMBERLING, CLARK, "A New Kind of Golden Triangle."

*KIMBERLING, CLARK, "Terms Common to Two Sequences Satisfying the Same Linear Recurrence."

*KOLB, RICKEY, A., (coauthors Joseph Arkin, David C. Arney and Frank R. Giordano) "A Note on Fundamental Properties of Recurring Series."

*LAHR, JOSEPH, "Recurrence Relations in Exponential Functions and in Damped Sinusoids and Their Applications in Electronics."

*LEE, JACK Y., "Some Basic Properties of the Fibonacci Line-Sequence."

*LIN, PIN-YEN, "De Moivre-Type Identities for the Tetrabonacci Numbers."

*LOH, R. P., (coauthors A. F. Horadam and A. G. Shannon) "Generalized Fibonacci and Lucas Factorizations."

*LONG, CALVIN T., (coauthor Shiro Ando) "Two Generalizations of Gould's Star of David Theorem."

*LUO, MING, "On Triangular Lucas Numbers."

*MAZZARELLA, FRANCESCO, (coauthors Odoardo Brugia and Piero Filipponi) "The Ring of Fibonacci (Fibonacci "Numbers" with Matrix Subscript)."

*NAGASAKA, KENJI, (coauthors Jau-Shyong Shiue and Chung-Wu Ho) "A Fast Algorithm of the Chinese Remainder Theorem and its Application to Fibonacci Numbers."

*PARISH, JAMES L., (coauthors Chung-wu Ho and Jau-shyong Shiue) "On the Sizes of Elements in the Complement of a Submonoid of Integers."

*PEELE, RHODES, (coauthor Daniel Flath) "A Carry Theorem for Rational Binomial Coefficients."

*PHILLIPS, GEORGE M., (coauthor Herta T. Freitag) "On Co-related Sequences Involving Generalized Fibonacci Numbers."

*ROBERTSON, E. F., (coauthors Colin M. Campbell, P. M. Heggie, and R. M. Thomas) "One-Relator Products of Cyclic Groups and Fibonacci-Like Sequences."

SATO, DAIHACHIRO, (coauthor Shiro Ando) "Mutually Exclusive Sets of Binomial Coefficients Each Pair of Which Gives Equal Product, Equal GCD and Equal LCM Properties Simultaneously."

*SATO, DAIHACHIRO, (coauthor Shiro Ando) "On the Proof of GCD and LCM Equalities Concerning the Generalized Binomial and Multinomial Coefficients."

*SCHAAKE, A. G., (coauthor J. C. Turner) "Generating the Pythagorean Triples via Simple Continued Fractions."

*SCHAAKE, A. G., (coauthor J. C. Turner) "On the Moebius Knot Tree and Euclid's Algorithm."

*SHANNON, A. G., (coauthors A. F. Horadam and R. P. Loh) "Generalized Fibonacci and Lucas Factorizations."

*SHIUE, JAU-SHYONG, (coauthors Chung-Wu Ho and Kenji Nagasaka) "A Fast Algorithm of the Chinese Remainder Theorem and its Application to Fibonacci Numbers."

*SHIUE, JAU-SHYONG, (coauthors Chung-wu Ho and James L. Parish) "On the Sizes of Elements in the Complement of a Submonoid of Integers."

*SOMER, LAWRENCE, "On Even Fibonacci Pseudoprimes."

*SOMER, LAWRENCE, (coauthor David Banks) "Period Patterns of Certain Second-Order Linear Recurrences Modulo A Prime."

*SOMER, LAWRENCE, "Possible Restricted Periods of Certain Lucas Sequences Modulo p."

*THOMAS, R. M., (coauthors Colin M. Campbell, P. M. Heggie, and E. F. Robertson) "One-Relator Products of Cyclic Groups and Fibonacci-Like Sequences."

*TURNER, J. C., (coauthor A. G. Schaake) "Generating the Pythagorean Triples via Simple Continued Fractions."

*TURNER, J. C., (coauthor A. G. Schaake) "On the Moebius Knot Tree and Euclid's Algorithm."

*WADDILL, MARCELLUS E., "Using Matrix Techniques to Establish Properties of a Generalized Tribonacci Sequence."

WILLIAMS, PAUL, "Resolving Fibonacci Patterns in Random Data."

INTRODUCTION

The numbers

$$1, 1, 2, 3, 5, 8, 13, 21, 34, 55, 89, \ldots,$$

known as the Fibonacci numbers, have been named by the nineteenth-century French mathematician Edouard Lucas after Leonard Fibonacci of Pisa, one of the best mathematicians of the Middle Ages, who referred to them in his book *Liber Abaci* (1202) in connection with his rabbit problem.

The astronomer Johann Kepler rediscovered the Fibonacci numbers, independently, and since then several renowned mathematicians have dealt with them. We only mention a few: J. Binet, B. Lamé, and E. Catalan. Edouard Lucas studied Fibonacci numbers extensively, and the simple generalization

$$2, 1, 3, 4, 7, 11, 18, 29, 47, 76, 123, \ldots,$$

bears his name.

During the twentieth century, interest in Fibonacci numbers and their applications rose rapidly. In 1961 the Soviet mathematician N. Vorobyov published *Fibonacci Numbers*, and Verner E. Hoggatt, Jr., followed in 1969 with his *Fibonacci and Lucas Numbers*. Meanwhile, in 1963, Hoggatt and his associates founded The Fibonacci Association and began publishing *The Fibonacci Quarterly*. They also organized a Fibonacci Conference in California, U.S.A., each year for almost sixteen years until 1979. In 1984, the First International Conference on Fibonacci Numbers and Their Applications was held in Patras, Greece, and the proceedings from this conference have been published. It was anticipated at that time that this conference would set the beginning of international conferences on the subject to be held every two or three years in different countries. With this intention as a motivating force, The Second International Conference on Fibonacci Numbers and Their Applications was held in San Jose, California, U.S.A., August 13-16, 1986. In order to carry on this new tradition, The Third International Conference on Fibonacci Numbers and Their Applications was held in Pisa, Italy, July 25-29, 1988. The proceedings from both of these conferences have also been published. Because of the continuous success of the preceeding three conferences, The Fourth International Conference on Fibonacci Numbers and Their Applications was held at Winston-Salem, North Carolina, July 30-August 3, 1990, and a Fifth Conference is scheduled for July 1992 in St. Andrews, Scotland.

It is impossible to overemphasize the importance and relevance of the Fibonacci numbers to the mathematical and physical sciences as well as other areas of study. The Fibonacci numbers appear in almost every branch of mathematics, like number theory, differential equations, probability, statistics, numerical analysis, and linear algebra. They also occur in physics, biology, chemistry, and electrical engineering.

It is believed that the contents of this book, like its predecessors, will prove useful to everyone interested in this important branch of mathematics and that this material may lead to additional results on Fibonacci numbers both in mathematics and in their applications to science and engineering.

The Editors

A FIBONACCI-BASED PSEUDO-RANDOM NUMBER GENERATOR

Peter G. Anderson

1. INTRODUCTION AND DISCLAIMER

Throughout this paper, we will speak of a (pseudo-)random number generator. This is a bit of a misnomer, since our number generator will definitely fail most of the tests that random number generators are usually expected to pass. However, our sequence does pass one important test — it is uniform — plus a few other tests that are usually not explicitly required of random number generators: it is very inexpensive to compute, it is easy to remember and to program, and it is easy to analyze; hence, it is trustworthy.

2. THE GENERATOR

If A and B are relatively prime integers, then the sequence of integers

$$S_k = kA \bmod B, k = 0, 1, 2, \ldots, B - 1$$

is a permutation of $\{0, 1, 2, \ldots, B - 1\}$. For our purposes, we choose as A and B two adjacent Fibonacci numbers,

$$A = F_n, \quad B = F_{n+1}$$

For example, if $A = 8$, $B = 13$ then our pseudo-random permutation of $\{0, \ldots, 12\}$ is

$$(0, 8, 3, 11, 6, 1, 9, 4, 12, 7, 2, 10, 5)$$

1

G. E. Bergum et al. (eds.), Applications of Fibonacci Numbers Volume 4, 1–8.
© 1991 *Kluwer Academic Publishers.*

The above sequence of integers repeats after B numbers are generated, and it could be used to generate a sequence of real numbers in the interval $[0,1)$ by using the value S_k/B. However for real numbers we propose the limiting case of the integer sequence:

$$s_k = k\phi - [k\phi], k = 0, 1, 2, \cdots$$

i.e., use the fractional part of multiples of the golden mean,

$$\phi = \frac{\sqrt{5}+1}{2}$$

Both of these sequences are simple, easy to remember, and easy to code. Whether they are analyzably uniform will be covered below; but first we present some applications.

3. SOME APPLICATIONS

The first application (beyond the impetus of needing to pull a quick and dirty sequence out of thin air for programming class examples in sequences to sort, and so forth) for which this family of sequences was originally developed was for the rendering of expensive computer graphics; e.g., fractals (Mandelbrot sets, Julia sets) and synthetic scene generation using ray tracing, which can be very expensive per pixel. Computer graphics software development is among the most frustrating: the programmer has to patiently wait to determine what the last change to the program might have done (is the picture what we wanted? is it centered? scaled? is the object even on the screen?) Graphics software developers have to relive the days of overnight processing. By painting the graphics screen's scan lines in a good random order (meaning in a jumbled order rather than the traditional left to right order), one can very quickly see a low resolution version and be able to make decisions early (including whether to let the process continue rendering to completion). Similarly, imagine scanning an image data base looking for some particular picture. High resolution images require millions of bytes of data which can be incredibly slow, especially for the users who must timeshare networks. If one can see a 5%-resolution image in 5% of the time that it would take to transmit the entire image (instead of a useless narrow high-resolution strip down the side of the screen), then one can successfully search a remote image date bank interactively.

The two ways, old and new, for rendering an image on a graphics screen with M rows and B columns are shown below. The function f is used to determine the color or brightness value for each pixel.

Traditional; painful; slow:	Proposed incremental resolution:
```for (x = 0; x < B; x = x + 1)```	```for (x = A; x ! = 0; x = (x+A) % B)```
```{```	```{```
``` for (y = 0; y < M; y = y+1)```	``` for (y = 0; y < M; y = y+1)```
``` {```	``` {```
``` plot f(x,y) at (x,y)```	``` plot f(x,y) at (x,y)```
``` }```	``` }```
```}```	```}```

Typical high resolution graphics screens provide approximately 1000 by 1000 pixels, so we can conveniently set A and B to the Fibonacci numbers 610 and 987, respectively.

One disadvantage of the improved program above is that early in the processing it illuminates a very small fraction of the display, producing a faint as well as low resolution image. An easy solution to this is to paint scan lines that are fat — fat enough to fill the gap between the current scan line and the one to its left without overpainting any previously correctly painted picture. A benefit of such a simple random number sequence as we are using is that no scan line is any closer to a line to its left than any scan line ever was to scan line 0 (zero). The width of the fat scan line is *width*:

```
width = 1000000;
for ( x = A; x != 0; x = (x+A) % B )
{
        if ( x < width ) width = x;
        for ( y = 0; y < M; y = y+1 )
        {
                plot a line:
                        from (x-width+1,y)
                        to (x,y)
                        of color f(x,y)
        }
}
```

There are several other applications for a uniform random number generator within computer graphics, such as shading and line and curve plotting. In shading (or "digital half-toning") an area, one wishes to randomly plot a certain fraction of the pixels black and the remainder of the pixels white, or, stated in a more continuous sense, plot a pixel black with a specified probability. Almost any random number generator can be used for such application (simply simulate the right roulette wheel), but we can guarantee that our random will not repeat itself, causing undesirable texturing, if we simply choose our B to be near the computer's largest integer.

A uniform random number generator can be used to draw straight lines as follows (without loss of too much generality, suppose that the slope of the line is between 0.0 and 1.0). To draw a line from (0,0) to (x,y), (where $0 \leq y \leq x$) go NORTHEAST y steps; go EAST $x - y$ steps; at random (each "step" entails painting the visited pixel black). "Go EAST" means move one pixel to the right; "go NORTHEAST" means move one pixel to the right and one pixel up. There are good straight line drawing algorithms, and the present one is not intended seriously to replace them; however, the present algorithm could serve as a visual *test* of uniformity of a proposed random number generator. Our generator passes this test well.

Curve drawing algorithms, however, could be of use. Generalize the line drawing process to integrate a differential equation. Specifically, assume that the curve passes through a given pixel, (x,y). Travel to and plot the next pixel that the curve passes through by moving one step along the curve's tangent line using the line drawing process which we described above.

4. UNIFORMITY

By "uniform" we mean that subsequences of our random sequence are smoothly distributed; i.e., in the subsequence

$$\{s_\alpha, s_{\alpha+1}, \cdots, s_\beta\}$$

for some range of the indices, $\alpha \cdots \beta$, gets no more crowded than it must. For the real number sequence, the random numbers are in the interval $[0,1)$, so optimally their minimum distance to a neighbor would be $(\beta - \alpha)^{-1}$; for the integers, the range is $[0, F_{n+1})$, giving an optimal crowding of $\frac{F_{n+1}}{\beta - \alpha}$. These optimal wishes are impossible to obtain with sequences developed as ours are, or with random number sequences in general: the next number added to such a set needs to bisect one of the intervals; if the optimum intervals are ever obtained for some interval, then simply adding one more number to the set would reduce the minimum distance between two numbers by at least a factor of two. We will see that 38% of the above so-called optimal is achievable (0.38196601125 is the distance between 1.0 and the golden mean).

The crowding analysis proceeds straightforwardly. Because of the simplicity of our construction, we will not lose generality if we assume that $\alpha = 0$, since the elements are in an arithmetic progression; a sequence with a non-zero starting point is simply a shift of the sequence with a zero starting point. We see that two elements in the sequence are close together only if a previous element was close to zero. Specifically, suppose that $| s_{p+q} - s_q | = a$; then, from the properties of modular arithmetic, $| s_p | = a$. (We will take a small liberty with absolute value notation here. The absolute value of a non-negative real number, x, considered reduced modulo 1 is the minimum of x and $1 - x$; the absolute value of an integer, k, considered reduced modulo B is the minimum of k and $B - k$. The notation "$| \, | x \, | \, |$" represents the distance between x and the nearest integer.) The preceding considers the real and the integer sequences together; below, we will discuss them separately (although connections will be evident to Fibonacci number connoisseurs).

If the sequence of real random numbers, $\{s_0, s_1, \cdots, s_\beta\}$ contains two numbers that are close together, then some previous element must be close to zero. Suppose that $| s_p | = \varepsilon$, so $| \, | p\phi | \, | = \varepsilon$. However, the best approximation to ϕ by rational numbers — say $\phi \approx \frac{q}{p}$, where p and q are integers — is when p and q are two successive Fibonacci numbers (see, e.g., [Cassels] or [Hardy and Wright]), in which case $| \phi - \frac{q}{p} | \approx \frac{1}{p^2\sqrt{5}}$. So our random number becomes small, $| \, | p\phi | \, | = | s_p | \approx \frac{1}{p\sqrt{5}}$, only when p is a Fibonacci Number. The observation that the irrational ϕ (and numbers equivalent to it) are the hardest numbers to approximate by rationals, translates, for us, to the observation that ϕ is the best possible choice to construct our sequence of real numbers. (The set, $\{p \, \theta \, mod \, 1 \mid p = 0, 1, 2, \cdots\}$ is dense in the unit interval for any irrational θ, but ϕ is the *best* irrational to use for uniform sequences.)

Now, we will analyze the integer sequence, $S_k = k \times F_n \, mod \, F_{n+1}$, with the help of the well-known identity which measures the discrepancy between the Fibonacci sequence and a geometric sequence:

$$F_n \times F_p - F_{n+1} \times F_{p-1} = (-1)^{p-1} F_{n-p}, \text{ for } n \geq p \geq 0.$$

(This is easily established by mathematical induction.) We interpret that identity as:

$$F_n \times F_p \equiv (-1)^{p-1} F_{n-p} \, mod \, F_{n+1}, \text{ for } n \geq p \geq 0,$$

This implies S_{F_p} is closer to zero than S_j for any $j < F_p$. In particular, $| S_{B-A} | = 1$ for the first time. For instance, using $A = 610$ and $B = 987$, we will generate 377 elements (38%) of the permutation of $\{0, 1, \ldots, 986\}$ before we generate two adjacent integers, and 233 elements (23.6%) before generating a pair that differ by 2. (The Fibonacci choice of parameters A and B to ensure good uniformity can also be justified by the well-known observation that the number of steps in Euclid's algorithm to compute the **gcd** of two numbers, A and B, is maximized (as a function of $max\,(A, B)$) when A and B are two successive Fibonacci numbers.)

5. RELATED WORK

[Knuth, vol. 3], in discussions of hash table algorithms (e.g., for the structure of a compiler's symbol table), presents a technique he calls "Fibonacci hashing" in which the entries in a table are searched in the order suggested by our sequences. Knuth points out that any sequence of the form $\{k\,\theta\,mod\,1 \mid k = 0, \cdots, n\}$, where θ is an irrational number always divides the unit interval into subintervals of only three different lengths. Furthermore, the next number, $(n + 1)\,\theta\,mod\,1$, goes into one of the largest of the existing intervals (in fact, it subdivides the *oldest* of the largest subintervals). When $\theta = \phi$, the subdivision is in a golden mean ratio.

A related set of numbers, the fractions whose denominators are powers of 2, is effectively used to locate approximate roots of continuous functions by "bisection searching." Our real number sequence is similarly effective at subdividing an interval to search for the maximum of a unimodular function. One needs to examine the function at four points, eliminate one of the three intervals, bisect one of the remaining two intervals, and iterate. The most effective way to do this is by golden mean subdivisions [Orr].

[Knuth, vol. 2] discusses, and quickly dismisses, sequences such as ours from the point of view of pseudo-randomness. According to him, most successful pseudo-random sequences are developed by the ("linear congruential") recursion,

$$X_{n+1} = (aX_n + c)\,mod\,m, \;\; n \leq 1,$$

where the parameters, $X_0, a, c,$ and m are appropriately chosen. However, we stand by our sequences for their uniformity and simplicity. Incremental image development is highly successful, and other applications should be as well.

6. SOME BACKGROUND COMMENTS

The integer random number sequence came to me while I was teaching an introductory computer programming class, when I needed a quick and dirty sequence of numbers suitable for exercising a sorting program. Later, for the same course, we needed a data stream to supply values for building a binary tree. The principle that "the more jumbled (random?) the input data, the more balanced and shallow the tree" clearly demonstrated these numbers to be very well jumbled.

Later, I found myself involved in a project to display some experimental data using computer graphics. This situation was indeed the mother of invention: the structure of the data was very unclear, I was new to this type of graphics programming, and the computer was not equipped to do floating-point arithmetic, so it was terribly slow. Rapid, low resolution rendering (with high resolution for the patient) was demanded. The result, "plywood" looking contour plots, are snapshot and shown in Figures 1, 2, and 3.

7. NEXT PROBLEMS TO SOLVE

7.1 Study the Binary Tree Structure:

Let θ be an irrational, and $s_n = n\, \theta \bmod 1$. Build a binary tree, with k nodes, using the data, $s_0, s_1, s_2, \cdots, s_{k-1}$, in that order. Study the depth of that tree as a function of k and θ. Elucidate and verify the conjecture that $\theta = \phi$ is optimal.

What is the relationship between θ's binary tree, as described above, and θ's continued fraction expansion?

7.2 Generalize to Multiple Dimensions:

What two irrationals, θ_1 and θ_2 optimally uniformly fill the unit square with

$$\vec{s}_n = (n\, \theta_1 \bmod 1, \, n\, \theta_2 \bmod 1)$$

This generalization is needed for image processing, Monte Carlo integration, and computer graphics.

BIBLIOGRAPHY

[1]   Cassels, J. W. S. An Introduction to Diophantine Approximation, Cambridge University Press, 1957.

[2]   Hardy, G. H. and Wright, E. M., The Theory of Numbers, Oxford, 1954.

[3]   Knuth, Donald E. The Art of Computer Programming, Volumes I, II, and III, Addison-Wesley, 1973.

[4]   Orr, Richard, Private correspondence.

Figures

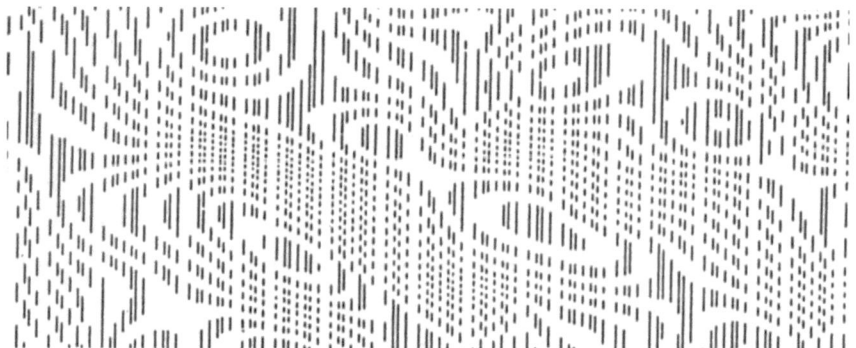

Figure 1. Plywood: 122 out of 610 scan lines rendered.

Figure 2. Plywood: 366 out of 610 scan lines rendered.

Figure 3. Plywood: all scan lines rendered.

ON THE PROOF OF GCD AND LCM EQUALITIES CONCERNING THE GENERALIZED BINOMIAL AND MULTINOMIAL COEFFICIENTS

Shiro Ando and Daihachiro Sato

1. INTRODUCTION

A strong divisibility sequence (or SDS) is a sequence of nonzero integers $\{a_n\}$ (n=1, 2, 3, ...) that satisfies

$$(a_m, a_n) = a_{(m, n)} \tag{1}$$

for any positive integers m, n, where (a, b) stands for the greatest common divisor of a and b. This terminology was named by Kimberling [7], although this concept had been studied before by Ward [9] and others.

The SDS, which involves the Fibonacci sequence as its typical example, is used to define the generalized binomial coefficients as

$$\binom{n}{h} = \prod_{i=1}^{n} a_i \Big/ \prod_{i=1}^{h} a_i \prod_{i=1}^{k} a_i, \text{ where h+k=n,} \tag{2}$$

and the generalized multinomial coefficients as

$$\left(\begin{smallmatrix} & n & \\ k_1, & ..., & k_m \end{smallmatrix} \right) = \prod_{i=1}^{n} a_i \Big/ \prod_{i=1}^{k_1} a_i \cdots \prod_{i=1}^{k_m} a_i, \tag{3}$$

where $k_1 + \cdots + k_m = n$.

The purpose of this note is to show that the basic property of the SDS given in the next section is useful to give non-p-adic proofs of GCD and LCM equalities concerning the generalized binomial and multinomial coefficients and their modified versions by giving some examples.

9

G. E. Bergum et al. (eds.), Applications of Fibonacci Numbers Volume 4, 9–16.

2. BASIC PROPERTY OF SDS

It is well known that the SDS is a divisibility sequence (or DS) that is defined by the condition

$$a_m \mid a_n \text{ (if } m \mid n) \tag{4}$$

and that the generalized binomial and multinomial coefficients are integers.

Basic Property: Let m, n and p be positive integers. If there are integers s and t such that $p=sm+tn$, then $(a_m, a_n) \mid a_p$.

Proof: Since $(m, n) \mid p$, $a_{(m,n)} \mid a_p$. From this and (1), we get the result.

Corollary: Let $n_1, \cdots, n_s$ and p be positive integers. If $p=t_1 n_1 + \cdots + t_s n_s$ for some integers $t_1, \cdots, t_s$, then we have $(a_{n_1}, \cdots, a_{n_s}) \mid a_p$, and therefore, we have

$$(a_p, a_{n_1}, \cdots, a_{n_s}) = (a_{n_1}, \cdots, a_{n_s}).$$

Remark 1: As the SDS is a DS, every term of the SDS is divisible by its initial term a_1. Hence, we can assume $a_1=1$ as long as we stay with the GCD or LCM equalities concerning the generalized binomial or multinomial coefficients and their modifications. Therefore, we always assume $a_1=1$ in the following.

Remark 2: The condition:

$$(a_m, a_n) \mid a_{m+n} \text{ and } (a_m, a_n) \mid a_{m-n} \text{ (if } m>n) \tag{5}$$

which was introduced in Hillman-Hoggatt [4] is equivalent to the condition (1) as we have mentioned in [2] before. In fact, the condition (5) is an immediate consequence of our basic property of the SDS. Conversely, repeated use of the first part of (5) leads to divisibility $a_m \mid a_{sm}$, which gives $a_{(m,n)} \mid (a_m, a_n)$, while using the expression $(m, n)=sm-tn$, where s and t are positive integers, the second part of (5) gives $(a_m, a_n) \mid (a_{sm}, a_{tn}) \mid a_{sm-tn} = a_{(m,n)}$.

Remark 3: We can not replace an SDS with a DS. For example, if we use a DS defined by $a_n = 2^{n-1}$, we can not define the generalized binomial or multinomial coefficients which satisfy the same GCD or LCM equalities as the ordinary ones do. On the other hand, the condition that $\{a_n\}$ is an SDS is not necessary. For example, for a DS defined by $a_1=1$, $a_n=2$ (for $n>1$), the same equalities hold.

3. APPLICATIONS TO GCD EQUALITIES

Example 1: Consider a generalized Pascal triangle which has the generalized binomial coefficients as its entries. Let

$$X = \prod_{i=1}^{n} a_i \Big/ \prod_{i=1}^{h} a_i \prod_{i=1}^{k} a_i, \text{ where } h+k=n. \qquad (6)$$

be any element inside the triangle, and let A, B, C, D, E, and F be six entries surrounding it. Then we have

$$(A, C, E) = (B, D, F). \qquad (7)$$

A p-adic proof was given by Hillman-Hoggatt [4], while a simple proof was given by Hitotumatu-Sato [5] in the case of ordinary Pascal's triangle. In [5], (B, D, F) | A, for instance, is shown by the equality

$$A = -(k+1)B+(h+1)D-(n+1)F. \qquad (8)$$

Now, we will modify this proof to apply to the generalized Pascal triangle. Since

$$A=\frac{a_h}{a_n}X, \quad B=\frac{a_h}{a_{k+1}}X, \quad D=\frac{a_{n+1}}{a_{h+1}}X, \quad F=\frac{a_k}{a_n}X \qquad (9)$$

in this case, we compare the two expressions

$$-a_{k+1}B+a_{h+1}D-a_nF-F=(-a_h+a_{n+1}-a_k-a_k a_n^{-1})X \text{ and } A=a_h a_n^{-1}X$$

instead of using the equality (8). By the relations (9) and the basic property of the SDS, we have

$$(B, D, F) \mid (a_h, a_{n+1}, a_k) \; X \mid a_1 X = X$$

and

$$(X, F) = (X, a_k a_n^{-1}X) = (a_n, a_k)a_n^{-1}X \mid a_h a_n^{-1}X = A$$

as $-h+(n+1)-k=1$ and $n-k=h$. Thus we can conclude that $(B, D, F) \mid A$.

The rest of the proof can be completed in a similar manner by referring the equalities given by the matrices in [5] and will be omitted here.

Example 2: In our former paper, Ando-Sato [2], we studied a generalization of the equality (7) to the m-dimensional generalized Pascal pyramid consisting of the entries (3). We first considered there the ordinary m-nomial coefficient $A=n!/k_1!k_2!\cdots k_m!$, and the set S of $m(m+1)$ entries surrounding it: A_{ij} (where i=1, 2, $\cdots$, m+1; j=1, 2, $\cdots$, m). They satisfy the equalities $nA_{ij}=k_jA$, $(k_{j-i+1}+1)A_{ij}=k_{j+i-1}A$ for j=1, 2, $\cdots$, m, and $(k_{j-m+1})A_{m+1\,j}=(n+1)A$, where the suffixes of k and the second suffixes of A are supposed to be considered mod m. We decomposed S into m-sets $S_j = \{A_{1j}, A_{2j}, \cdots, A_{m+1\ j}\}$, (where $j=1, 2, \cdots, m$) following Hoggatt-Alexanderson [6]. The conclusion was $\gcd S_1' = \gcd S_2' = \cdots = \gcd S_m'$, where $S_j' = S - S_j$. For the proof we showed $\gcd S_j' = \gcd S$ for $j=1$ and m odd, as an example, using the equalities

$$A = ((n+1) - k_1 - k_2 - \cdots - k_m)A$$

$$= (k_2+1)A_{m+1\ 2} - (k_3+1)A_{m2} - nA_{12} - \cdots - (k_4+1)A_{m-1\ 2}, \tag{10}$$

$$\frac{A}{n} = \frac{A_{11}}{k_1} = \frac{A_{12}}{k_2} = \cdots = \frac{A_{1m}}{k_m} \tag{11}$$

and

$$\frac{A}{k_{m-r}+1} = \frac{A_{2\ 1-r}}{k_{2-r}} = \cdots = \frac{A_{m\ m-1-r}}{k_{2m-2-r}} = \frac{A_{m+1\ m-r}}{n+1}, \text{ for } r=1, 2, \cdots, m, \tag{12}$$

where the denominators of (11) and (12) satisfy the equalities

$$n = k_1 + k_2 + \cdots + k_m \tag{13}$$

and

$$(k_{m-r}+1) + k_{2-r} + \cdots + k_{2m-2-r} = n+1, \tag{14}$$

respectively.

For the generalized version, we gave a p-adic proof in [2]. Now, we will give a non-p-adic proof using the basic property of the SDS. In this case, we use

$$a_{n+1}A = a_{k_2+1}A_{m+1\ 2}, \ a_{k_1}A = a_{k_3+1}A_{m2}, \ a_{k_2}A = a_n A_{12}, \cdots, \ a_{k_m}A = a_{k_4+1}A_{m-1\ 2}$$

and the equality $1 = (n+1) - k_1 - \cdots - k_m$ instead of (10) to get

$$\gcd S_2 = (A_{12}, A_{22}, \cdots, A_{m+1\ 2}) \mid (a_{n+1}, a_{k_1}, \cdots, a_{k_m})A \mid a_1 A = A. \tag{15}$$

The denominators of (11) and (12) will be replaced with the corresponding terms of the SDS. If we consider (13) and (14), then we see that each denominator of (11) and (12) is divisible by the GCD of other denominators in the same formula, the same is true for the numerators so that we can conclude $(A, \gcd S_1') \mid \gcd S_1$. As $S_2 \subset S_1'$, from this relation and (15), we have $\gcd S_1' \mid \gcd S_1$ to prove $\gcd S_1 = \gcd S$, which is our goal.

4. APPLICATIONS TO LCM EQUALITIES

Before we present examples of applying the basic property of SDS to LCM equalities, we will define the GCD and the LCM on a set of rational numbers.

Let $A = \{a_1, a_2, \cdots, a_n\}$ be a set of positive rational numbers. Denote $D = \{p_1, p_2, \cdots, p_s\}$ the set of all primes that divide a denominator or a numerator of some of a_i

in A. Then we can express a_i as $a_i = p_1^{v_{i1}} p_2^{v_{i2}} \cdots p_s^{v_{is}}$, where v_{ij} ($j=1, 2, \cdots, s$) are integers. Let

$$u_j = \min_{1 \leq i \leq n} v_{ij}, \quad v_j = \max_{1 \leq i \leq n} v_{ij},$$

for $j=1, 2, \cdots, s$ and define the GCD and the LCM of A by $\gcd A = p_1^{u_1} p_2^{u_2} \cdots p_s^{u_s}$, $\operatorname{lcm} A = p_1^{v_1} p_2^{v_2} \cdots p_2^{v_s}$. If $a_1, a_2, \cdots, a_n$ are integers, then these definitions coincide with the ordinary ones. We can prove easily the following formulas:

$$\gcd \{aa_1, aa_2, \cdots, aa_n\} = a \gcd A,$$
$$\operatorname{lcm} \{aa_1, aa_2, \cdots, aa_n\} = a \operatorname{lcm} A,$$

for any rational number a, and

$$\gcd \{a_1^{-1}, a_2^{-1}, \cdots, a_n^{-1}\} = (\operatorname{lcm} A)^{-1}. \tag{16}$$

In the following we use the simple notation for the GCD of rational numbers as well as the GCD of integers: $\gcd \{a_1, a_2, \cdots, a_n\} = (a_1, a_2, \cdots, a_n)$. We will also use the notation a | b for two positive rational numbers to mean that $a^{-1}b$ is an integer. Then we have

$$a \mid b \Leftrightarrow (a, b) = a \tag{17}$$

where a and b are positive rational numbers.

Example 3: Let $a_1, a_2, a_3, \cdots$ be a SDS, and let

$$X = \left[\begin{array}{c} n \\ h \end{array}\right] = \prod_{i=1}^{n+1} a_i \Big/ \prod_{i=1}^{h} a_i \prod_{i=1}^{k} a_i, \text{ where } h+k=n,$$

be any entry inside the generalized modified Pascal triangle defined in Ando [1]. If A, B, C, D, E, and F are six entries surrounding it as in example 1, then we have an LCM equality

$$\operatorname{lcm} \{A, C, E\} = \operatorname{lcm} \{B, D, F\} \tag{18}$$

instead of GCD equality (7). A p-adic proof of (18) was shown in [1]. We will give here a non-p-adic proof. If we use (16), we can equivalently express this relation as

$$(A^{-1}, C^{-1}, E^{-1}) = (B^{-1}, D^{-1}, F^{-1}),$$

in terms of the GCD of rational numbers.

For the case $a_n = n$ ($n=1, 2, 3, \cdots$), Sato [8] gave a simple proof by showing the linear relations

$$\begin{pmatrix} A^{-1} \\ C^{-1} \\ E^{-1} \end{pmatrix} = \begin{pmatrix} h+1 & n+2 & -k \\ h & n+1 & -k \\ -h & -n-2 & k+1 \end{pmatrix} \begin{pmatrix} B^{-1} \\ D^{-1} \\ F^{-1} \end{pmatrix}, \quad \begin{pmatrix} B^{-1} \\ D^{-1} \\ F^{-1} \end{pmatrix} = \begin{pmatrix} h+1 & -n-2 & -k \\ -h & n+1 & k \\ -h & n+2 & k+1 \end{pmatrix} \begin{pmatrix} A^{-1} \\ C^{-1} \\ E^{-1} \end{pmatrix}.$$

In the case of generalized version, we have the relations:

$$a_h A^{-1} = a_{n+1} X^{-1}, \quad a_{n+2} C^{-1} = A_{k+1} X^{-1}, \quad a_k E^{-1} = a_{h+1} X^{-1},$$
$$a_h B^{-1} = a_{k+1} X^{-1}, \quad a_{n+2} D^{-1} = a_{h+1} X^{-1}, \quad a_k F^{-1} = a_{n+1} X^{-1}.$$

Using these relations, we have

$$\gcd \{B^{-1}, D^{-1}, F^{-1}\} \mid \gcd \{a_{k+1}, a_{h+1}, a_{n+1}\} X^{-1} \mid a_1 X^{-1} = X^{-1}$$

and

$$a_h (B^{-1}, X^{-1}) \mid (a_{h+1}, a_h) X^{-1} \mid a_{n+1} X^{-1} = a_h A^{-1},$$

from which we have $(B^{-1}, X^{-1}) \mid A^{-1}$, and arrive at the conclusion: $(B^{-1}, D^{-1}, F^{-1}) = A^{-1}$.

The rest of the proof can be accomplished in a similar manner and will be omitted.

Example 4: Let X be any element in the Pascal triangle which has six elements A, B, C, D, E and F surrounding it and other six elements H, I, J, L, R and T next to them.

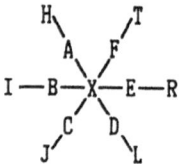

Then we have

$$\text{lcm } \{X, A, B, H, I\} = \text{lcm } \{A, B, H, I\}, \tag{19}$$

as is stated in Ando-Sato [3]. If we notice the equality (16), this is equivalent to

$$(1, A', B', H', I') = (A', B', H', I'), \tag{20}$$

where $A', B', \cdots$ denote $XA^{-1}, XB^{-1}, \cdots$, respectively.

In Pascal's triangle where $a_n = n$ $(n=1, 2, 3, \cdots)$, as

$$A' = \frac{n}{h}, \quad B' = \frac{k+1}{h}, \quad H' = \frac{n(n-1)}{h(h-1)}, \quad I' = \frac{(k+1)(k+2)}{h(h-1)}, \tag{21}$$

we have the equality $hA' + 2(n-1)B' - (n+1)H' + (n-1)I' = 1$, from which (20) is clear.

In the generalized case, we use

$$A' = \frac{a_n}{a_h}, \quad B' = \frac{a_{k+1}}{a_h}, \quad H' = \frac{a_n a_{n-1}}{a_h a_{h-1}}, \quad I' = \frac{a_{k+1} a_{k+2}}{a_h a_{h-1}} \tag{22}$$

instead of (21).

Considering n=h+k to use the basic property of the SDS, we have

$$(a_{k+2}, a_{h-1}) \mid a_{n+1}, \quad (a_{k+1}, a_n) \mid a_{h-1}, \quad (a_{n+1}, a_{k+1}) \mid a_h.$$

Using these relations and (22), we see that

$$a_h a_{h-1}(B', I') = a_{k+1}(a_{h-1}, a_{k+2}) \mid a_{k+1} a_{n+1},$$

$$a_h a_{h-1}(B', I', H') \mid a_h a_{h-1}(a_{n-1}B', a_{n-1}(B', I'), a_{n+1}H')$$

$$\mid (a_{h-1} a_{n-1} a_{k+1}, (a_{n-1} a_{k+1} a_{n+1}, a_n a_{n-1} a_{n+1}))$$

$$= a_{n-1}(a_{h-1} a_{k+1}, a_{n+1}(a_{k+1}, a_n)) \mid a_{n-1}(a_{h-1} a_{k+1}, a_{n+1} a_{h-1})$$

$$= a_{n-1} a_{h-1}(a_{k+1}, a_{n+1}) \mid a_{n-1} a_{h-1} a_h$$

Therefore, $(B', I', H') \mid a_{n-1}$. $(A', B', I', H') \mid (a_h A', (B', I', H')) \mid (a_n, a_{n-1}) = a_{(n, n-1)} = a_1 = 1.$
Thus we have (20) by (17), completing the proof.

Example 5: In the same situation as in example 4, we have

$$\text{lcm } \{X, A, C, E, H, J, R\} = \text{lcm } \{A, C, E, H, J, R\} \tag{23}$$

Using the same notations as in example 4, we have the following expressions along with (22):

$$C' = \frac{a_{k+1}}{a_{n+1}}, \quad E' = \frac{a_{h+1}}{a_k}, \quad J' = \frac{a_{k+1} a_{k+2}}{a_{n+1} a_{n+2}}, \quad R' = \frac{a_{h+1} a_{h+2}}{a_k a_{k-1}} \ .$$

First, we assume that n is odd. Then we have

$$(a_n, a_{h+1}, a_{k+1}) \mid a_{(h+1)+(k+1)-n} = a_2, \quad (a_n, a_{h+1}, a_{k+1}) \mid (a_n, a_2) = a_{(n,2)} = a_1 = 1.$$

Hence,

$$(A', C', E') \mid 1 \ . \tag{24}$$

If h or k is even, we also have (24) in a similar manner.

In the remaining case, n is even, and h and k are odd. Hence, let n=2m, h+1=2i and k+1=2j. Then we have $(a_{2m}, a_{2i}, a_{2j}) = a_{(2m,2i,2j)} = a_2$ as $i+j-m=1$. Considering $(a_2, a_{2m-1}) = a_1 = 1$ and $a_2 \mid a_{2(i-1)}$, we have $(A', H') \mid (a_h A', a_{2(i-1)} a_2^{-1} a_h H') = (a_{2m}, a_{2m} a_2^{-1} a_{2m-1}) = a_{2m} a_2^{-1} (a_2, a_{2m-1}) = a_{2m} a_2^{-1}$. In a similar manner, we also have $(C', H') \mid a_{2j} a_2^{-1}$ and $(E', R') \mid a_{2i} a_2^{-1}$. Hence we have $(A', C', E', H', J', R') \mid a_2^{-1}(a_{2m}, a_{2i}, a_{2j}) = 1$, establishing (23).

REFERENCES

[1]   Ando, S. "A Triangular Array with Hexagon Property, Dual to Pascal's Triangle." Applications of Fibonacci Numbers, Edited by A.N. Philippou, A.F. Horadam and G.E. Bergum, Kluwer Academic Publishers, (1988) pp. 61-67.

[2]   Ando, S. and Sato, D. "A GCD Property on Pascal's Pyramid and the Corresponding LCM Property of the Modified Pascal Pyramid." Applications of Fibonacci Numbers, Volume 3, Edited by A.N. Philippou, A.F. Horadam and G.E. Bergum, Kluwer Academic Publishers, (1990) pp. 7-14.

[3]   Ando, S. and Sato, D. "Translatable and Rotatable Configurations which Give Equal Product, Equal GCD and Equal LCM Properties Simultaneously." Applications of Fibonacci Numbers Volume 3, Edited by A.N. Philippou, A.F. Horadam and G.E. Bergum, Kluwer Academic Publishers, (1990) pp. 15-26.

[4]   Hillman, A. P. and Hoggatt, V. E. Jr. "A Proof of Gould's Pascal Hexagon Conjecture." The Fibonacci Quarterly, 10 (1972): pp. 565-568, 598.

[5]   Hitotumatu, S. and Sato, D. "The Star of David Theorem (I)." The Fibonacci Quarterly 13 (1975): p. 70.

[6]   Hoggatt, V. E. Jr. and Alexanderson, G. L. "A Property of Multinomial Coefficients." The Fibonacci Quarterly, 9 (1971): pp. 351-356, 420-421.

[7]   Kimberling, C. "Strong Divisibility Sequences with Nonzero Initial Term." The Fibonacci Quarterly 16 (1978): pp. 541-544.

[8]   Sato, D. "Star of David Theorem (II)-A Simple Proof of Ando's Theorem." Personal Communication.

[9]   Ward, M. "Note on Divisibility Sequences." Bull. Amer. Math. Soc. 42 (1936) pp. 843-845.

SUPERCUBE

Joseph Arkin, David C. Arney, Lee S. Dewald and Frank R. Giordano

A FEW BRIEF HISTORICAL NOTES ON PERFECT CUBES

(1) In 1888, the first <u>perfect</u> magic cube ever constructed was of order 8, and was placed in *"The Memoirs of the National Academy of Science"* [3].

(2) Martin Gardner defines a perfect magic cube as follows: "A perfect magic cube is a cubical array of positive integers from 1 to N^3 such that every straight line of N cells adds up to a constant. These lines include the orthogonals (the lines parallel to an edge), the two main diagonals of every orthogonal cross section and the four space diagonals. The constant is $(1+2+3+...+N^3)/N^2 = \frac{1}{2}(N^4+N)$" [4].

(3) E. G. Straus, in 1976, in a private letter to Arkin, described how he constructed a 7x7x7 perfect magic cube. This may be the lowest possible order of a perfect Latin 3-cube [5].

(4) In 1985, Arkin superimposed 6 orthogonal Latin cubes of order 7 to form 20 separate Latin 3-cubes [1].

(5) A perfect 4-dimensional hypercube of order 7 was constructed at West Point in 1989 [2].

LATIN K-CUBE OF ORDER N

A Latin square of order n is an nxn square in which each of the numbers 0, 1, . . . , n − 1 occurs exactly once in each row and exactly once in each column. For example

17

G. E. Bergum et al. (eds.), Applications of Fibonacci Numbers Volume 4, 17–32.

01	012	0123
10	120	1230
	201	2301
		3012

are Latin squares of orders 2, 3, 4, respectively. Two Latin squares of order n are orthogonal, when one is superimposed on the other, every ordered pair 00, 01, . . . , $(n-1)(n-1)$ occurs. Thus

012		012		00	11	22
120	and	201	superimpose to	12	20	01
201		120		21	02	10

and therefore are orthogonal squares of order 3. A set of Latin squares of order n is orthogonal if every two of them are orthogonal. As an example the 4x4 square of triples

000	111	222	333
123	032	301	210
231	320	013	102
312	203	130	021

represents three mutually orthogonal squares of order 4 since each of the 16 pairs 00, 01, . . . , 33 occurs in each of the three possible positions among the 16 triples.

We can generalize all these concepts to nxnxn cubes and cubes of higher dimensions. A Latin cube of order n is an nxnxn cube (n rows, n columns and n files) in which the numbers 0, 1, . . . n − 1 are entered so that each number occurs exactly once in each row, column and file. If we list the cube in terms of the n squares of order n which form its different levels we can list the cubes

01	10		012	120	201
10	01	and	120	201	012
			201	012	120

as Latin cubes of order 2 and 3, respectively.

Orthogonality of Latin cubes is the following relation among three Latin cubes: three Latin cubes of order n are orthogonal if, when superimposed, each ordered triple 000, 001, . . ., $(n-1)(n-1)(n-1)$ will occur. For example the pair of 3x3 Latin squares

00	11	22
12	20	01
21	02	10

leads to the four 3x3x3 cubes

a:	012	120	201
	120	201	012
	201	012	120

b:	012	120	201
	201	012	120
	120	201	012

c:	021	102	210
	210	021	102
	102	210	021

d:	021	102	210
	102	210	021
	210	021	102

Superimposed these lead to a cube of quadruples in three levels with I over II over III, abdc:

0000	1122	2211	1111	2200	0022	2222	0011	1100
1221	2010	0102	2002	0121	1210	0110	1202	2021
2112	0201	1020	0220	1012	2101	1001	2120	0212

I		II		III

where each ordered triple occurs in every one of the four possible positions in the quadruples.
Note: We define a cube of triples (say abc) where each ordered triple occurs in some order in the 27 cells of the three levels (I over II over III) of the cube abc as a *Latin 3-cube of order 3.*

SUPERCUBE

In this paper we have constructed an orthogonal 4-cube of order 8. This <u>Supercube</u> consists of eight Latin 4-cubes.

To individually construct each one of the eight cubes, we superimpose 4 orthogonal Latin cubes of order 8. Now, each one of the resulting 4096 cells throughout the Supercube contains four digits, where each ordered quadruple (0000, 0001, . . . , 7777) occurs only once in every cell.

We believe this construction to be the smallest possible perfect 4-dimensional array (found to date) where <u>each one of the eight cubes</u> that make up the supercube, is <u>perfect</u>.

Each of the eight Latin 4-cubes is perfect in the following way (we consider only one of the eight cubes at a time): the sum (31108) of the elements in each minor diagonal is equal to the sum of the elements of a row in each of the 2 directions in each of the respective squares (layers) that make up each Latin 4-cube of order 8. The sum (31108) of the elements of a row in each direction of a cube is equal to the sum of the elements in each of the 4 major diagonals and the sum on all the diagonals of the cube is the same (namely 31108). The sum (31108) of each of the eight major space diagonals throughout the supercube is the same. The construction of the cube is based on the 3 orthogonal cubes

$$A_{ijk}^{(2)} = x_i + 2x_j - 3x_k,$$

$$A_{ijk}^{(-2)} = x_i - 2x_j - 3x_k,$$

$$A_{ijk}^{(3)} = x_i + 3x_j + 2x_k,$$

where $(x_1, \ldots, x_8) = (0, 1, \ldots, y)$ and arithmetic is (mod 8).

DOUBLY-MAGIC PROPERTIES

Designate each cell by the coordinates in the array as E(k, s, r, c). The indices -- k, s, r, and c -- correspond to cube, square, row, and column of the entry in the array and take on values from 1 through 8.

By labeling the four digits in each quadruple ABCD, new entries for the 4-cube can be formed, *in columns*, by the following four three-digit combinations - CDC, DCD, ADA, and DAD. New entries, *in rows*, by the following two three digit combinations - BCB and CBC. With these new entries, the 4-cube not only has a new magic sum of 3108 in base 10, but also a doubly-magic sum. That is by squaring each entry in base 10 and summing the eight appropriate numbers, we obtain 1,640,100.

Using the row entries in the cubes BCB, or CBC, we have the doubly-magic sums

$$\sum_{j=1}^{8} E(k, s, r, j)^2 = 1{,}640{,}100,$$

In what follows, we consider the squares and cubes that are made by using the column entries ADA, DAD, CDC, or DCD instead of ABCD. We have for the 512 sums

$$\sum_{i=1}^{8} E(k, s, i, c)^2 = 1{,}640{,}100,$$

and

$$\sum_{i=1}^{8} E(i, s, r, c)^2 = 1{,}640{,}100,$$

This same doubly-magic sum holds for the major space diagonals with these new three digit entries.

BROKEN DIAGONALS

We now explain the uppercase and the lowercase respectively over the columns A and C throughout the 4-Cube.

Let I(n, 1, k, s), under some column labeled I, represent the number ADA, on the particular row where it is found, and as well let I(n, 2, k, s) represent DAD, where the indices k and s correspond to cube and square and n = 0, 1, . . . , 7 ranges in some order from 0 through 7). For example, I(4, 1, 1, 1) under column I, in Cube 1, Square 1, represents 454 or I(4, 2, 1, 1) represents 545. In the same way, in Cube 3, Square 8, we have M(7, 1, 3, 8) = 767 or M(7, 2, 3, 8) = 676.

Although there are many other ways of routing the doubly-magic broken diagonals through this 4th dimensional cube, we will limit our discussion to routing our diagonals, from top to bottom, that is, from cube 1, square 1, through cube 8, square 8.

It is easy to show that

$$\sum_{n=1}^{8} I(n, f, k, s)^2 = 1,640,100,$$

where f=1 or 2, and n=0, 1, . . . , 7 ranges in some order from 0 through 7, since the digits in A and D are the same under each and every I. Thus, from cube 1, square 1, there are 2x7! broken diagonals (with a sum of 3108 and a double sum of 1, 640,100) emanating from each digit in A (to give ADA and DAD) that is under I, and so for all 8 digits we have 2x8! broken diagonals in ADA and DAD.

In the exact way we found the broken diagonals under I, using digits in A and D, we find the broken diagonals under the remaining 7 uppercase letters, J, K, L, M, N, P, and Q.

We get the same results for the lower case letters, r, s, t, u, v, w, x, and y, except that the results contain the digits in C and D, and we have the arrangements for CDC and DCD.

In Figure 1 each of the eight 8x8 squares is orthogonal, the 64 lowercase letters are superimposed with the 64 uppercase letters.

To see how this chart works, it will suffice to use, say Ir, as an example. Since, in Ir, we have both an upper and lower case at the same time, we form, in columns, the following four doubly-magic three-digit combinations - ADA, DAD, CDC, and DCD. It is evident that by starting in cube 1 and tracing Ir through cube 1, cube 2, . . . , cube 7, cube 8, we have traced the route of a major space diagonal. Let us display one of the possible four routes of this Ir, namely -- ADA

$$(020)^2 + (606)^2 + (111)^2 + (737)^2 + (565)^2 + (343)^2 + (454)^2 + (272)^2 = 1,640,100.$$

In the same way, using any orthogonal pair of letters, starting in cube 1 and tracing the pair through cube 1, cube 2, . . . , cube 8, we can form 8! broken diagonals in the ADA, DAD, CDC, and DCD.

REFERENCES

[1] Arkin, Joseph. "An Extension of E. G. Straus' Perfect Latin 3-Cube of Order 7." *Pacific Journal of Mathematics*, Vol. 118, No. 2 (June 1985), pp. 277-280.

[2] Arkin, Joseph, Arney, David C., and Porter, Bruce J. "A Perfect 4-Dimensional Hypercube of Order 7 (the Cameron Cube)." *Journal of Recreational Mathematics*, Vol. 21, No. 2 (1989), pp. 81-88.

[3] Barnard, F. A. P. "Theory of Magic Squares and of Magic Cubes." in The Memoires of *the National Academy of Science*, 4 (1988): pp. 209-270.

[4] Gardner, Martin. "Mathematical Games." *Scientific American* (January 1976), pp. 120, 122.

[5] Straus, E. G., Personal Correspondence to Joseph Arkin, January 1976, giving a detailed construction of a 7x7x7 perfect magic cube written to base 7 with digits 000 to 666.

BIBLIOGRAPHY

a) Arkin, J. and Straus, E. G., "Latin k-Cubes," *Fibonacci Quarterly*, Vol. 12, No. 3 (October 1974), pp. 288-292.

b) Arkin, J., Hoggatt, V. E. Jr., and Strauss, E. G. "Systems of Magic Latin k-Cubes." *Canadian Journal of Mathematics*, 28:6, pp. 1153-1161, 1976.

c) Arkin, J., Smith, P., "Trebly Magic Systems in a Latin 3-Cube of Order Eight," *Fibonacci Quarterly*, Vol. 14, No. 2 (April 1976), pp. 167-170.

Ir	Js	Kt	Lu	Mv	Nw	Px	Qy
Nv	Mw	Qx	Py	Jr	Is	Lt	Ku
Qw	Pv	Ny	Mx	Ls	Kr	Ju	It
Ks	Lr	Iu	Jt	Pw	Qv	My	Nx
Ly	Kx	Jw	Iv	Qu	Pt	Ns	Mr
Pu	Qt	Ms	Nr	Ky	Lx	Iw	Jv
Mt	Nu	Pr	Qs	Ix	Jy	Kv	Lw
Jx	Iy	Lv	Kw	Nt	Mu	Qr	Ps

CUBE 1

Mw	Nv	Py	Qx	Is	Jr	Ku	Lt
Js	Ir	Lu	Kt	Nw	Mv	Qy	Px
Lr	Ks	Jt	Iu	Qv	Pw	Nx	My
Pv	Qw	Mx	Ny	Kr	Ls	It	Ju
Qt	Pu	Nr	Ms	Lx	Ky	Jv	Iw
Kx	Ly	Iv	Jw	Pt	Qu	Mr	Ns
Iy	Jx	Kw	Lv	Mu	Nt	Ps	Qr
Nu	Mt	Qs	Pr	Jy	Ix	Lw	Kv

CUBE 2

Ny	Mx	Qw	Pv	Ju	It	Ls	Kr
Iu	Jt	Ks	Lr	My	Nx	Pw	Qv
Kt	Lu	Ir	Js	Px	Qy	Mv	Nw
Qx	Py	Nv	Mw	Lt	Ku	Jr	Is
Pr	Qs	Mt	Nu	Kv	Lw	Ix	Jv
Lv	Kw	Jx	Iy	Qr	Ps	Nt	Mu
Jw	Iv	Ly	Kx	Ns	Mr	Qu	Pt
Ms	Nr	Pu	Qt	Iw	Jv	Ky	Lx

CUBE 3

Jt	Iu	Lr	Ks	Nx	My	Qv	Pw
Mx	Ny	Pv	Qw	It	Ju	Kr	Ls
Py	Qx	Mw	Nv	Ku	Lt	Is	Jr
Lu	Kt	Js	Ir	Qy	Px	Nw	Mv
Kw	Lv	Iy	Jx	Ps	Qr	Mu	Nt
Qs	Pr	Nu	Mt	Lw	Kv	Jy	Ix
Nr	Ms	Qt	Pu	Jv	Iw	Lx	Ky
Iv	Jw	Kx	Ly	Mr	Ns	Pt	Qu

CUBE 4

Qu	Pt	Ns	Mr	Ly	Kx	Jw	Iv
Ky	Lx	Iw	Jv	Pu	Qt	Ms	Nr
Ix	Jy	Kv	Lw	Mt	Nu	Pr	Qs
Nt	Mu	Qr	Ps	Jx	Iy	Lv	Kw
Mv	Nw	Px	Qy	Ir	Js	Kt	Lu
Jr	Is	Lt	Ku	Nv	Mw	Qx	Py
Ls	Kr	Ju	It	Qw	Pv	Ny	Mx
Pw	Qv	My	Nx	Ks	Lr	Iu	Jt

CUBE 5

Lx	Ky	Jv	Iw	Qt	Pu	Nr	Ms
Pt	Qu	Mr	Ns	Kx	Ly	Iv	Jw
Mu	Nt	Ps	Qr	Iy	Jx	Kw	Lv
Jy	Ix	Lw	Kv	Nu	Mt	Qs	Pr
Is	Jr	Ku	Lt	Mw	Nv	Py	Qx
Nw	Mv	Qy	Px	Js	Ir	Lu	Kt
Qv	Pw	Nx	My	Lr	Ks	Jt	Iu
Kr	Ls	It	Ju	Pv	Qw	Mx	Ny

CUBE 6

Kv	Lw	Ix	Jy	Pr	Qs	Mt	Nu
Qr	Ps	Nt	Mu	Lv	Kw	Jx	Iy
Ns	Mr	Qu	Pt	Jw	Iv	Ly	Kx
Iw	Jv	Ky	Lx	Ms	Nr	Pu	Qt
Ju	It	Ls	Kr	Ny	Mx	Qw	Pv
My	Nx	Pw	Qv	Iu	Jt	Ks	Lr
Px	Qy	Mv	Nw	Kt	Lu	Ir	Js
Lt	Ku	Jr	Is	Qx	Py	Nv	Mw

CUBE 7

Ps	Qr	Mu	Nt	Kw	Lv	Iy	Jx
Lw	Kv	Jy	Ix	Qs	Pr	Nu	Mt
Jv	Iw	Lx	Ky	Nr	Ms	Qt	Pu
Mr	Ns	Pt	Qu	Iv	Jw	Kx	Ly
Nx	My	Qv	Pw	Jt	Iu	Lr	Ks
It	Ju	Kr	Ls	Mx	Ny	Pv	Qw
Ku	Lt	Is	Jr	Py	Qx	Mw	Nv
Qy	Px	Nw	Mv	Lu	Kt	Js	Ir

CUBE 8

Figure 1

I r	J s	K t	L u	M v	N w	P x	Q y
0722	7201	3067	4544	6375	1656	5430	2113
4365	3646	7420	0103	2732	5211	1077	6554
5676	2355	6133	1410	3221	4702	0564	7047
1231	6712	2574	5057	7666	0345	4123	3400
7153	0470	4616	3335	1504	6027	2241	5762
3514	4037	0251	7772	5143	2460	6606	1325
2007	5524	1742	6261	4450	3173	7315	0636
6440	1163	5305	2626	0017	7534	3752	4271

CUBE 1 SQUARE I

N v	M w	Q x	P y	J r	I s	L t	K u
2450	5173	1315	6636	4007	3524	7742	0261
6017	1534	5752	2271	0440	7163	3305	4626
7504	0027	4241	3762	1153	6470	2616	5335
3143	4460	0606	7325	5514	2037	6251	1772
5221	2702	6564	1047	3676	4355	0133	7410
1666	6345	2123	5400	7231	0712	4574	3057
0375	7656	3430	4113	6722	1201	5067	2544
4732	3211	7077	0554	2365	5646	1420	6103

CUBE 1 SQUARE II

Q w	P v	N y	M x	L s	K r	J u	I t
4211	3732	7554	0077	2646	5365	1103	6420
0656	7375	3113	4430	6201	1722	5544	2067
1345	6666	2400	5123	7712	0231	4057	3574
5702	2221	6047	1564	3355	4676	0410	7133
3460	4143	0325	7606	5037	2514	6772	1251
7027	0504	4762	3241	1470	6153	2335	5616
6534	1017	5271	2752	0163	7440	3626	4305
2173	5450	1636	6315	4524	3007	7261	0742

CUBE 1 SQUARE III

K s	L r	I u	J t	P w	Q v	M y	N x
6163	1440	5626	2305	0534	7017	3271	4752
2524	5007	1261	6742	4173	3450	7636	0315
3037	4514	0772	7251	5460	2143	6325	1606
7470	0153	4335	3616	1027	6504	2762	5241
1712	6231	2057	5574	7345	0666	4400	3123
5355	2676	6410	1133	3702	4221	0047	7564
4646	3365	7103	0420	2211	5732	1554	6077
0201	7722	3544	4067	6656	1375	5113	2430

CUBE 1 SQUARE IV

L y	K x	J w	I v	Q u	P t	N s	M r
5047	2564	6702	1221	3410	4133	0355	7676
1400	6123	2345	5666	7057	0574	4712	3231
0113	7430	3656	4375	6544	1067	5201	2722
4554	3077	7211	0732	2103	5420	1646	6365
2636	5315	1173	6450	4261	3742	7524	0007
6271	1752	5534	2017	0626	7305	3163	4440
7762	0241	4027	3504	1335	6616	2470	5153
3325	4606	0460	7143	5772	2251	6037	1514

CUBE 1 SQUARE V

P u	Q t	M s	N r	K y	L x	I w	J v
7335	0616	4470	3513	1762	6241	2027	5504
3772	4251	0037	7514	5325	2606	6460	1143
2261	5742	1524	6007	4636	3315	7173	0450
6626	1305	5163	2440	0271	7752	3534	4017
0544	7067	3201	4722	6113	1430	5656	2375
4103	3420	7646	0365	2554	5077	1211	6732
5410	2133	6355	1676	3047	4564	0702	7221
1057	6574	2712	5231	7400	0123	4345	3666

CUBE 1 SQUARE VI

M t	N u	P r	Q s	I x	J y	K v	L w
1574	6057	2231	5712	7123	0400	4666	3345
5133	2410	6676	1355	3564	4047	0221	7702
4420	3103	7365	0646	2077	5554	1732	6211
0067	7544	3722	4201	6430	1113	5375	2656
6305	1626	5440	2163	0752	7271	3017	4534
2742	5261	1007	6524	4315	3636	7450	0173
3251	4772	0514	7037	5606	2325	6143	1460
7616	0335	4153	3470	1241	6762	2504	5027

CUBE 1 SQUARE VII

J x	I y	L v	K w	N t	M u	Q r	P s
3606	4325	0143	7460	5251	2772	6514	1037
7241	0762	4504	3027	1616	6335	2153	5470
6752	1271	5017	2534	0305	7626	3440	4163
2315	5636	1450	6173	4742	3261	7007	0524
4077	3554	7732	0211	2420	5103	1365	6646
0430	7113	3375	4656	6067	1544	5722	2201
1123	6400	2666	5345	7574	0057	4231	3712
5564	2047	6221	1702	3133	4410	0676	7355

CUBE 1 SQUARE VIII

M w	N v	P y	G x	I s	J r	K u	L t
7356	0675	4413	3130	1701	6222	2044	5567
3711	4232	0054	7577	5346	2665	6403	1120
2202	5721	1547	6064	4655	3376	7110	0433
6645	1366	5100	2423	0212	7731	3557	4074
0527	7004	3262	4741	6170	1453	5635	2316
4160	3443	7625	0306	2537	5014	1272	6751
5473	2150	6336	1615	3024	4507	0761	7242
1034	6517	2771	5252	7463	0140	4326	3605

CUBE 2 SQUARE I

J s	I r	L u	K t	N w	M v	Q y	P x
5024	2507	6761	1242	3473	4150	0336	7615
1463	6140	2326	5605	7034	0517	4771	3252
0170	7453	3635	4316	6527	1004	5262	2741
4537	3014	7272	0751	2160	5443	1625	6306
2655	5376	1110	6433	4202	3721	7547	0064
6212	1731	5557	2074	0645	7366	3100	4423
7701	0222	4044	3567	1356	6675	2413	5130
3346	4665	0403	7120	5711	2232	6054	1577

CUBE 2 SQUARE II

L r	K s	J t	I u	Q v	P w	N x	M y
3665	4346	0120	7403	5232	2711	6577	1054
7222	0701	4567	3044	1675	6356	2130	5413
6731	1212	5074	2557	0366	7645	3423	4100
2376	5655	1433	6110	4721	3202	7064	0547
4014	3537	7751	0272	2443	5160	1306	6625
0453	7170	3316	4635	6004	1527	5741	2262
1140	6463	2605	5326	7517	0034	4252	3771
5507	2024	6242	1761	3150	4473	0615	7336

CUBE 2 SQUARE III

P v	Q w	M x	N y	K r	L s	I t	J u
1517	6034	2252	5771	7140	0463	4605	3326
5150	2473	6615	1336	3507	4024	0242	7761
4443	3160	7306	0625	2014	5537	1751	6272
0004	7527	3741	4262	6453	1170	5316	2635
6366	1645	5423	2100	0731	7212	3074	4557
2721	5202	1064	6547	4376	3655	7433	0110
3232	4711	0577	7054	5665	2346	6120	1403
7675	0356	4130	3413	1222	6701	2567	5044

CUBE 2 SQUARE IV

Q t	P u	N r	M s	L x	K y	J v	I w
2433	5110	1376	6655	4064	3547	7721	0202
6074	1557	5731	2212	0423	7100	3366	4645
7567	0044	4222	3701	1130	6413	2675	5356
3120	4403	0665	7346	5577	2054	6232	1711
5242	2761	6507	1024	3615	4336	0150	7473
1605	6326	2140	5463	7252	0771	4517	3034
0316	7635	3453	4170	6741	1262	5004	2527
4751	3272	7014	0537	2306	5625	1443	6160

CUBE 2 SQUARE V

K x	L y	I v	J w	P t	Q u	M r	N s
0741	7262	3004	4527	6316	1635	5453	2170
4306	3625	7443	0160	2751	5272	1014	6537
5615	2336	6150	1473	3242	4761	0507	7024
1252	6771	2517	5034	7605	0326	4140	3463
7130	0413	4675	3356	1567	6044	2222	5701
3577	4054	0232	7711	5120	2403	6665	1346
2064	5547	1721	6202	4433	3110	7376	0655
6423	1100	5366	2645	0074	7557	3731	4212

CUBE 2 SQUARE VI

I y	J x	K w	L v	M u	N t	P s	Q r
6100	1423	5645	2366	0557	7074	3212	4731
2547	5064	1202	6721	4110	3433	7655	0376
3054	4577	0711	7232	5403	2120	6346	1665
7413	0130	4356	3675	1044	6567	2701	5222
1771	6252	2034	5517	7326	0605	4463	3140
5336	2615	6473	1150	3761	4242	0024	7507
4625	3306	7160	0443	2272	5751	1537	6014
0262	7741	3527	4004	6635	1316	5170	2453

CUBE 2 SQUARE VII

N u	M t	Q s	P r	J y	I x	L w	K v
4272	3751	7537	0014	2625	5306	1160	6443
0635	7316	3170	4453	6262	1741	5527	2004
1326	6605	2463	5140	7771	0252	4034	3517
5761	2242	6024	1507	3336	4615	0473	7150
3403	4120	0346	7665	5054	2577	6711	1232
7044	0567	4701	3222	1413	6130	2356	5675
6557	1074	5212	2731	0100	7423	3645	4366
2110	5433	1655	6376	4547	3064	7202	0721

CUBE 2 SQUARE VIII

N y	M x	Q w	P v	J u	I t	L s	K r
3213	4730	0556	7075	5644	2367	6101	1422
7654	0377	4111	3432	1203	6720	2546	5065
6347	1664	5402	2121	0710	7233	3055	4576
2700	5223	1045	6566	4357	3674	7412	0131
4462	3141	7327	0604	2035	5516	1770	6253
0025	7506	3760	4243	6472	1151	5337	2614
1536	6015	2273	5750	7161	0442	4624	3307
5171	2452	6634	1317	3526	4005	0263	7740

CUBE 3 SQUARE I

I u	J t	K s	L r	M y	N x	P w	Q v
1161	6442	2624	5307	7536	0015	4273	3750
5526	2005	6263	1740	3171	4452	0634	7317
4035	3516	7770	0253	2462	5141	1327	6604
0472	7151	3337	4614	6025	1506	5760	2243
6710	1233	5055	2576	0347	7664	3402	4121
2357	5674	1412	6131	4700	3223	7045	0566
3644	4367	0101	7422	5213	2730	6556	1075
7203	0720	4546	3065	1654	6377	2111	5432

CUBE 3 SQUARE II

K t	L u	I r	J s	P x	Q y	M v	N w
7720	0203	4065	3546	1377	6654	2432	5111
3367	4644	0422	7101	5730	2213	6075	1556
2674	5357	1131	6412	4223	3700	7566	0045
6233	1710	5576	2055	0664	7347	3121	4402
0151	7472	3614	4337	6506	1025	5243	2760
4516	3035	7253	0770	2141	5462	1604	6327
5005	2526	6740	1263	3452	4171	0317	7634
1442	6161	2307	5624	7015	0536	4750	3273

CUBE 3 SQUARE III

Q x	P y	N v	M w	L t	K u	J r	I s
5452	2171	6317	1634	3005	4526	0740	7263
1015	6536	2750	5273	7442	0161	4307	3624
0506	7025	3243	4760	6151	1472	5614	2337
4141	3462	7604	0327	2516	5035	1253	6770
2223	5700	1566	6045	4674	3357	7131	0412
6664	1347	5121	2402	0233	7710	3576	4055
7377	0654	4432	3111	1720	6203	2065	5546
3730	4213	0075	7556	5367	2644	6422	1101

CUBE 3 SQUARE IV

P r	Q s	M t	N u	K v	L w	I x	J v
6576	1055	5233	2710	0121	7402	3664	4347
2131	5412	1674	6357	4566	3045	7223	0700
3422	4101	0367	7644	5075	2556	6730	1213
7065	0546	4720	3203	1432	6111	2377	5654
1307	6624	2442	5161	7750	0273	4015	3536
5740	2263	6005	1526	3317	4634	0452	7171
4253	3770	7516	0035	2604	5327	1141	6462
0614	7337	3151	4472	6243	1760	5506	2025

CUBE 3 SQUARE V

L v	K w	J x	I y	Q r	P s	N t	M u
4604	3327	7141	0462	2253	5770	1516	6035
0243	7760	3506	4025	6614	1337	5151	2472
1750	6273	2015	5536	7307	0624	4442	3161
5317	2634	6452	1171	3740	4263	0005	7526
3075	4556	0730	7213	5422	2101	6367	1644
7432	0111	4377	3654	1065	6546	2720	5203
6121	1402	5664	2347	0576	7055	3233	4710
2566	5045	1223	6700	4131	3412	7674	0357

CUBE 3 SQUARE VI

J w	I v	L y	K x	N x	M r	Q u	P t
2045	5566	1700	6223	4412	3131	7357	0674
6402	1121	5347	2664	0055	7576	3710	4233
7111	0432	4654	3377	1546	6065	2203	5720
3556	4075	0213	7730	5101	2422	6644	1367
5634	2317	6171	1452	3263	4740	0526	7005
1273	6750	2536	5015	7624	0307	4161	3442
0760	7243	3025	4506	6337	1614	5472	2151
4327	3604	7462	0141	2770	5253	1035	6516

CUBE 3 SQUARE VII

M s	N r	P u	Q t	I w	J v	K y	L x
0337	7614	3472	4151	6760	1243	5025	2506
4770	3253	7035	0516	2327	5604	1462	6141
5263	2740	6526	1005	3634	4317	0171	7452
1624	6307	2161	5442	7273	0750	4536	3015
7546	0065	4203	3720	1111	6432	2654	5377
3101	4422	0644	7367	5556	2075	6213	1730
2412	5131	1357	6674	4045	3566	7700	0223
6055	1576	5710	2233	0402	7121	3347	4664

CUBE 3 SQUARE VIII

J t	I u	L r	K s	N x	M y	Q v	P w
4667	3344	7122	0401	2230	5713	1575	6056
0220	7703	3565	4046	6677	1354	5132	2411
1733	6210	2076	5555	7364	0647	4421	3102
5374	2657	6431	1112	3723	4200	0066	7545
3016	4535	0753	7270	5441	2162	6304	1627
7451	0172	4314	3637	1006	6525	2743	5260
6142	1461	5607	2324	0515	7036	3250	4773
2505	5026	1240	6763	4152	3471	7617	0334

CUBE 4 SQUARE I

M x	N y	P v	Q w	I t	J u	K r	L s
6515	1036	5250	2773	0142	7461	3607	4324
2152	5471	1617	6334	4505	3026	7240	0763
3441	4162	0304	7627	5016	2535	6753	1270
7006	0525	4743	3260	1451	6172	2314	5637
1364	6647	2421	5102	7733	0210	4076	3555
5723	2200	6066	1545	3374	4657	0431	7112
4230	3713	7575	0056	2667	5344	1122	6401
0677	7354	3132	4411	6220	1703	5565	2046

CUBE 4 SQUARE II

P y	Q x	M w	N y	K u	L t	I s	J r
0354	7677	3411	4132	6703	1220	5046	2565
4713	3230	7056	0575	2344	5667	1401	6122
5200	2723	6545	1066	3657	4374	0112	7431
1647	6364	2102	5421	7210	0733	4555	3076
7525	0006	4260	3743	1172	6451	2637	5314
3162	4441	0627	7304	5535	2016	6270	1753
2471	5152	1334	6617	4026	3505	7763	0240
6036	1515	5773	2250	0461	7142	3324	4607

CUBE 4 SQUARE III

L u	K t	J s	I r	Q y	P x	N w	M v
2026	5505	1763	6240	4471	3152	7334	0617
6461	1142	5324	2607	0036	7515	3773	4250
7172	0451	4637	3314	1525	6006	2260	5743
3535	4016	0270	7753	5162	2441	6627	1304
5657	2374	6112	1431	3200	4723	0545	7066
1210	6733	2555	5075	7647	0364	4102	3421
0703	7220	3046	4565	6354	1677	5411	2132
4344	3667	7401	0122	2713	5230	1056	6575

CUBE 4 SQUARE IV

K w	L v	I y	J x	P s	Q r	M u	N t
1102	6421	2647	5364	7555	0076	4210	3733
5545	2066	6200	1723	3112	4431	0657	7374
4056	3575	7713	0230	2401	5122	1344	6667
0411	7132	3354	4677	6046	1565	5703	2220
6773	1250	5036	2515	0324	7607	3461	4142
2334	5617	1471	6152	4763	3240	7026	0505
3627	4304	0162	7441	5270	2753	6535	1016
7260	0743	4525	3006	1637	6314	2172	5451

CUBE 4 SQUARE V

Q s	P r	N u	M t	L w	K v	J y	I x
3270	4753	0535	7016	5627	2304	6162	1441
7637	0314	4172	3451	1260	6743	2525	5006
6324	1607	5461	2142	0773	7250	3036	4515
2763	5240	1026	6505	4334	3617	7471	0152
4401	3122	7344	0667	2056	5575	1713	6230
0046	7565	3703	4220	6411	1132	5354	2677
1555	6076	2210	5733	7102	0421	4647	3364
5112	2431	6657	1374	3545	4066	0200	7723

CUBE 4 SQUARE VI

N r	M s	Q t	P u	J v	I w	L x	K y
5431	2112	6374	1657	3066	4545	0723	7200
1076	6555	2733	5210	7421	0102	4364	3647
0565	7046	3220	4703	6132	1411	5677	2354
4122	3401	7667	0344	2575	5056	1230	6713
2240	5763	1505	6026	4617	3334	7152	0471
6607	1324	5142	2461	0250	7773	3515	4036
7314	0637	4451	3172	1743	6260	2006	5525
3753	4270	0016	7535	5304	2627	6441	1162

CUBE 4 SQUARE VII

I v	J w	K x	L y	M r	N s	P t	Q u
7743	0260	4006	3525	1314	6637	2451	5172
3304	4627	0441	7162	5753	2270	6016	1535
2617	5334	1152	6471	4240	3763	7505	0026
6250	1773	5515	2036	0607	7324	3142	4461
0132	7411	3677	4354	6565	1046	5220	2703
4575	3056	7230	0713	2122	5401	1667	6344
5066	2545	6723	1200	3431	4112	0374	7657
1421	6102	2364	5647	7076	0555	4733	3210

CUBE 4 SQUARE VIII

Q u	P t	N s	M r	L y	K x	J w	I v
1135	6416	2670	5353	7562	0041	4227	3704
5572	2051	6237	1714	3125	4406	0660	7343
4061	3542	7724	0207	2436	5115	1373	6650
0426	7105	3363	4640	6071	1552	5734	2217
6744	1267	5001	2522	0313	7630	3456	4175
2303	5620	1446	6165	4754	3277	7011	0532
3610	4333	0155	7476	5247	2764	6502	1021
7257	0774	4512	3031	1600	6323	2145	5466

CUBE 5 SQUARE I

K y	L x	I w	J v	P u	Q t	M s	N r
3247	4764	0502	7021	5610	2333	6155	1476
7600	0323	4145	3466	1257	6774	2512	5031
6313	1630	5456	2175	0744	7267	3001	4522
2754	5277	1011	6532	4303	3620	7446	0165
4436	3115	7373	0650	2061	5542	1724	6207
0071	7552	3734	4217	6426	1105	5363	2640
1562	6041	2227	5704	7135	0416	4670	3353
5125	2406	6660	1343	3572	4051	0237	7714

CUBE 5 SQUARE II

I x	J y	K v	L w	M t	N u	P r	Q s
5406	2125	6343	1660	3051	4572	0714	7237
1041	6562	2704	5227	7416	0135	4353	3670
0552	7071	3217	4734	6105	1426	5640	2363
4115	3436	7650	0373	2542	5061	1207	6724
2277	5754	1532	6011	4620	3303	7165	0446
6630	1313	5175	2456	0267	7744	3522	4001
7323	0600	4466	3145	1774	6257	2031	5512
3764	4247	0021	7502	5333	2610	6476	1155

CUBE 5 SQUARE III

N t	M u	Q r	P s	J x	I y	L v	K w
7774	0257	4031	3512	1323	6600	2466	5145
3333	4610	0476	7155	5764	2247	6021	1502
2620	5303	1165	6446	4277	3754	7532	0011
6267	1744	5522	2001	0630	7313	3175	4456
0105	7426	3640	4363	6552	1071	5217	2734
4542	3061	7207	0724	2115	5436	1650	6373
5051	2572	6714	1237	3406	4125	0343	7660
1416	6135	2355	5670	7041	0562	4704	3227

CUBE 5 SQUARE IV

M v	N w	P x	Q y	I r	J s	K t	L u
4650	3373	7115	0436	2207	5724	1542	6061
0217	7754	3552	4071	6640	1363	5105	2426
1704	6227	2041	5562	7353	0670	4416	3135
5343	2660	6406	1125	3714	4237	0051	7572
3021	4502	0764	7247	5476	2155	6333	1610
7466	0145	4323	3600	1031	6512	2774	5257
6175	1456	5630	2313	0522	7001	3267	4744
2532	5011	1277	6754	4165	3446	7620	0303

CUBE 5 SQUARE V

J r	I s	L t	K u	N v	M w	Q x	P y
6522	1001	5267	2744	0175	7456	3630	4313
2165	5446	1620	6303	4532	3011	7277	0754
3476	4155	0333	7610	5021	2502	6764	1247
7031	0512	4774	3257	1466	6145	2323	5600
1353	6670	2416	5135	7704	0227	4041	3562
5714	2237	6051	1572	3343	4660	0406	7125
4207	3724	7542	0061	2650	5373	1115	6436
0640	7363	3105	4426	6217	1734	5552	2071

CUBE 5 SQUARE VI

L s	K r	J u	I t	Q w	P v	N y	M x
0363	7640	3426	4105	6734	1217	5071	2552
4724	3207	7061	0542	2373	5650	1436	6115
5237	2714	6572	1051	3660	4343	0125	7406
1670	6353	2135	5416	7227	0704	4562	3041
7512	0031	4257	3774	1145	6466	2600	5323
3155	4476	0610	7333	5502	2021	6247	1764
2446	5165	1303	6620	4011	3532	7754	0277
6001	1522	5744	2267	0456	7175	3313	4630

CUBE 5 SQUARE VII

P w	Q v	M y	N x	K s	L r	I u	J t
2011	5532	1754	6277	4446	3165	7303	0620
6456	1175	5313	2630	0001	7522	3744	4267
7145	0466	4600	3323	1512	6031	2257	5774
3502	4021	0247	7764	5155	2476	6610	1333
5660	2343	6125	1406	3237	4714	0572	7051
1227	6704	2562	5041	7670	0353	4135	3416
0734	7217	3071	4552	6363	1640	5426	2105
4373	3650	7436	0115	2724	5207	1061	6542

CUBE 5 SQUARE VIII

L x	K y	J v	I w	Q t	P u	N r	M s
6541	1062	5204	2727	0116	7435	3653	4370
2106	5425	1643	6360	4551	3072	7214	0737
3415	4136	0350	7673	5042	2561	6707	1224
7052	0571	4717	3234	1405	6126	2340	5663
1330	6613	2475	5156	7767	0244	4022	3501
5777	2254	6032	1511	3320	4603	0465	7146
4264	3747	7521	0002	2633	5310	1176	6455
0623	7300	3166	4445	6274	1757	5531	2012

CUBE 6 SQUARE I

P t	Q u	M r	N s	K x	L y	I v	J w
4633	3310	7176	0455	2264	5747	1521	6002
0274	7757	3531	4012	6623	1300	5166	2445
1767	6244	2022	5501	7330	0613	4475	3156
5320	2603	6465	1146	3777	4254	0032	7511
3042	4561	0707	7224	5415	2136	6350	1673
7405	0126	4340	3663	1052	6571	2717	5234
6116	1435	5653	2370	0541	7062	3204	4727
2551	5072	1214	6737	4106	3425	7643	0360

CUBE 6 SQUARE II

M u	N t	P s	Q r	I y	J x	K w	L v
2072	5551	1737	6214	4425	3106	7360	0643
6435	1116	5370	2653	0062	7541	3727	4204
7126	0405	4663	3340	1571	6052	2234	5717
3561	4042	0224	7707	5136	2415	6673	1350
5603	2320	6146	1465	3254	4777	0511	7032
1244	6767	2501	5022	7613	0330	4156	3475
0757	7274	3012	4531	6300	1623	5445	2166
4310	3633	7455	0176	2747	5264	1002	6521

CUBE 6 SQUARE III

J y	I x	L w	K v	N u	M t	Q s	P r
0300	7623	3445	4166	6757	1274	5012	2531
4747	3264	7002	0521	2310	5633	1455	6176
5254	2777	6511	1032	3603	4320	0146	7465
1613	6330	2156	5475	7244	0767	4501	3022
7571	0052	4234	3717	1126	6405	2663	5340
3136	4415	0673	7350	5561	2042	6224	1707
2424	5106	1360	6643	4072	3551	7737	0214
6062	1541	5727	2204	0435	7116	3370	4653

CUBE 6 SQUARE IV

I s	J r	K u	L t	M w	N v	P y	Q x
3224	4707	0561	7042	5673	2350	6136	1415
7663	0340	4126	3405	1234	6717	2571	5052
6370	1653	5435	2116	0727	7204	3062	4041
2737	5214	1072	6551	4360	3643	7425	0106
4455	3176	7310	0633	2002	5521	1747	6264
0012	7531	3757	4274	6445	1166	5300	2623
1501	6022	2244	5767	7156	0475	4613	3330
5146	2465	6603	1320	3511	4032	0254	7777

CUBE 6 SQUARE V

N w	M v	Q y	P x	J s	I r	L u	K t
1156	6475	2613	5330	7501	0022	4244	3767
5511	2032	6254	1777	3146	4465	0603	7320
4002	3521	7747	0264	2455	5176	1310	6633
0445	7166	3300	4623	6012	1531	5757	2274
6727	1204	5062	2541	0370	7653	3435	4116
2360	5643	1425	6106	4737	3214	7072	0551
3673	4350	0136	7415	5224	2707	6561	1042
7234	0717	4571	3052	1663	6340	2126	5405

CUBE 6 SQUARE VI

Q v	P w	N x	M y	L r	K s	J t	I u
7717	0234	4052	3571	1340	6663	2405	5126
3350	4673	0415	7136	5707	2224	6042	1561
2643	5360	1106	6425	4214	3737	7551	0072
6204	1727	5541	2062	0653	7370	3116	4435
0166	7445	3623	4300	6531	1012	5274	2757
4521	3002	7264	0747	2176	5455	1633	6310
5032	2511	6777	1254	3465	4146	0320	7603
1475	6156	2330	5613	7022	0501	4767	3244

CUBE 6 SQUARE VII

K r	L s	I t	J r	P v	Q w	M x	N y
5465	2146	6320	1603	3032	4511	0777	7254
1022	6501	2767	5244	7475	0156	4330	3613
0531	7012	3274	4757	6166	1445	5623	2300
4176	3455	7633	0310	2521	5002	1264	6747
2214	5737	1551	6072	4643	3360	7106	0425
6653	1370	5116	2435	0204	7727	3541	4062
7340	0663	4405	3126	1717	6234	2052	5571
3707	4224	0042	7561	5350	2673	6415	1136

CUBE 6 SQUARE VIII

K v	L w	I x	J y	P r	Q s	M t	N u
2404	5127	1341	6662	4053	3570	7716	0235
6043	1560	5706	2225	0414	7137	3351	4672
7550	0073	4215	3736	1107	6424	2642	5361
3117	4434	0652	7371	5540	2063	6205	1726
5275	2756	6530	1013	3622	4301	0167	7444
1632	6311	2177	5454	7265	0746	4520	3003
0321	7602	3464	4147	6776	1255	5033	2510
4766	3245	7023	0500	2331	5612	1474	6157

CUBE 7 SQUARE I

Q r	P s	N t	M u	L v	K w	J x	I y
0776	7255	3033	4510	6321	1602	5464	2147
4331	3612	7474	0157	2766	5245	1023	6500
5622	2301	6167	1444	3275	4756	0530	7013
1265	6746	2520	5003	7632	0311	4177	3454
7107	0424	4642	3361	1550	6073	2215	5736
3540	4063	0205	7726	5117	2434	6652	1371
2053	5570	1716	6235	4404	3127	7341	0662
6414	1137	5351	2672	0043	7560	3706	4225

CUBE 7 SQUARE II

N s	M r	Q u	P t	J w	I v	L y	K x
6137	1414	5672	2351	0560	7043	3225	4706
2570	5053	1235	6716	4127	3404	7662	0341
3063	4540	0726	7205	5434	2117	6371	1652
7424	0107	4361	3642	1073	6550	2736	5215
1746	6265	2003	5520	7311	0632	4454	3177
5301	2622	6444	1167	3756	4275	0013	7530
4612	3331	7157	0474	2245	5766	1500	6023
0255	7776	3510	4033	6602	1321	5147	2464

CUBE 7 SQUARE III

I w	J v	K y	L x	M s	N r	P u	Q t
4245	3766	7500	0023	2612	5331	1157	6474
0602	7321	3147	4464	6255	1776	5510	2033
1311	6632	2454	5177	7746	0265	4003	3520
5756	2275	6013	1530	3301	4622	0444	7167
3434	4117	0371	7652	5063	2540	6726	1205
7073	0550	4736	3215	1424	6107	2361	5642
6560	1043	5225	2706	0137	7414	3672	4351
2127	5404	1662	6341	4570	3053	7235	0716

CUBE 7 SQUARE IV

J u	I t	L s	K r	N y	M x	Q w	P v
7361	0642	4424	3107	1736	6215	2073	5550
3726	4205	0063	7540	5371	2652	6434	1117
2235	5716	1570	6053	4662	3341	7127	0404
6672	1351	5137	2414	0225	7706	3560	4043
0510	7033	3255	4776	6147	1464	5602	2321
4157	3474	7612	0331	2500	5023	1245	6766
5444	2167	6301	1622	3013	4530	0756	7275
1003	6520	2746	5265	7454	0177	4311	3632

CUBE 7 SQUARE V

M y	N x	P w	Q v	I u	J t	K s	L r
5013	2530	6756	1275	3444	4167	0301	7622
1454	6177	2311	5632	7003	0520	4746	3265
0147	7464	3602	4321	6510	1033	5255	2776
4500	3023	7245	0766	2157	5474	1612	6331
2662	5341	1127	6404	4235	3716	7570	0053
6225	1706	5560	2043	0672	7351	3137	4414
7736	0215	4073	3550	1361	6642	2424	5107
3371	4652	0434	7117	5726	2205	6063	1540

CUBE 7 SQUARE VI

P x	Q y	M v	N w	K t	L u	I r	J s
3652	4371	0117	7434	5205	2726	6540	1063
7215	0736	4550	3073	1642	6361	2107	5424
6706	1225	5043	2560	0351	7672	3414	4137
2341	5662	1404	6127	4716	3235	7053	0570
4023	3500	7766	0245	2474	5157	1331	6612
0464	7147	3521	4602	6033	1510	5776	2255
1177	6454	2632	5311	7520	0003	4265	3746
5530	2013	6275	1756	3167	4444	0622	7301

CUBE 7 SQUARE VII

L t	K u	J r	I s	Q x	P y	N v	M w
1520	6003	2265	5746	7177	0454	4632	3311
5167	2444	6622	1301	3530	4013	0275	7756
4474	3157	7331	0612	2023	5500	1766	6245
0033	7510	3776	4255	6464	1147	5321	2602
6351	1672	5414	2137	0706	7225	3043	4560
2716	5235	1053	6570	4341	3662	7404	0127
3205	4726	0540	7063	5652	2371	6117	1434
7642	0361	4107	3424	1215	6736	2550	5073

CUBE 7 SQUARE VIII

P s	Q r	M u	N t	K w	L v	I y	J x
5070	2553	6735	1216	3427	4104	0362	7641
1437	6114	2372	5651	7060	0543	4725	3206
0124	7407	3661	4342	6573	1050	5236	2715
4563	3040	7226	0705	2134	5417	1671	6352
2601	5322	1144	6467	4256	3775	7513	0030
6246	1765	5503	2020	0611	7332	3154	4477
7755	0276	4010	3533	1302	6621	2447	5164
3312	4631	0547	7174	5745	2266	6000	1523

CUBE 8 SQUARE I

L w	K v	J y	I x	Q s	P r	N u	M t
7302	0621	4447	3164	1755	6276	2010	5533
3745	4266	0000	7523	5312	2631	6457	1174
2256	5775	1513	6030	4601	3322	7144	0467
6611	1332	5154	2477	0246	7765	3503	4020
0573	7050	3236	4715	6124	1407	5661	2342
4134	3417	7671	0352	2563	5040	1226	6705
5427	2104	6362	1641	3070	4553	0735	7216
1060	6543	2725	5206	7437	0114	4372	3651

CUBE 8 SQUARE II

J v	I w	L x	K y	N r	M s	Q t	P u
1543	6060	2206	5725	7114	0437	4651	3372
5104	2427	6641	1362	3553	4070	0216	7735
4417	3134	7352	0671	2040	5563	1705	6226
0050	7573	3715	4236	6407	1124	5342	2661
6332	1611	5477	2154	0765	7246	3020	4503
2775	5256	1030	6513	4322	3601	7467	0144
3266	4745	0523	7000	5631	2312	6174	1457
7621	0302	4164	3447	1276	6755	2533	5010

CUBE 8 SQUARE III

M r	N s	P t	Q u	I v	J w	K x	L y
3631	4312	0174	7457	5266	2745	6523	1000
7276	0755	4533	3010	1621	6302	2164	5447
6765	1246	5020	2503	0332	7611	3477	4154
2322	5601	1467	6144	4775	3256	7030	0513
4040	3563	7705	0226	2417	5134	1352	6671
0407	7124	3342	4661	6050	1573	5715	2236
1114	6437	2651	5372	7543	0060	4206	3725
5553	2070	6216	1735	3104	4427	0641	7362

CUBE 8 SQUARE IV

N x	M y	Q v	P w	J t	I u	L r	K s
0715	7236	3050	4573	6342	1661	5407	2124
4352	3671	7417	0134	2705	5226	1040	6563
5641	2362	6104	1427	3216	4735	0553	7070
1206	6725	2543	5060	7651	0372	4114	3437
7164	0447	4621	3302	1533	6010	2276	5755
3523	4000	0266	7745	5174	2457	6631	1312
2030	5513	1775	6256	4467	3144	7322	0601
6477	1154	5332	2611	0020	7503	3765	4246

CUBE 8 SQUARE V

I t	J u	K r	L s	M x	N y	P v	Q w
2467	5144	1322	6601	4030	3513	7775	0256
6020	1503	5765	2246	0477	7154	3332	4611
7533	0010	4276	3755	1164	6447	2621	5302
3174	4457	0631	7312	5523	2000	6266	1745
5216	2735	6553	1070	3641	4362	0104	7427
1651	6372	2114	5437	7206	0725	4543	3060
0342	7661	3407	4124	6715	1236	5050	2573
4705	3226	7040	0563	2352	5671	1417	6134

CUBE 8 SQUARE VI

K u	L t	I s	J r	P y	Q x	M w	N v
4226	3705	7563	0040	2671	5352	1134	6417
0661	7342	3124	4407	6236	1715	5573	2050
1372	6651	2437	5114	7725	0206	4060	3543
5735	2216	6070	1553	3362	4641	0427	7104
3457	4174	0312	7631	5000	2523	6745	1266
7010	0533	4755	3276	1447	6164	2302	5621
6503	1020	5246	2765	0154	7477	3611	4332
2144	5467	1601	6322	4513	3030	7256	0775

CUBE 8 SQUARE VII

Q y	P x	N w	M v	L u	K t	J s	I r
6154	1477	5611	2332	0503	7020	3246	4765
2513	5030	1256	6775	4144	3467	7601	0322
3000	4523	0745	7266	5457	2174	6312	1631
7447	0164	4302	3621	1010	6533	2755	5276
1725	6206	2060	5543	7372	0651	4437	3114
5362	2641	6427	1104	3735	4216	0070	7553
4671	3352	7134	0417	2226	5705	1563	6040
0236	7715	3573	4050	6661	1342	5124	2407

CUBE 8 SQUARE VIII

A NOTE ON FUNDAMENTAL PROPERTIES OF RECURRING SERIES

Joseph Arkin, David C. Arney, Frank R. Giordano and Rickey A. Kolb

In this paper we consider certain recurring series and find some new fundamental properties in these series.

For completeness, we list some results of Arkin [2], [3], where the following identities were developed. We put

$$(c_0 + c_1 x + c_2 x^2 + \cdots + c_n x^n + \cdots)^k = c_0^{(k)} + c_1^{(k)} x + c_1^{(k)} x^2 + \cdots + c_n^{(k)} x^n + \cdots, \qquad (1.1)$$

where the c_n are integers and $k \geq 1$. We define $c_n^{(k)}$ by means of (1.1) for all integral k. (The series considered are purely formal power series, and the question of their convergence is irrelevant). We then have:

$$c_n^{(k)} \equiv 0 \ (\mathrm{mod}\ k/(n,k))\ n = 1, 2, 3, \ldots, \qquad (1.2)$$

and when $c_0 = 1$,

$$c_n^{(-k)} \equiv 0 \ (\mathrm{mod}\ k/(n,k)). \qquad (1.3)$$

The congruences above then led to the following supplementary results [3].

If for a prime P and for some integers t and $r(r \geq 1)$, the congruence

33

G. E. Bergum et al. (eds.), Applications of Fibonacci Numbers Volume 4, 33–36.
© 1991 Kluwer Academic Publishers.

$c_v^{(d)} \equiv 0 \pmod{P^{r-s}}$, $(v=mP+t)$, holds for $m = 0, 1, 2, \ldots$ and for $0 \le s \le r$, then

$$c_v^{(w+d)} \equiv 0 \pmod{P^{r-s}} \ (w=P^r a) \text{ for any integer a.} \tag{1.4}$$

If $c_0 = 1$, and if $c_v^{(\pm d)} \equiv 0 \pmod{P^{r-s}}$, P being a prime, and t is any integer for which the congruence holds, then

$$c_v^{(\pm w \pm d)} \equiv 0 \pmod{P^{r-s}}.$$

In this paper, we put

$$(f(x))^k = C(k,0) + C(k,1)x + C(k,2)x^2 + \ldots + C(k,n)x^n + \ldots \tag{2.1}$$

such that

$$f(x) = \sum_{v=0}^{\infty} (2av+b)x^{v(av+b)}. \tag{2.2}$$

Theorem: Let $P = 4am+3$ be a prime. Then

$$C(2,n) \equiv 0 \pmod{P^2} \tag{2.3}$$

for $n \equiv -2m-d \pmod{P}$ where d is an integer and $d = (b^2+3)/2a$. Further, since b is odd, b $= 2w+1$ makes $2(w^2 + w + 1) = ad$.

Proof: Squaring the $f(x)$ in (2.2), we get

$$(f(x))^2 = \sum_{v=0}^{\infty} \sum_{t=0}^{\infty} (2av+b)(2at+b)x^{v(av+b)+t(at+b)}.$$

Now, let $n=uP-2m-d$; then $C(2,n)$ is the coefficient of x^{uP} in the expansion of x^{d+2m} $(f(x))^2$. Hence,

$$C(2,n) = \sum \sum (2av+b)(2at+b), \tag{2.4}$$

where the sum extends over all nonnegative v, t satisfying $v(av+b)+t(at+b)+2m+d=uP$.

The condition under the summation can be written, after multiplication by 4a,

$$4a(av+b)+4at(at+b)+8am+4ad=4auP,$$

so that

$$4a(av+b)+4at(at+b)+8am+2b^2+6=4auP,$$

which leads to

$$(2av+b)^2 + (2at+b)^2 = (4au-2)P. \tag{2.5}$$

Since -1 is not a quadratic residue of P, P can divide the left side (2.5) only by dividing each term. Now the Theorem follows from (2.4).

Of the many special cases possible using the Theorem above, we state some results of Arkin, Jacobi, and Ramanujan. In the Theorem above, let m=m/2, v=s/2, a=2, b=1, d=1, and w=0, so that (2.2) becomes

$$f(x)= \sum_{s=0}^{\infty} (2s+1)x^{s(s+1)/2}. \tag{3.1}$$

Let

$$g(x) = \prod_{n=1}^{\infty}(1-x^n) \tag{3.2}$$

and

$$(g(x))^k = \sum_{n=0}^{\infty} p^{(k)}(n)x^n. \tag{3.3}$$

A well-known identity of Jacobi [4] is

$$(g(x))^3 = \sum_{s=0}^{\infty} (-1)^s(2s+1)x^{s(s+1)/2} \tag{3.4}$$

which is of course (after including the (-1)) identical to (3.1).

Arkin [1], has proved that, if $P = 4m+3$ is a prime, then

$$p^{(6)}(n) \equiv 0 \pmod{P^2} \text{ for } n \equiv -1-m \pmod{P}, \tag{4.1}$$

and among other things that

$$p^{(-1)}(7m+5) \equiv 0 \pmod{7}. \tag{4.2}$$

The above (4.2) is a well-known congruence of Ramanujan [5], but proved in a different way.

ACKNOWLEDGEMENTS

The authors wish to thank the referee for his very valuable help in setting up this paper and we thank the editor for his patience and interest.

REFERENCES

[1] Arkin, J. "An Extension of a Theorem of Ramanujan." *American Mathematical Monthly,* 73 (1966): pp. 1087-1090.

[2] Arkin, J. "Congruences for the kth Power of a Power Series." *American Mathematical Monthly, 71* (1964): pp. 899-900.

[3] Arkin, J. "Congruences for the kth Power of a Power Series (II)." *American Mathematical Monthly, 72* (1965): pp. 390-393.

[4] Jacobi, C. G. J. *Fundamenta Nova,* (1829):, 66, (7); Werke, 1, p. 237.

[5] Ramanujan, S. Proc. Cambridge Phil. Soc., 19 (1919): pp. 207-210.

PERIOD PATTERNS OF CERTAIN SECOND-ORDER LINEAR RECURRENCES MODULO A PRIME

David Banks and Lawrence Somer

1. INTRODUCTION

Throughout this paper, p will denote a fixed prime. Let a and b be integers and let the Lucas sequence $u(a,b)$ denote the second-order linear recurrence satisfying

$$u_{n+2} = au_{n+1} + bu_n \tag{1}$$

with initial terms $u_0 = 0$, $u_1 = 1$. Let $\mu(a,b)$ denote the period of $u(a,b)$ modulo p. It is known (see [2, pages 344-345]) that if $b \not\equiv 0 \ (mod \ p)$, then $u(a,b)$ is purely periodic modulo p. If $b \not\equiv 0$ $(mod \ p)$, define $ord(b)$ to be the exponent of b modulo p. It was shown by Somer in [5, Theorem 11] and [6, Theorem 4.5.1] that if $ord(-b) = ord(-b')$, then the set of periods modulo p appearing among the Lucas sequences $u(a,b)$ is the same as the set of periods modulo p appearing among the Lucas sequences $u(a',b')$, where both a and a' vary over all the residues modulo p. However, it was not shown in [5] and [6] that the number of recurrences $u(a,b)$ modulo p having a given period equals the number of recurrences $u(a',b')$ modulo p having that same period. This will be shown in Theorem 1 by means of a period-preserving map between the recurrences modulo p.

Before stating Theorem 1, we will need the following lemma.

Lemma 1: Let b and b' be non-zero residues modulo p such that $ord(-b) = ord(-b')$. Then b'/b is a quadratic residue modulo p. Let r be such that $b' \equiv r^2 b (mod \ p)$ with $ord(r)$ minimal. Then

$$ord(r) \mid ord(-b).$$

37

G. E. Bergum et al. (eds.), Applications of Fibonacci Numbers Volume 4, 37–40.
© 1991 *Kluwer Academic Publishers.*

We now state the main result.

Theorem 1: Let b, b', and r satisfy the conditions of Lemma 1. Then there exists a one-to-one period-preserving map f between the set of all recurrences $u(a,b)$ *modulo* p and the set of all recurrences $u(a',b')$ *modulo* p, given by

$$f(u(a,b)) = u(ra,b'),$$

where a varies over all the residues *modulo* p.

2. PRELIMINARIES

Associated with the Lucas sequence $u(a,b)$ is the characteristic polynomial

$$x^2 - ax - b$$

with characteristic roots r_1 and r_2. Also associated with $u(a,b)$ is the companion matrix

$$M = \begin{pmatrix} 0 & 1 \\ b & a \end{pmatrix}.$$

Note that the determinant of M is $-b$ and the eigenvalues of M are the characteristic roots r_1 and r_2. Given the vector (u_n, u_{n+1}), it follows from the recursion relation defining $u(a,b)$ that

$$M(u_n, u_{n+1})^{\mathrm{T}} = (u_{n+1}, u_{n+2})^{\mathrm{T}}. \tag{2}$$

Applying (2) n times to the vector $(u_0, u_1) = (0, 1)$, one sees that

$$M^n(0, 1)^{\mathrm{T}} = (u_n, u_{n+1})^{\mathrm{T}}. \tag{3}$$

It further follows by induction that

$$M^n = \begin{pmatrix} bu_{n-1} & u_n \\ bu_n & u_{n+1} \end{pmatrix}. \tag{4}$$

It is known (see [4, page 620]) and follows from (4) that

$$\mu(a,b) = ord(M), \tag{5}$$

where $ord(M)$ denotes the order of M modulo p. It is also known (see [7, Theorem 3]) that

$$ord(-b) \mid \mu(a,b). \tag{6}$$

3. PROOFS OF THE LEMMA AND THE MAIN THEOREM

Proof of Lemma 1: Since $ord(-b) = ord(-b')$, it follows that $-b$ and $-b'$ are both quadratic residues modulo p if $ord(-b) \mid (p-1)/2$ and are both quadratic non-residues modulo p if $ord(-b) \nmid (p-1)/2$. Thus, b'/b is a quadratic residue modulo p.

Let $t = ord(-b)$. It suffices to prove that

$$r^t \equiv (mod\ p). \tag{7}$$

We have that

$$r^{2t} \equiv [(-b')/(-b)]^t \equiv 1 \ (mod\ p).$$

Hence, if t is odd, then

$$r^t \equiv -(-r)^t \equiv \pm 1 \ (mod\ p)$$

and

$$r^t \equiv 1 \ (mod\ p)$$

by the definition of r. Now suppose that t is even. Since

$$ord(-b') = ord(-b) = t,$$

we have that

$$-b' \equiv (-b)^{2n+1} \ (mod\ p),$$

where $1 \leq 2n+1 \leq t$. Hence,

$$r^t \equiv (r^2)^{t/2} \equiv [(-b)^{2n+1}/(-b)]^{t/2} \equiv (-b)^{nt} \equiv 1 \ (mod\ p).$$

Proof of Theorem 1: Clearly, the map f is one-to-one. It now remains to prove that $\mu(ra,b') = \mu(a,b)$. Since $ord(-b)$ divides both $\mu(ra,b')$ and $\mu(a,b)$ by (6), we have by Lemma 1 that

$$r^{\mu(ra,b')} \equiv r^{\mu(a,b)} \equiv 1 \ (mod\ p).$$

Consequently, since

$$\begin{pmatrix} 0 & 1 \\ r^2 b & ra \end{pmatrix} \equiv r \begin{pmatrix} 0 & 1 \\ b & a \end{pmatrix} \begin{pmatrix} r & 0 \\ 0 & 1 \end{pmatrix} (mod\ p),$$

then

$$\mu(ra, b') = ord \begin{pmatrix} 0 & 1 \\ r^2 b & ra \end{pmatrix} \equiv ord \begin{pmatrix} 0 & 1 \\ b & a \end{pmatrix} = \mu(a, b).$$

4. ACKNOWLEDGMENT

We wish to thank the referee for his helpful suggestions modifying the statement of Theorem 1, shortening the proof of Theorem 1, and making the paper more concise.

REFERENCES

[1]   Banks, D. "Bounds on the Periods of Second-Order Linear Recursions Over Cyclic Groups." Master's Thesis. University of North Carolina at Chapel Hill, 1987.

[2]   Carmichael, R. D. "On Sequences of Integers Defined by Recurrence Relations." *Quart. J. Pure Appl. Math. 48* (1920): pp. 343-372.

[3]   Robinson, D. W. "The Fibonacci Matrix Modulo *m*." *The Fibonacci Quarterly 1*, No. 2 (1963): pp. 29-37.

[4]   Robinson, D. W. "A Note on Linear Recurrent Sequences Modulo *m*." *Amer. Math. Monthly 73* (1966): pp. 619-621.

[5]   Somer, L. "Possible Periods of Primary Fibonacci-Like Sequences with Respect to a Fixed Odd Prime." *The Fibonacci Quarterly 20*, No. 4 (1982): pp. 311-333.

[6]   Somer, L. "The Divisibility and Modular Properties of *k*th-Order Linear Recurrences Over the Ring of Integers of an Algebraic Number Field With Respect to Prime Ideals." Ph.D. Thesis. The University of Illinois at Urbana-Champaign, 1985.

[7]   Wyler, O. "On Second-Order Recurrences." *Amer. Math. Monthly 72* (1965): pp. 500-506.

NEARLY ISOSCELES TRIANGLES WHERE THE VERTEX ANGLE IS A MULTIPLE OF THE BASE ANGLE

Marjorie Bicknell-Johnson

1. $\alpha - n\alpha$ TRIANGLES

Let a triangle in which the vertex angle is a positive integral multiple n of a base angle be called an $\alpha - n\alpha$ triangle. We find integral solutions for the lengths of the sides by a recursive method. We note that, for any particular α for which there is an integral $\alpha - n\alpha$ triangle, $\cos \alpha$ must be a rational number by the law of cosines.

Draw $\triangle QCA$ such that $\angle C = n\alpha$, $\angle A = \alpha$. Draw CB so that $\angle QCB = \alpha$. Label $AC = b$, $AB = c$, $BC = a$, $QB = q$, and $QC = p$. Then $\triangle ABC$ is an $\alpha - (n-1)\alpha$ triangle.

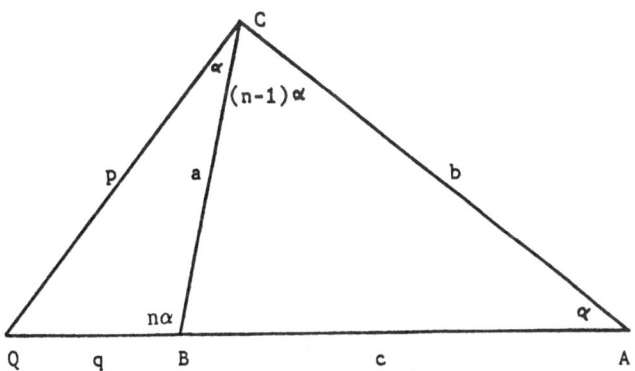

Figure 1

G. E. Bergum et al. (eds.), Applications of Fibonacci Numbers Volume 4, 41–50.

Since $\triangle QBC$ is similar to $\triangle QCA$, $p/q = b/a = (q+c)/p$, yielding

$$p = qb/a \quad \text{and} \quad p^2 = q(q+c).$$

Solving these together gives us

$$q = \frac{ca^2}{b^2 - a^2}, \qquad p = \frac{abc}{b^2 - a^2}, \qquad \text{and} \qquad q+c = \frac{cb^2}{b^2 - a^2}.$$

Inspecting these equations, note that p, b, and (q+c) are sides of an $\alpha - n\alpha$ triangle. Multiply by $(b^2 - a^2)$ to write simplified expressions for the sides in a similar $\alpha - n\alpha$ triangle and divide by the common factor b, and the integral sides a'_n, b'_n, c'_n of an $\alpha - n\alpha$ triangle where c'_n is the side opposite $n\alpha$ and a'_n is the side opposite α, are given by

$$a'_n = a_{n-1}c_{n-1}, \qquad b'_n = b^2_{n-1} - a^2_{n-1}, \qquad c'_n = c_{n-1}b_{n-1} \qquad (1.1)$$

so we have a recursive solution, except for possible common factors. In actually using (1.1), a'_n, b'_n, and c'_n will develop the common factor b_{n-2}, $n > 2$, which will have to be removed to keep them relatively prime, making primitive solutions. For now, we can quickly write solutions for small values of n successively by removing common factors by inspection. The final recursive solution will be

$$a_n = \frac{a_{n-1}c_{n-1}}{b_{n-2}}, \qquad b_n = \frac{b^2_{n-1} - a^2_{n-1}}{b_{n-2}}, \qquad c_n = \frac{c_{n-1}b_{n-1}}{b_{n-2}} \qquad (1.2)$$

where a_n, b_n, and c_n are sides of an $\alpha - n\alpha$ triangle with c_n opposite $n\alpha$ and a_n opposite angle α, $(a_n, b_n, c_n) = 1$, which is proved in Section 4.

Now, in an $\alpha - n\alpha$ triangle, we must have $(n+1)\alpha < 180°$ because we have to leave room for the third angle. We can easily obtain $\cos \alpha = s/2r$ from the $\alpha - \alpha$ triangle, $a_1 = c_1 = r$, $b_1 = s$ since $(r, s) = 1$. Thus, $180°/(n+1) > \alpha > 0$, so writing the cosines of these three quantities gives us $\cos \dfrac{180°}{n+1} < \cos \alpha = \dfrac{s}{2r} < 1$, or, multiplying through by 2r,

$$2r \cos \frac{180°}{n+1} < s < 2r \qquad (1.3)$$

Now we examine some special cases, using only (1.1) and (1.3).

2. $\alpha - 2\alpha$, $\alpha - 3\alpha$, $\alpha - 4\alpha$, and $\alpha - 5\alpha$ TRIANGLES

We write primitive integral solutions for the sides of $\alpha - 2\alpha$ triangles as

$$a = r^2, \qquad b = s^2 - r^2, \qquad c = rs \qquad (2.1)$$

for integers r and s where $(r, s) = 1$, $1 < r < s < 2r$, $\alpha < 60°$, by substituting $a_1 = c_1 = r$ and $b_1 = s$ into (1.1) and $n = 2$ into (1.3), agreeing with the results in [3], [5]. All other $\alpha - 2\alpha$

triangles are similar to these by an integral scale factor. Applying (2.1), there are 192 values of a, b and c in primitive $\alpha - 2\alpha$ triangles where all three lengths are less than 1000. The first values for $r \leq 8$ are given below. Several cases are "nearly isosceles", which we will investigate in the next section.

Vertex Angle 2α

r	s	$a = r^2$	$b = s^2 - r^2$	$c = rs$
2	3	4	5	6
3	4	9	7	12
3	5	9	16	15
4	5	16	9	20
4	7	16	33	28
5	6	25	11	30
5	7	25	24	35
5	8	25	39	40
5	9	25	56	45
6	7	36	13	42
6	11	36	85	66
7	8	49	15	56
7	9	49	32	63
7	10	49	51	70
7	11	49	72	77
7	12	49	95	84
7	13	49	120	91
8	9	64	17	72
8	11	64	57	88
8	13	64	105	104
8	15	64	161	120

Now return to (1.1) and substitute the results of (2.1). We can write $a'_3 = a_2 c_2 = r^3 s$, $b'_3 = (b_2)^2 - (a_2)^2 = (s^2 - r^2)^2 - (r^2)^2 = s^4 - 2s^2 r^2$, and $c'_3 = c_2 b_2 = rs(s^2 - r^2)$. Factor out the common factor $b_1 = s$, and write the integral sides a_3, b_3, and c_3 of the $\alpha - 3\alpha$ triangles where c_3 is opposite 3α and a_3 is opposite α as

$$a_3 = r^3, \qquad b_3 = s^3 - 2sr^2, \qquad c_3 = r(s^2 - r^2). \tag{2.2}$$

In the table below, we list the sides of primitive $\alpha - 3\alpha$ triangles for $2 \leq r \leq 8$, where $(r, s) = 1$, and $\sqrt{2}\, r < s < 2r$ from Eq. (1.3). Now, there are 12 solutions where a, b, and c are all less than 1000 (the twelfth occurs when $r = 9$, $s = 13$), and one triangle with sides 125, 279, 280 which is nearly isosceles. As one last observation, the smallest $\alpha - 3\alpha$ triangle, 8, 3, 10, and the smallest $\alpha - 2\alpha$ triangle, 4, 5, 6, have the same area, which we will investigate in Section 4.

Vertex Angle 3α

r	s	$a = r^3$	$b = s^3 - 2sr^2$	$c = r(s^2 - r^2)$
2	3	8	3	10
3	5	27	35	48
4	7	64	119	132
5	8	125	112	195
5	9	125	279	280
6	11	216	539	510
7	10	343	20	357
7	11	343	253	504
7	12	343	552	665
7	13	343	923	840
8	13	512	533	840
8	15	512	1455	1288

Similarly, by substituting formulas for the sides of $\alpha - 3\alpha$ triangles given in (2.2) into (1.1) and dividing by the common factor $b_2 = s^2 - r^2$, the sides of the $\alpha - 4\alpha$ triangles are

$$a_4 = r^4, \qquad b_4 = s^4 - 3s^2r^2 + r^4, \qquad c_4 = r(s^3 - 2sr^2). \tag{2.3}$$

We compute the primitive $\alpha - 4\alpha$ triangles below for $3 \leq r \leq 8$, where we recall that, by (1.3), $2r \cos 36° < s < 2r$, and $2 \cos 36° = (1 + \sqrt{5})/2 = \Phi$, so $\Phi r < s < 2r$.

Vertex Angle 4α

r	s	$a = r^4$	$b = s^4 - 3s^2r^2 + r^4$	$c = r(s^3 - 2sr^2)$
3	5	81	31	105
4	7	256	305	476
5	9	625	1111	1395
6	11	1296	2869	3234
7	12	2401	1969	3864
7	13	2401	6119	6461
8	13	4096	209	4264
8	15	4096	11521	11640

Now there are only two primitive solutions where all sides are less than 1000. For $\alpha - 5\alpha$ triangles, there are only two primitive solutions for which all three sides are less than 10,000, and none for all three sides less than 1000. The first three solutions are given below, letting $(r, s) = 1$ and $\sqrt{3} \, r < s < 2r$.

Vertex Angle 5α

r	s	$a = r^5$	$b = s^5 - 4s^3r^2 + 3sr^4$	$c = r(s^4 - 3s^2r^2 + r^4)$
4	7	1024	231	1220
5	9	3125	3024	5555
6	11	7776	12155	17214

Now we turn our attention to nearly isosceles triangles.

3. NEARLY ISOSCELES $\alpha - n\alpha$ TRIANGLES, n = 2, 3, 4

Notice that, if $\triangle ABC$ is an $\alpha - 2\alpha$ triangle, $\triangle ABC$ cannot be isosceles and have all three sides with integral length. There are two ways that $\triangle ABC$ could be isosceles. If $\angle B = \angle A = \alpha$, then $a = b$ and $\triangle ABC$ is a 45-45-90 triangle where $\cos \alpha = \sqrt{2}/2$. If $\angle B = \angle C = 2\alpha$, then $b = c$ and $\triangle ABC$ is a 36-72-72 "golden triangle" [1], where $\cos \alpha = (1 + \sqrt{5})/4$. Note that $\cos \alpha$ is half of the golden ratio $\Phi = (1 + \sqrt{5})/2$ and that $b/a = \Phi$.

Let an $\alpha - 2\alpha$ triangle where the lengths of a pair of sides differ by one be called a "nearly isosceles" triangle. There are two cases, $b - c = \pm 1$ and $a - b = \pm 1$.

Consider the case $b - c \pm 1$. Then $b - c = \pm 1 = s^2 - rs - r^2$, which equation is known to have solutions $r = F_n$ and $s = F_{n+1}$. Then $\cos \alpha = F_{n+1}/2F_n$, and the limiting value of $\cos \alpha$ as $n \to \infty$ is $\Phi/2$, giving us the golden triangle with angles 36-72-72. The first eight values of the Fibonacci case $b - c = \pm 1$ are given below.

Nearly Isosceles, Vertex 2α, Fibonacci Case

r	s	$a = r^2$	$b = s^2 - r^2$	$c = rs$
2	3	4	5	6
3	5	9	16	15
5	8	25	39	40
8	13	64	105	104
13	21	169	272	273
21	34	441	715	714
34	55	1156	1869	1870
55	89	3025	4896	4895
F_n	F_{n+1}	F_n^2	$F_{n+2}F_{n-1}$	F_nF_{n+1}

Consider the case $a - b = \pm 1$. Then $s^2 - 2r^2 = \pm 1$, where it is known that the solutions are related to the Pell numbers 1, 2, 5, 12, ..., $P_n = 2P_{n-1} + P_{n-2}$. If $r = P_n$, then $s = P_n + P_{n-1}$, which give the values of the Pellian case $a - b = \pm 1$ listed below.

Nearly Isosceles, Vertex 2α, Pellian Case

r	s	$a = r^2$	$b = s^2 - r^2$	$c = rs$
2	3	4	5	6
5	7	25	24	35
12	17	144	145	204
29	41	841	840	1189
70	99	4900	4901	6930
169	239	28561	28560	40391
408	577	166464	166465	235416
985	1393	970225	970224	1372105
P_n	$P_n + P_{n-1}$	P_n^2	$P_{n-1}P_{n+1}$	$P_n(P_n + P_{n-1})$

Here $\cos\,\alpha = \frac{s}{2r} = \frac{P_n + P_{n-1}}{2P_n}$, and the limiting value of $\cos\,\alpha$ as $n\to\infty$ is $\sqrt{2}/2$ (see Bicknell [2]). If $a = b$, then $s^2 = 2r^2$, and $s/2r = \sqrt{2}/2 = \cos 45°$, the isosceles right triangle.

Compare the lengths of the sides of the nearly isosceles $\alpha - 2\alpha$ triangles with Pythagorean triples a, b, c, $a^2 + b^2 = c^2$, given parametrically by $a = m^2 - n^2$, $b = 2mn$, and $c = m^2 + n^2$, $(m, n) = 1$, $1 < n < m$, m and n not both odd. The solutions of $u^2 - 2v^2 = \pm 1$ give $m = u + v$ and $n = v$, the values of m and n which make $a - b = \pm 1$ [Olds, 7], again the equation related to the Pell numbers. Taking m and n as successive integers provides $c = b + 1$, while in the $\alpha - 2\alpha$ triangles, we used successive Fibonacci numbers.

In the $\alpha - 3\alpha$ triangles, there appears to be only one nearly isosceles triangle, with sides 125, 279, 280. The case $b - a = \pm 1$ gives $s^3 - 2sr^2 - r^3 = (s + r)(s^2 - sr - r^2) = \pm 1$, but there is no solution because $(s + r) > 1$. The case $b - c = \pm 1$ gives $s^3 - 2sr^2 - rs^2 + r^3 = \pm 1$. In trials for $r < 100$ and all suitable values for s, only the solution $r = 5$, $s = 9$ arises.

In the $\alpha - 4\alpha$ triangles, there are no nearly isosceles triangles. The case $b - a = \pm 1$ has no solution, since $b - a = s^4 - 3s^2r^2 + r^4 - r^4 = s^2(s^2 - 3r^2) = \pm 1$ cannot be solved because $s > 1$. Now, the case $b - c = s^4 - 3s^2r^2 + r^4 - rs^3 + 2r^3s = \pm 1$ cannot be solved because the polynomial factors into $(s - r)(s^3 - 3r^2s - r^3)$ where $s - r > 1$.

Now, we return to the solutions for the $\alpha - n\alpha$ triangles to study their form.

4. $\alpha - n\alpha$ TRIANGLES

Earlier, we wrote expressions for sides a, b, and c for $\alpha - n\alpha$ triangles for $n = 2$, 3, 4, and 5. Similarly, we can apply Eq. (1.1) as long as we like to list solutions for a_n, b_n, and c_n in $\alpha - n\alpha$ triangles. We list the first few solutions in the table below to study their form.

Vertex Angle $n\alpha$

n	a_n	b_n	c_n
1	r	s	r
2	r^2	$s^2 - r^2$	rs
3	r^3	$s^3 - 2sr^2$	$r(s^2 - r^2)$
4	r^4	$s^4 - 3s^2r^2 + r^4$	$r(s^3 - 2sr^2)$
5	r^5	$s^5 - 4s^3r^2 + 3sr^4$	$r(s^4 - 3s^2r^2 + r^4)$
6	r^6	$s^6 - 5s^4r^2 + 6s^2r^4 - r^6$	$r(s^5 - 4s^3r^2 + 3sr^4)$
7	r^7	$s^7 - 6s^5r^2 + 10s^3r^4 - 4sr^6$	$r(s^6 - 5s^4r^2 + 6\ s^2r^4 - r^6)$
...	...	...	...

The table suggests the identities

$$a_n = ra_{n-1} = r^n \qquad \text{and} \qquad c_n = rb_{n-1}. \qquad (4.1)$$

Eq. (4.1) can be proved by mathematical induction, using Figure 1, the law of cosines, and Equation (1.1). At the same time, we prove (1.2) and find a recursive relationship for the b_n. Now, for n = 1, 2, ..., 7, (4.1) holds, as well as $(b_n, r) = 1$, for r, s positive relatively prime integers with $\cos 180/(n+1) < \cos \alpha = s/2r < 1$ and $n > 1$. The induction assumption is that for all $j < n$ there is a rational-sided, hence primitive integral, $\alpha - j\alpha$ triangle, that $a_j = r^j$, that $c_j = rb_{j-1}$, and that $(b_j, r) = 1$.

By (1.1) we deduce that

$$a'_n = a_{n-1}c_{n-1} = r^n b_{n-2}$$

$$c'_n = c_{n-1}b_{n-1} = rb_{n-1}b_{n-2}$$

and, with the law of cosines,

$$\begin{aligned}
b'_n &= (b_{n-1})^2 - (a_{n-1})^2 \\
&= (2\cos \alpha\ b_{n-1} - c_{n-1})c_{n-1} \\
&= [2(s/2r)b_{n-1} - rb_{n-2}]rb_{n-2} \\
&= (sb_{n-1} - r^2b_{n-2})b_{n-2}
\end{aligned}$$

Since $(r, s) = 1$ and we have assumed $(r, b_j) = 1$ for $j < n$, it follows that a'_n/b_{n-2}, b'_n/b_{n-2}, and c'_n/b_{n-2} are relatively prime integers, which makes the associated triangle $a_n = a'_n/b_{n-2}$, $b_n = b'_n/b_{n-2}$, $c_n = c'_n/b_{n-2}$ primitive and establishes (1.2) as well as (4.1), and the recursion (4.2),

$$b_n = sb_{n-1} - r^2b_{n-2}. \qquad (4.2)$$

48   M. BICKNELL-JOHNSON

Now, return to the table at the beginning of this section. Notice Pascal's triangle forming in the coefficients of the forms given for the b_n. If we consider each b_n as a polynomial in s and $-r^2$, then the coefficients of the b_n lie on the rising diagonals of Pascal's triangle, as

$$b_n = \sum_{k=0}^{[n/2]} (-1)^k \binom{n-k}{k} s^{n-2k} r^{2k} \qquad (4.3)$$

where [x] is the greatest integer in x. This can be proved by induction now that we have (4.2), but it is proved by Hoggatt and Long [4] since the b_n are a special case of their generalized Fibonacci polynomials evaluated at integers s and $-r^2$. Hoggatt and Long [4] define general Fibonacci polynomials as

$$u_{n+2}(x, y) = xu_{n+1}(x, y) + yu_n(x, y) \qquad (4.4)$$

where $u_0 = 0$, $u_1 = 1$, $u_2 = x$, $u_3 = x^2 + y$. Note that, if we formally let $x = s$ and $y = -r^2$, then $b_n = u_{n+1}(s, -r^2)$ for each positive integer n, because $b_1 = s = u_2(s, -r^2)$, $b_2 = s^2 - r^2 = u_3(s, -r^2)$, and (4.2) written as $b_{n+1} = sb_n - r^2 b_{n-1}$ becomes $u_{n+2}(s, -r^2) = su_{n+1}(s, -r^2) - r^2 u_n(s, -r^2)$ which is (4.4) for $x = s$ and $y = -r^2$. Of course, (4.3) means that we find Fibonacci numbers if we sum the absolute values of the coefficients in the expressions for b_n. We note that the expressions for b_n also agree with Carroll and Yanosko [3].

Equations (4.1) and (4.2) could also have been proved by looking at a simple diagram in which $a_{n-1} = ka_n$ and $b_{n-1} = kc_n$, which we give as Figure 2 below. We omit a second proof, but instead show some area relationships. Draw an $\alpha - (n-1)\alpha$ triangle ABC in which the sides a_{n-1}, b_{n-1}, and c_{n-1} are given in the table. Note that angle $B = \pi - n\alpha$ is the supplement of $n\alpha$, so extend AB to A' to draw an $\alpha - n\alpha$ triangle with scale factor k so that $a_{n-1} = ka_n$ and $BA' = kb_n$. Then $A'C = kc_n$, and angle $A' = \alpha$, so that $b_{n-1} = kc_n$ as well as the given $a_{n-1} = ka_n$. By (4.1), we see that $k = 1/r$.

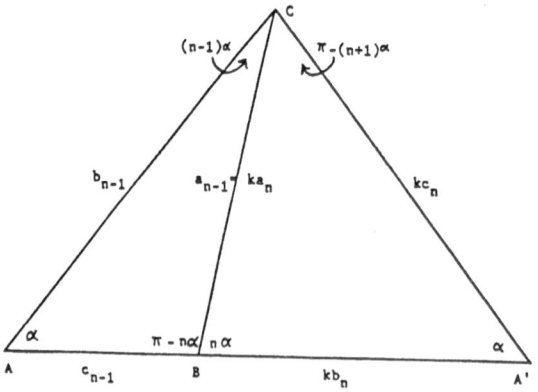

Figure 2

Let A_n be the area of an $\alpha - n\alpha$ triangle given by Equation (1.2). Then $k^2 A_n$ is the area of $\triangle A'BC$ and A_{n-1} is the area of $\triangle ABC$. Since those triangles have the same altitude from C, $A_{n-1}/k^2 A_n = c_{n-1}/kb_n$. Let $k = 1/r$, and apply (4.1) to write $r^2 A_{n-1}/A_n = rc_{n-1}/b_n = r^2 b_{n-2}/b_n$, or,

$$A_{n-1}/A_n = b_{n-2}/b_n. \tag{4.5}$$

In particular, for $n = 3$, $A_2/A_3 = s/(s^3 - 2sr^2) = 1/(s^2 - 2r^2)$. From Section 3, $s^2 - 2r^2 = 1$ for $r = P_{2n}$ and $s = P_{2n-1} + P_{2n-1}$ for Pell numbers P_n. So $A_2 = A_3$ for $r = 2$, $s = 3$; $r = 12$, $s = 17$; ...; $r = P_{2n}$, $s = P_{2n} + P_{2n-1}$.

5. NEARLY ISOSCELES $\alpha - n\alpha$ TRIANGLES

Nearly isosceles $\alpha - n\alpha$ triangles would either approximate an $\alpha - n\alpha - n\alpha$ triangle where $\alpha = \pi/(2n+1)$ with $b_n - c_n = \pm 1$, or an $\alpha - n\alpha - \alpha$ triangle where $\alpha = \pi/(n+2)$ with $b_n - a_n = \pm 1$.

First, consider the case $b_n - a_n = \pm 1$. Recall the form of b_n from (4.3). If the sum of the coefficients of the terms of the polynomial expression for b_n is 1, then the sum of the coefficients of the terms of $b_n - a_n$ is 0, so that $b_n - a_n$ will have a factor $s - r > 1$, since $b_n - a_n = 0$ is homogeneous. Let k_n be the sum of the coefficients of the terms of b_n, which is the value of b_n when $r = s = 1$. Since $k_n = b_n$ when $r = s = 1$, k_n must fulfill the same recursion as b_n, or $k_n = sk_{n-1} - r^2 k_{n-2}$, but $r = s = 1$, so $k_n = k_{n-1} - k_{n-2}$. By inspection, $k_1 = 1$, $k_2 = 0$, $k_3 = -1$, ..., and we can prove that $k_{6m} = k_{6m+1} = 1$, $k_{6m+2} = 0$, $k_{6m+3} = k_{6m+4} = -1$, and $k_{6m+5} = 0$. Thus, $b_n - a_n = \pm 1$ is impossible if $n \equiv 0, 1$ (mod 6). Also, by inspecting (4.3), $s \mid (b_n - a_n)$ whenever $n = 4k$, so $b_n - a_n = \pm 1$ is impossible if $n \equiv 0$ (mod 4).

As to the case $b_n - c_n = \pm 1$, we need some divisibility properties for b_n. We show that $b_2 = s^2 - r^2$ divides b_2, b_5, b_8, ..., b_{3m-1}. By observation, b_2 divides b_2 and b_5. We can show by induction that b_2 divides b_{3m-1}. Assume that b_2 divides b_{3m-1} for $m = k - 1$, or that b_2 divides $b_{3(k-1)-1} = b_{3k-4}$. For $m = k$, by (4.2) we have

$$
\begin{aligned}
b_{3k-1} &= sb_{3k-2} - r^2 b_{3k-3} \\
&= s(sb_{3k-3} - r^2 b_{3k-4}) - r^2 b_{3k-3} \\
&= (s^2 - r^2)b_{3k-3} - sr^2 b_{3k-4}
\end{aligned}
$$

Whenever $(s^2 - r^2)$ divides b_{3k-4}, then $(s^2 - r^2)$ divides b_{3k-1}, so $(s_2 - r^2)$ divides b_{3m-1} for all $m \geq 1$. By (4.1) and (4.2)

$$b_{3m+1} - c_{3m+1} = b_{3m+1} - rb_{3m} = (s-r)b_{3m} - r^2 b_{3m-1}$$

Now, $(s^2 - r^2)$ divides b_{3m-1}, so $(s-r)$ divides both terms on the right. Since $(s-r) > 1$ divides $(b_{3m+1} - c_{3m+1})$, $b_{3m+1} - c_{3m+1} = \pm 1$ is impossible, so the case $b_n - c_n = \pm 1$ is impossible for $n = 3m + 1$.

These methods will not yield a complete solution. The only nearly isosceles triangles found to date are given in this paper. Perhaps someone will find a way to show whether or not other nearly isosceles triangles are possible.

The author is grateful for the referee's many helpful suggestions, including the proof used for (4.1), (1.2) and (4.2) as well as a proof using Figure 2. The referee gave very generously of his/her time, and this paper would not have been finished without his/her help.

REFERENCES

[1]   Bicknell, Marjorie and Hoggatt, Verner E., Jr. "Golden Triangles, Rectangles and Cuboids," *The Fibonacci Quarterly, 7.1* (1969): pp. 73-91.

[2]   Bicknell, Marjorie. "A Primer on the Pell Sequence and Related Sequences," *The Fibonacci Quarterly, 13.4* (1975): pp. 345-349.

[3]   Carroll, Joseph E. and Yanosko, Ken, "The Determination of a Class of Primitive Integral Triangles," *The Fibonacci Quarterly*, to appear.

[4]   Hoggatt, Verner E. Jr., and Long, Calvin T. "Divisibility Properties of Generalized Fibonacci Polynomials," *The Fibonacci Quarterly, 12.2* (1974): pp. 113-120.

[5]   Luthar, R. S. "Integer-sided Triangles with One Angle Twice Another," *The College Mathematics Journal*, 15:3 (1984): pp. 55-6.

[6]   Leonard, Bill, Schwartzman, Jan and Shultz, Harris. "Double-Angle Triangles, Pentagons, Paradoxes, and Gold," *AMATYC Review, 10:1*, Fall, 1988.

[7]   Olds, C. D. Continued Fractions, New Mathematical Library, Random House, New York, 1963.

THE RING OF FIBONACCI
(FIBONACCI "NUMBERS" WITH MATRIX SUBSCRIPT)

Odoardo Brugia, Piero Filipponi and Francesco Mazzarella

1. INTRODUCTION

Several authors (e.g., see [8]) have considered the Fibonacci numbers F_x where the subscript x is an arbitrary real number and showed that these (complex) numbers enjoy most of the properties of the usual Fibonacci numbers F_m (m integral). A quite natural extension of the numbers F_x leads to the definition of the Fibonacci numbers F_z and Lucas numbers L_z

$$F_z = (\alpha^z - \beta^z)/\sqrt{5} \tag{1.1}$$

$$L_z = \alpha^z + \beta^z, \tag{1.2}$$

where the subscript z is an arbitrary complex number and $\alpha = -1/\beta = (1 + \sqrt{5})/2$.

In this paper, after giving some basic definitions, we further extend (1.1) and (1.2) by considering the Fibonacci and Lucas *numbers* F_A and L_A, where A is a (complex) square matrix of arbitrary order so that the quantities F_A and L_A are square (complex) matrices of the same order as that of A (Section 2).

Then, we show that *mutatis mutandis* F_A and L_A have many properties of the usual numbers F_m and L_m. More precisely, we show that, replacing the alphabetical (variable) part, a, and the numerical (constant) part, k, of the subscript of a Fibonacci (and/or Lucas) number by a matrix A of order n and a scalar matrix kI (I the identity matrix of order n), respectively,

51

G. E. Bergum et al. (eds.), Applications of Fibonacci Numbers Volume 4, 51–62.
© 1991 *Kluwer Academic Publishers.*

a Fibonacci identity changes into a matrix identity. If the subscript involves more than one alphabetical quantity (say, $a, b, c, \ldots$) then the matrix identity holds subject to the condition that the matrices $A, B, C, \ldots$ commute (Sections 3 and 4).

In Section 5 several numerical examples are shown whence some theoretical properties of F_A and L_A can be checked on, while a possible future investigation in this field is outlined in Section 6. Further properties of these matrices are demonstrated *passim*.

2. DEFINITIONS AND PRELIMINARIES

Following [5, Ch. XII], let $z \neq 0$ be an arbitrary complex number and define the principal value of its logarithm as

$$\log z = \ln |z| + i \arg z, \tag{2.1}$$

where $i = \sqrt{-1}$. In particular, for $z = x$ (real) we have

$$\log x = \ln |x| + i \arg x = \begin{cases} \ln x \text{ if } x > 0 \\ \ln |x| + i\pi \text{ if } x < 0. \end{cases} \tag{2.2}$$

Moreover, for $z = a + ib$, let us define the exponential function x^z as

$$x^z = e^{z \log x} = e^{(a + ib)(\ln|x| + i \arg x)}$$
$$= |x|^a e^{-b \arg x}[\cos (b \ln |x| + a \arg x) + i \sin (b \ln |x| + a \arg x)]. \tag{2.3}$$

Before introducing matrices and their functions, let us mention *en passant* that, as a direct consequence of (2.2), the successive derivatives of the function $f(z) = x^z$ with respect to z are given by

$$f^{(k)}(z) = x^z (\log x)^k, \tag{2.4}$$

where the superscript (k) denotes the kth derivative.

Let A_n (or simply A, if there is not fear of confusion) be a generic (complex) square matrix of order n and let I_n and 0_n be the identity matrix and the zero matrix, respectively. The definition of the matrix exponential function $f(A) = x^A$ follows directly from the previous definitions and the general theory of the functions of matrices [4, Ch. 5]. In fact, it is well known [4] that there exists at least one non-singular matrix T such that

$$A = TJT^{-1}, \tag{2.5}$$

where J is the so-called *Jordan normal form* of the matrix A, and the function x^A can be expressed as

$$x^A = Tx^J T^{-1}. \tag{2.6}$$

The structure of the matrix x^J is available in [4, p. 100] and, once z is seen as a generic eigenvalue of A, the calculation of the entries of x^J can be readily carried out with the aid of (2.2) and (2.4).

We recall that the function x^A can be represented also as a polynomial in A

$$x^A = c_0 I + c_1 A + c_2 A^2 + \cdots + c_{n-1} A^{n-1}, \tag{2.7}$$

where coefficients c_j $(j = 0, 1, \ldots, n-1)$ (some of which may vanish) are given by the solution of a suitable system of n equations and n unknowns [4]. Throughout the paper, we sometimes shall make use of this representation.

Let us point out some properties of the function x^A which will be utilized in the sequel. Let x and y be arbitrary real numbers.

Proposition 1: $x^{yI} = x^y I$

Proposition 2: $x^A y^A = y^A x^A$

Proposition 3: $x^A y^A = (xy)^A$.

Now, we are in a position to define what is to be meant by the Fibonacci and Lucas *numbers* with matrix subscript, thus extending the definitions (1.1) and (1.2) which can be thought of as particular cases of Fibonacci and Lucas *numbers* whose subscript is an arbitrary 1-by-1 complex matrix. Let

$$F_A = (\alpha^A - \beta^A)/\sqrt{5}, \tag{2.8}$$
$$L_A = \alpha^A + \beta^A. \tag{2.9}$$

These quantities are clearly square matrices of the same order as that of A.

From the previous definitions it follows that:

(i)   If the real part a_j of the eigenvalues $\lambda_j = a_j + ib_j$ of A $(j = 1, 2, \ldots, n)$ is an integer and $b_j = k_j/\ln\alpha$ $(k_j \in \mathbb{Z})$, and A is a real matrix of *simple structure*, that is it can be reduced to the diagonal form

$$D = \text{diag}\,\{\lambda_1, \lambda_2, \ldots, \lambda_n\} = T^{-1} A\, T,$$

then F_A and L_A are real.

(ii) F_A is singular iff at least one eigenvalue of A has the form

$$\lambda_j = \frac{2k\pi(\pi - i2 \ln \alpha)}{4 \ln^2\alpha + \pi^2} \quad (k \in \mathbb{Z}). \tag{2.10}$$

L_A is singular iff at least one eigenvalue of A has the form

$$\lambda_j = \frac{(2k+1)\pi(\pi - i2 \ln \alpha)}{4 \ln^2\alpha + \pi^2} \quad (k \in \mathbb{Z}). \tag{2.11}$$

(iii)

$$\det(F_A) = \prod_{j=1}^{n} F_{\lambda_j}, \tag{2.12}$$

$$\det(L_A) = \prod_{j=1}^{n} L_{\lambda_j}. \tag{2.13}$$

(iv)   For any given square matrix A, there exist infinitely many matrices X and Y such that

$$F_X = A \text{ and } L_Y = A.$$

(v)   If all eigenvalues of A have the form

$$\lambda_j = \frac{(2k+1)\pi^2 + 4 \ln^2\alpha - i4\pi \ln \alpha}{4 \ln^2\alpha + \pi^2} \quad (j = 1, 2, \ldots, n; k \in \mathbb{Z}) \tag{2.14}$$

and A is of simple structure, then $F_A = L_A$. We point out that the matrix I is the only real matrix fulfilling the above conditions.

The proofs of the statements (ii), (iv) and (v) are deducible from some results established in [9].

Remark

     The matrices F_A and L_A can also be expressed (cf. [4, p. 113]) by the following expansion

$$F_A = \frac{1}{\sqrt{5}} \sum_{k=0}^{\infty} \frac{x_k - y_k}{k!} A^k \tag{2.15}$$

$$L_A = \sum_{k=0}^{\infty} \frac{x_k + y_k}{k!} A^k, \tag{2.16}$$

where

$$x_k = (\ln \alpha)^k \quad , \quad y_k = (\ln|\beta| + i\pi)^k. \tag{2.17}$$

It must be noted that, even though x_k and y_k are complex quantities, the above expansions lead to a real matrix whenever A fulfills the requirements of statement (i). Letting $A = I$ in (2.15) and (2.16), we have clearly (cf. (3.1) and (3.2))

$$\frac{1}{\sqrt{5}} \sum_{k=0}^{\infty} \frac{x_k - y_k}{k!} = \sum_{k=0}^{\infty} \frac{x_k + y_k}{k!} = 1. \tag{2.18}$$

3. ELEMENTARY PROPERTIES OF F_A AND L_A

In this section some properties of the matrices F_A and L_A are shown which are related to some analogous elementary properties of the usual Fibonacci and Lucas numbers.

First, let us state the following theorem.

Theorem 1: If $D = \text{diag } \{d_1, d_2, \ldots, d_n\}$ is an arbitrary diagonal matrix, then

$$F_D = \text{diag}\{F_{d_1}, F_{d_2}, \ldots, F_{d_n}\} \ .$$

Proof: $F_D = (\alpha^D - \beta^D)/\sqrt{5} = \text{diag}\{\alpha^{d_1} - \beta^{d_1}, \ldots, \alpha^{d_n} - \beta^{d_n}\}/\sqrt{5} = \text{diag}\{F_{d_1}, \ldots, F_{d_n}\}.$ Q.E.D.

Analogously, we can state

Theorem 2: $L_D = \text{diag } \{L_{d_1}, L_{d_2}, \ldots, L_{d_n}\}.$

Corollary 1: $F_{zI} = F_z I$ and $L_{zI} = L_z I$ for z an arbitrary complex number.

As particular examples, we have

$$F_0 = 0, \ F_I = F_{2I} = I, \ F_{3I} = 2I, \ldots \tag{3.1}$$

$$L_0 = 2I, \ L_I = I, \ L_{2I} = 3I, \ldots \ . \tag{3.2}$$

Theorem 3: $F_{A+2I} = F_{A+I} + F_A$, A being an arbitrary square matrix.

Proof: We have

$$
\begin{aligned}
F_{A+2I} - F_{A+I} - F_A &= \tfrac{1}{\sqrt{5}} \ (\alpha^{A+2I} - \beta^{A+2I} - \alpha^{A+I} + \beta^{A+I} - \alpha^A + \beta^A) \\
&= \tfrac{1}{\sqrt{5}} \ [\alpha^A(\alpha^{2I} - \alpha^I - I) - \beta^A(\beta^{2I} - \beta^I - I)] \\
&= \tfrac{1}{\sqrt{5}} \ [\alpha^A(\alpha^2 - \alpha - 1) - \beta^A(\beta^2 - \beta - 1)] \\
&= \tfrac{1}{\sqrt{5}} \ (\alpha^A \cdot 0 - \beta^A \cdot 0) = 0. \qquad \text{Q.E.D.}
\end{aligned}
$$

Analogously, we have

Theorem 4: $L_{A+2I} = L_{A+I} + L_A.$

Then, after recalling that $x^{-A} = (x^A)^{-1} = (x^{-1})^A$ [4, p. 114], we state

Theorem 5: $F_{-A} = -(-1)^A F_A.$

Proof: We can write

$$F_{-A} = [\alpha^{-A}(\beta^{-A}\beta^A) - \beta^{-A}(\alpha^{-A}\alpha^A)]/\sqrt{5},$$

whence, by using the associative property of the matrix multiplication and Proposition 3, we get

$$F_{-A} = (\alpha\beta)^{-A}(\beta^A - \alpha^A)/\sqrt{5} = -(-1)^{-A}F_A = -(-1)^A F_A. \qquad \text{Q.E.D.}$$

Analogously, we have

Theorem 6: $L_{-A} = (-1)^A L_A.$

Finally, let us state the following:

Theorem 7: $L_A = F_{A-I} + F_{A+I}.$

Proof: $F_{A-I} + F_{A+I} = [\alpha^A(\alpha^{-I} + \alpha^I) - \beta^A(\beta^{-I} + \beta^I)]/\sqrt{5}$

$$= [\alpha^A(\alpha^{-1} + \alpha) - \beta^A(\beta^{-1} + \beta)]/\sqrt{5} = (\sqrt{5}\alpha^A + \sqrt{5}\beta^A)/\sqrt{5} = L_A. \qquad \text{Q.E.D.}$$

The above properties established for matrices F_A and L_A are analogous to those enjoyed by the usual numbers F_{a+k} and L_{a+k}. We simply replaced the alphabetical part of the subscript (say, a) by a generic matrix A, and the numerical part (say, k) by the scalar matrix kI.

4. SOME MATRIX FIBONACCI IDENTITIES

Firstly, let us prove the following matrix identity which is analogous to the *Simson formula* for the usual Fibonacci numbers.

Theorem 8: $F_{A-I}F_{A+I} = F_A^2 + (-1)^A.$

Proof: By using Propositions 1, 2, and 3, we can write

$$F_{A-I}F_{A+I} = (\alpha^{A-I} - \beta^{A-I})(\alpha^{A+I} - \beta^{A+I})/5 = [L_{2A} - \alpha^A\beta^A(\alpha\beta^{-I} + \alpha^{-I}\beta)]/5$$

$$= [L_{2A} - (-1)^A(\alpha\beta^{-I} + \alpha^{-I}\beta)]/5 = [L_{2A} - (-1)^A(\alpha/\beta + \beta/\alpha)]/5$$

$$= [L_{2A} + 3(-1)^A]/5 = [\alpha^{2A} + \beta^{2A} - 2(\alpha\beta)^A + 5(-1)^A]/5$$

$$= [(\alpha^A - \beta^A)/\sqrt{5}]^2 + (-1)^A = F_A^2 + (-1)^A. \qquad \text{Q.E.D.}$$

Then, we state the following:

Theorem 9: $F_{2A} = F_A L_A.$

Proof: $F_A L_A = (\alpha^A - \beta^A)(\alpha^A + \beta^A)/\sqrt{5} = [\alpha^{2A} - \beta^{2A} + (\alpha\beta)^A - (\beta\alpha)^A]/\sqrt{5} = F_{2A}.$

$$\text{Q.E.D.}$$

Now, let us consider identities involving Fibonacci and Lucas numbers whose subscripts consist of two alphabetical parts (say, a and b) and a possible constant part (say, k). The usual substitutions $a \leftarrow A$, $b \leftarrow B$ and $k \leftarrow kI$ lead to a correct matrix Fibonacci identity provided A and B commute. In fact, it can be proved that, if A and B commute, then $x^A x^B = x^B x^A = x^{A+B}$ [1, p. 167] and $\alpha^A \beta^B = \beta^B \alpha^A = (-1)^B \alpha^{A-B}$. This implies that $F_A F_B = F_B F_A$, $L_A L_B = L_B L_A$ and $F_A L_B = L_B F_A$.

As an example, let us consider the identities I_{27}, I_{21} and I_{23} [6, p. 59] and state the following theorems.

Theorem 10: If A and B commute, then $L_{A+B+I} = F_{A+I} L_{B+I} + F_A L_B$.

Proof: Using Theorem 7, we can write

$$F_{A+I} L_{B+I} + F_A L_B = (\alpha^{A+I} - \beta^{A+I})(\alpha^{B+I} + \beta^{B+I})/\sqrt{5} + (\alpha^A - \beta^A)(\alpha^B + \beta^B)/\sqrt{5}$$

$$= F_{A+B+2I} + F_{A+B} + \{\alpha^A \beta^B [I + (\alpha\beta)^I] - \beta^A \alpha^B [I + (\beta\alpha)^I]\}/\sqrt{5}$$

$$= L_{A+B+I} + [I + (-1)^I](\alpha^A \beta^B - \beta^A \alpha^B)/\sqrt{5}$$

Since $I + (-1)^I = 0$, the theorem is proved.   Q.E.D.

Theorem 11: If A and B commute, then $F_{A+B} + (-1)^B F_{A-B} = F_A L_B$.

Proof: Recalling that also B and $A - B$ commute and using Proposition 3, we can write

$$F_A L_B = (\alpha^A - \beta^A)(\alpha^B + \beta^B)/\sqrt{5} = F_{A+B} + (\alpha^A \beta^B - \beta^A \alpha^B)/\sqrt{5}$$

$$= F_{A+B} + \alpha^B \beta^B (\alpha^{A-B} - \beta^{A-B})/\sqrt{5} = F_{A+B} + (-1)^B F_{A-B}.$$ Q.E.D.

5. NUMERICAL EXAMPLES

In this section a number of numerical examples are given, whence some of the theoretical properties of F_A and L_A established in previous sections can be investigated.

Example 1: A is real, of simple structure and has integral eigenvalues: F_A and L_A are real.

$$A = \begin{bmatrix} 2 & 1 \\ 1 & 2 \end{bmatrix} \quad \begin{cases} \lambda_1 = 1 \\ \lambda_2 = 3 \end{cases}. \tag{5.1}$$

By using (2.8), (2.9) and [7, formulae (2.24)-(2.27)], we get

$$F_A = \frac{1}{2} \begin{bmatrix} 3 & 1 \\ 1 & 3 \end{bmatrix}, \quad L_A = \frac{1}{2} \begin{bmatrix} 5 & 3 \\ 3 & 5 \end{bmatrix}. \tag{5.2}$$

Example 2: A is real with integral (namely, zero) eigenvalues but it is not of simple structure: F_A and L_A are complex.

$$A = \begin{bmatrix} 0 & 1 \\ 0 & 0 \end{bmatrix} \quad \begin{cases} \lambda_1 = 0 \\ \lambda_2 = 0 \end{cases}. \tag{5.3}$$

By using [4, p. 100], (2.4) and (2.2), we get

$$\alpha^A = \begin{bmatrix} 1 & \ln \alpha \\ 0 & 1 \end{bmatrix}, \quad \beta^A = \begin{bmatrix} 1 & \ln|\beta| + i\pi \\ 0 & 1 \end{bmatrix},$$

whence

$$F_A = \begin{bmatrix} 0 & (2\ln \alpha - i\pi)/\sqrt{5} \\ 0 & 0 \end{bmatrix}, \quad L_A = \begin{bmatrix} 2 & i\pi \\ 0 & 2 \end{bmatrix}. \tag{5.4}$$

Example 3: A is real, of simple structure but its eigenvalues are not integral: F_A and L_A are complex. Let us consider the well-known matrix Q [6, p. 65]. Using [7, formulae (2.24)-(2.27)], after some manipulations we get

$$F_Q = \frac{1}{\sqrt{5}} \begin{bmatrix} \alpha F_\alpha - \beta F_\beta & F_\alpha - F_\beta \\ F_\alpha - F_\beta & \alpha F_\beta - \beta F_\alpha \end{bmatrix} \approx \begin{bmatrix} 0.803 + i0.294 & 0.156 - i0.165 \\ 0.156 - i0.165 & 0.647 + i0.459 \end{bmatrix},$$

$$L_Q = \frac{1}{\sqrt{5}} \begin{bmatrix} \alpha L_\alpha - \beta L_\beta & L_\alpha - L_\beta \\ L_\alpha - L_\beta & \alpha L_\beta - \beta L_\alpha \end{bmatrix} \approx \begin{bmatrix} 1.767 - i0.656 & 0.935 + i0.370 \\ 0.935 + i0.370 & 0.833 - i1.026 \end{bmatrix}.$$

Example 4: The statement (iv) (Section 2) implies that, for a given matrix A, there exist infinitely many matrices B and C ($B \neq A$, $C \neq A$) such that $F_A = F_B$ and $L_A = L_C$. A trivial example of this fact is given by the equalities $F_I = F_{2I}$ (see (3.1)) and $L_{-2I} = L_{2I}$ (see Theorem 6). A further interesting example results by considering the matrix $xH = x[h_{ij}]$ of order n, where $h_{ij} = 1$ (0) if $i + j = n + 1$ ($\neq n + 1$) and x is an arbitrary quantity. The eigenvalues of xH are

$$\begin{cases} \lambda_1 = x \text{ (with multiplicity } \mu = \lfloor (n+1)/2 \rfloor \text{ on the characteristic polynomial)} \\ \lambda_2 = -x \ (\mu = \lfloor n/2 \rfloor) \end{cases}$$

and xH is of simple structure as it is symmetric. Consequently, any function $f(xH)$ defined at x and $-x$ can be expressed in the form $f(xH) = c_0 I + c_1 xH$ [4], where the coefficients c_0 and c_1

can be found by [7, formula (2.23)]. After some simple calculations, we get

$$\begin{cases} c_0 = (F_x + F_{-x})/2 \\ c_1 = (F_x - F_{-x})/(2x) \ \ (x \neq 0), \end{cases} \tag{5.5}$$

whence

$$F_{xH} = \frac{F_x + F_{-x}}{2} I + \frac{F_x - F_{-x}}{2} H, \tag{5.6}$$

$$L_{xH} = \frac{L_x + L_{-x}}{2} I + \frac{L_x - L_{-x}}{2} H. \tag{5.7}$$

As particular cases ($x = m$ an integer), we have

$$F_{mH} = \begin{cases} F_m I & (m \text{ odd}) \\ F_m H & (m \text{ even}) \end{cases}, \tag{5.8}$$

$$L_{mH} = \begin{cases} L_m H & (m \text{ odd}) \\ L_m I & (m \text{ even}) \end{cases}. \tag{5.9}$$

From Corollary 1, it is seen that $F_{mH} = F_{mI}$ (m odd) and $L_{mH} = L_{mI}$ (m even).

Example 5: Let x be an arbitrary quantity and U_n be the *universal* (all unity) matrix of order n. It can be readily proved (e.g., see [2]) that

$$F_{xU_n} = \frac{F_{nx}}{n} U_n, \quad L_{xU_n} = \frac{L_{nx} - 2}{n} U_n + 2I . \tag{5.10}$$

As a particular instance, letting $n = 5$ and $x = 1$ in (5.10), we get

$$F_{U_5} = U_5. \tag{5.11}$$

A question arises spontaneously: "Apart from the case (5.11) and the trivial cases $F_{mI} = F_m I = mI$ for $m = 0, 1, 5$, has the matrix equation $F_X = X$ further solutions?" The answer is in the affirmative. More precisely, we show that F_A enjoys the following cute property.

Theorem 12: If A is a matrix of simple structure with eigenvalues $\lambda_j \in \{0, 1, 5\}$, then $F_A = A$.

Proof: Let $D = \text{diag} \{\lambda_1, \lambda_2, ..., \lambda_n\}$. Since $A = TDT^{-1}$, we have $F_A = TF_D T^{-1} = T$ diag $\{F_{\lambda_1}, ..., F_{\lambda_n}\}T^{-1}$. Since, if $\lambda_j \in \{0, 1, 5\}$, then $F_{\lambda_j} = \lambda_j$, we can write $F_A = T$ diag $\{\lambda_1, \lambda_2, ..., \lambda_n\}T^{-1} = A$.   Q.E.D.

Thus the equalities $F_0 = 0$, $F_1 = 1$ and $F_5 = 5$ can be seen as a corollary of Theorem 12 ($A = [a_{11}] = 0$ or 1 or 5, is a 1-by-1 matrix).

Analogously, it can be proved that $L_A = A$ ($A = \pm I$), $F_A = A - I$ ($\lambda_j \in \{2, 3, 4\}$), and $L_A = A + I$ ($\lambda_j \in \{2, 3\}$).

Let us give two numerical examples of Theorem 12.

Example 5.1: Consider the following real M-matrix [3]

$$A = \frac{1}{4} \begin{bmatrix} -6 & -2 & 14 \\ -7 & 1 & 13 \\ -11 & -7 & 29 \end{bmatrix} \quad \begin{cases} \lambda_1 = 0 \\ \lambda_2 = 1 \\ \lambda_3 = 5 . \end{cases} \tag{5.12}$$

By using [3, formula (39)], the entries x_{ij} of x^A (x an arbitrary quantity) can be readily found

$$\begin{array}{lll}
x_{11} = (-x^5 - x + 6)/4 & x_{12} = (-x^5 + 3x - 2)/4 & x_{13} = (3x^5 - x - 2)/4 \\
x_{21} = (-x^5 - 2x - 2)/4 & x_{22} = (-x^5 + 6x - 1)/4 & x_{23} = (3x^5 - 2x - 2)/4 \\
x_{31} = (-2x^5 - x + 3)/4 & x_{32} = (-2x^5 + 3x - 1)/4 & x_{33} = (6x^5 - x - 1)/4.
\end{array} \tag{5.13}$$

Letting $x = \alpha$ and $x = \beta$ in the above expressions and using (2.8), the equality $F_A = A$ can be readily found.

Example 5.2: Consider the following complex matrix

$$A = \begin{bmatrix} 3 & 2i \\ -2i & 3 \end{bmatrix} \quad \begin{cases} \lambda_1 = 1 \\ \lambda_2 = 5 \end{cases} . \tag{5.14}$$

By using [7, formulae (2.24)-(2.27)], we obtain

$$x^A = \frac{x}{2} \begin{bmatrix} x^4 + 1 & i(x^4 - 1) \\ -i(x^4 - 1) & x^4 + 1 \end{bmatrix},$$

whence, after replacing x by α and β and using (2.8),

$$F_A = \frac{1}{2} \begin{bmatrix} F_5 + F_1 & i(F_5 - F_1) \\ -i(F_5 - F_1) & F_5 + F_1 \end{bmatrix} = A. \tag{5.15}$$

Finally, let us point out that the necessity of the condition in Theorem 12 does not hold. In fact, if ζ_k is the generic solution to the trascendental equation $\alpha^z - \beta^z = z\sqrt{5}$ and

$A = T$ diag $\{\zeta_1, \zeta_2, ..., \zeta_n\}$ T^{-1} is a matrix with complex entries, then $F_A = A$. As a particular instance, the following numerical values of the above solutions have been obtained by means of a pocket calculator (10^{-8} accuracy)

$$\begin{cases} \zeta_1 = -9.9352320 + i0.53423299 \\ \zeta_2 = -2.0037580 - i0.19744498 \end{cases} \qquad (5.16)$$

6. FURTHER WORK

Let us write

$$A = F_0(A), F_A = F_1(A), F_{F_A} = F_2(A),... \text{ and so on,} \qquad (6.1)$$

and ask ourselves what conditions the eigenvalues of A must be subject to for the limit

$$\lim_{k \to \infty} F_k(A) = \Lambda_F(A) \qquad (6.2)$$

to exist. The same question can obviously be posed about the existence of the analogous limit $\Lambda_L(A)$. The existence of these limits implies clearly the existence of the limits

$$l_F(z) = \lim_{k \to \infty} F_k(z), \quad l_L(z) = \lim_{k \to \infty} L_k(z), \qquad (6.3)$$

where $F_k(z)$ and $L_k(z)$ can be defined in the same way as that of (6.1). Solutions to (6.3) are obviously related to the solutions of the trascendental equations (cf. (5.16))

$$\begin{cases} \alpha^z - \beta^z = z\sqrt{5} \\ \alpha^z + \beta^z = z \end{cases} \qquad (6.4)$$

A deep investigation in this direction seems to be rather interesting in spite of its difficulty. Currently, we can only state (cf. Theorem 12) that, if $A = T$ diag $\{\lambda_1, \lambda_2, ..., \lambda_n\}$ T^{-1}, the eigenvalues λ_j of A are integral and $-5 \le \lambda_j \le 5$, then

$$\Lambda_F(A) = T \text{ diag } \{\xi_1, \xi_2, ..., \xi_n\} T^{-1}, \qquad (6.5)$$

where

$$\xi_j = \begin{cases} |\lambda_j|, \text{ if } \lambda_j = 0 \text{ or } \pm 5 \\ 1, \text{ otherwise,} \end{cases} \qquad (6.6)$$

and the convergence of $F_k(A)$ to $\Lambda_F(A)$ requires at most three steps, that is $F_k(A) = \Lambda_F(A)$ for $0 \le k \le 3$.

ACKNOWLEDGEMENT

This work has been carried out in the framework of an agreement between the Italian PT Administration and the Fondazione Ugo Bordoni.

REFERENCES

[1]   Barnett, S. Matrix Methods for Engineers and Scientists. London: McGraw-Hill Book Co., 1979.

[2]   Filipponi, P. "Functions of Matrices: A Set of Matrices for Testing Computer Programs." *Ricerche di Automatica, 8*, No. 2-3 (1977): pp. 210-216.

[3]   Filipponi, P. "Testing the Efficiency of Procedures for Calculating Matrix Functions." *Ricerche di Automatica, 11*, No. 1 (1980): pp. 32-50.

[4]   Gantmacher, F.R. The Theory of Matrices, Vol. I. New York: Chelsea, 1959.

[5]   Ghizzetti, A. Lezioni di Analisi Matematica, Vol. 1. Rome: Veschi, 1953.

[6]   Hoggatt, V.E. Jr. Fibonacci and Lucas Numbers. Boston: Houghton Mifflin, 1969.

[7]   Horadam, A.F. and Filipponi, P. "Cholesky Algorithm Matrices of Fibonacci Type and Properties of Generalized Sequences." Fond. U. Bordoni Tech. rept. 3T0988. *The Fibonacci Quarterly* (to appear).

[8]   Lahr, J. Theorie elektrischer Leitungen unter Anwendung und Erweiterung der Fibonacci-Funktion." Dissertation ETH No. 6958, Zurich, 1981.

[9]   Mazzarella, F. "Un Tipo di Equazione Trascendente Relativa ad una Estensione dei Numeri di Fibonacci." *Note Recensioni Notizie, 38*, No. 1-2 (1989): pp. 15-20.

ONE-RELATOR PRODUCTS OF CYCLIC GROUPS AND FIBONACCI-LIKE SEQUENCES

C. M. Campbell, P. M. Heggie, E. F. Robertson and R. M. Thomas

1. INTRODUCTION

In this paper, we mention some properties of certain generalized Fibonacci sequences we have looked at while investigating one-relator products of cyclic groups. The particular groups we have investigated are those defined by presentations of the form

$$< a, b : a^2 = b^n = R(a, b) = 1 >,$$

where $R(a, b)$ is a word of the form $ab^{j(1)}ab^{j(2)}...ab^{j(r)}$ with $r \geq 2$ and $0<j(i)<n$ for each i. Such a group is called a *one-relator product* of the cyclic groups C_2 and C_n of orders 2 and n respectively, in that it is formed from the free product of C_2 and C_n by imposing the single extra relation $R(a, b) = 1$. We denote this group by $G(n; j(1), j(2), ..., j(r))$.

One question of particular interest when dealing with group presentations is the following: given a presentation, is the corresponding group finite or infinite? It was shown in [10] that there is no general and effective method for resolving this question, but there do exist techniques for providing answers for certain classes of presentation. If the corresponding group is finite, then we would wish to know the order and structure of the group.

The structure of $G(n; j(1), ..., j(r))$ is easy to determine if $r = 2$, and the case $r = 3$ was dealt with in [1]. If $r = 4$ and $j(1) + j(2) + j(3) + j(4) = 0$, with each $j(i)$ in $\{ \pm 1, \pm 2 \}$, then the corresponding groups were determined in [5], and further results on groups with $r = 4$ may be found in [4] and [7]. In particular, the group $G(n; -1, 1, 2, -2)$ is metabelian of order $2ng_n$ if n is odd or $n \leq 4$, and infinite otherwise [3, 5], where (g_n) denotes the Lucas sequence. We

63

G. E. Bergum et al. (eds.), Applications of Fibonacci Numbers Volume 4, 63–68.
© 1991 *Kluwer Academic Publishers.*

now consider the groups $G = G(\alpha, n)$ defined by the presentations

$$< a, b : a^2 = b^n = ab^{-1}ab(abab^{-1})^{\alpha-1}ab^2ab^{-2} = 1 >$$

for $n \geq 1$ and $\alpha \geq 1$. We can show that $[G : G'] = 2n$ and $[G' : G''] = v_n(\alpha)$, where $v_n = v_n(\alpha)$ is defined by $v_0 = 0$, $v_1 = 1$, $v_n = \alpha v_{n-1} + v_{n-2} + 1 + (-1)^{n-1}$ $(n \geq 2)$. If $\alpha = 1$, then $v_n = g_n - 1 - (-1)^n$ (see Corollary 5 below), so that this is in accordance with the above result. Here G' denotes the smallest normal subgroup N of G such that G/N is abelian, $G^{(2)}$ or G'' denotes $(G')'$, $G^{(3)}$ or G''' denotes $(G'')'$, etc. We say that G is *soluble* of *derived length* n if $G^{(n)} = 1$ but $G^{(n-1)} \neq 1$. Clearly, G is soluble of derived length 1 if and only if G is abelian, and we say that G is *metabelian* if it is soluble of derived length 2.

The groups $G(\alpha, n)$ are particularly interesting for a variety of reasons. Apart from the general interest in one-relator products of cyclic groups, the finiteness of these groups, for suitable values of n but independently of α, shows that there is no upper bound on the length of the extra relator such that a one-relator product of cyclic groups is finite. In particular, if n is odd, then G is finite and soluble with derived length at most 3 for any value of α, and further to this, if n is odd and g_n is coprime to $\alpha - 1$, then G is a metabelian group of order $2nv_n(\alpha)$. There are also interesting connections between the $G(\alpha, n)$ and the *Fibonacci groups* $F(r, n)$; see [3] and [11] for recent surveys of the Fibonacci and related groups. We shall not prove any group-theoretic results here; a preliminary report [2] contains the proofs of those results, and they will be published officially elsewhere. Instead, we shall present some properties of the associated sequences. These are necessary for the group theoretic work, and may be of some independent interest.

If n is odd or $n \leq 7$, the groups $G(\alpha, n)$ are known to be finite and soluble of derived length at most 3, with the exception of the group $G(1, 6)$ (or $G(6; -1, 1, 2, -2)$ in the previous notation). However, in general, the groups need not be finite; for example, the groups $G(5, 12)$ and $G(6, 8)$ are infinite soluble groups. Also, not all the groups are soluble; for example, $G(6, 10)$ is a finite insoluble group involving PSU(3, 4) and $G(7, 10)$ an infinite insoluble group involving the Higman-Sims simple group. The group $G(6, 10)$ is particularly interesting, in that the related group $H(6, 10)$ with presentation

$$< a, b, z : a^2 = b^{10} = z, ab^{-1}ab(abab^{-1})^5ab^2ab^{-2} = 1 >$$

is then a finite insoluble group of deficiency zero of order at least 114,967,210,176,000.

2. THE SEQUENCES

It was shown in [8] that a sequence (u_n) satisfying the identity $u_n^2 - u_{n-1}u_{n+1} = (-1)^{n+1}$ must be defined by $u_0 = 0$, $u_1 = 1$, $u_n = \alpha u_{n-1} + u_{n-2}$ $(n \geq 2)$ for some value of α, and then $x = u_{n-1} + u_{n+1}$, $y = u_n$ is the general solution of the Diophantine equation $x^2 - (\alpha^2 + 4)y^2 = \pm 4$; in fact

$$(u_{n-1} + u_{n+1})^2 - (\alpha^2 + 4)u_n^2 = (-1)^n 4.$$

In particular, if $\alpha = 1$, we have the Fibonacci sequence (f_n). As part of our investigation into the groups $G(\alpha, n)$, we derived some properties of (u_n), and the related sequences (v_n), defined as in Section 1, and (w_n), defined by $w_0 = 0$, $w_1 = 1$, $w_n = \alpha w_{n-1} + w_{n-2} + 1$ $(n \geq 2)$. Writing u_{n+1} as $\alpha u_n + u_{n-1}$, we have $4\alpha u_n u_{n-1} + 4u_{n-1}^2 - 4u_n^2 = (-1)^n 4$, and hence

Proposition 1: $u_n^2 + \alpha u_n u_{n-1} + u_{n-1}^2 = 2u_n^2 + (-1)^n.$

In the special case that $\alpha = 2$, we have $[u_n(2) + u_{n-1}(2)]^2 - 2u_n(2)^2 = (-1)^n$. Rewriting x as $2z$, we see that the equation $x^2 - 8y^2 = \pm 4$ is equivalent to $z^2 - 2y^2 = \pm 1$, and so $z = u_{n-1}(2) + u_n(2)$, $y = u_n(2)$ is the general solution of the Diophantine equation $z^2 - 2y^2 = \pm 1$, which is confirmed by the fact that $[u_{n-1}(2) + u_n(2)] + u_n(2)\sqrt{2} = (1 + \sqrt{2})^n$. For general values of α, we note

Proposition 2: (i) $u_{2m+1} = u_m^2 + u_{m+1}^2$, (ii) $\alpha u_{2m+2} = u_{m+2}^2 - u_m^2$.

Proposition 2 is readily proved by induction, and we omit the details. We then have

Proposition 3: (i) $u_{2m+1} + u_{2m} = u_{m+1}^2 + 2u_m u_{m+1} - (\alpha - 1)u_m^2$,

(ii) $u_{2m+2} + u_{2m+1} = (\alpha + 1)u_{m+1}^2 + 2u_m u_{m+1} + u_m^2$.

To prove Proposition 3 (i), use Proposition 2 to give

$$\alpha(u_{2m+1} + u_{2m}) = \alpha(u_m^2 + u_{m+1}^2) + (u_{m+1}^2 - u_{m-1}^2) = \alpha u_{m+1}^2 + (u_{m+1}^2 + \alpha u_m^2 - u_{m-1}^2).$$

Rewriting u_{m+1}^2 as $(\alpha u_m + u_{m-1})^2$ gives

$$\alpha(u_{2m+1} + u_{2m}) = \alpha u_{m+1}^2 + [(\alpha u_m + u_{m-1})^2 + \alpha u_m^2 - u_{m-1}^2]$$
$$= \alpha u_{m+1}^2 + 2\alpha u_m(\alpha u_m + u_{m-1}) - (\alpha^2 - \alpha)u_m^2,$$

and rewriting $\alpha u_m + u_{m-1}$ as u_{m+1} yields the result. To prove (ii), use Proposition 2 to give

$$\alpha(u_{2m+2} + u_{2m+1}) = (u_{m+2}^2 - u_m^2) + \alpha(u_m^2 + u_{m+1}^2) = u_{m+2}^2 + \alpha u_{m+1}^2 + (\alpha - 1)u_m^2.$$

Rewriting u_{m+2} as $\alpha u_{m+1} + u_m$ gives

$$\alpha(u_{2m+2} + u_{2m+1}) = (\alpha u_{m+1} + u_m)^2 + \alpha u_{m+1}^2 + (\alpha - 1)u_m^2 = (\alpha^2 + \alpha)u_{m+1}^2 + 2\alpha u_m u_{m+1} + \alpha u_m^2$$

as required. We then have

Proposition 4: $\alpha v_n = u_{n+1} + u_{n-1} - 1 - (-1)^n$,

which can also be proved by induction. If $\alpha = 1$, then $u_{n+1} + u_{n-1} = f_{n+1} + f_{n-1} = g_n$, so we have

Corollary 5: $v_n(1) = g_n - 1 - (-1)^n$

as mentioned in Section 1. More generally, we have

Proposition 6: (i) $v_{2n+1} = (\alpha^2 + 2)v_{2n-1} - v_{2n-3}$;

(ii) $v_{2n+2} = (\alpha^2 + 2)v_{2n} - v_{2n-2} + 2\alpha$.

To prove (i), use Proposition 4 to get

$$\alpha[(\alpha^2+2)v_{2n-1} - v_{2n-3}] = (\alpha^2 + 2)u_{2n} + (\alpha^2 + 1)u_{2n-2} - u_{2n-4} .$$

Writing, in turn, u_{2n-4} as $u_{2n-2} - \alpha u_{2n-3}$ and u_{2n-3} as $u_{2n-1} - \alpha u_{2n-2}$ yields $(\alpha^2 + 2)u_{2n} + \alpha u_{2n-1}$, and then, writing u_{2n-1} as $u_{2n+1} - \alpha u_{2n}$ yields

$$\alpha[(\alpha^2 + 2)v_{2n-1} - v_{2n-3}] = \alpha u_{2n+1} + 2u_{2n} = u_{2n+2} + u_{2n} = \alpha v_{2n+1}$$

by Proposition 4 as required. To prove (ii), we again use Proposition 4 to get

$$\alpha[(\alpha^2 + 2)v_{2n} - v_{2n-2} + 2\alpha] = (\alpha^2 + 2)u_{2n+1} + (\alpha^2 + 1)u_{2n-1} - u_{2n-3} - 2.$$

Writing, in turn, u_{2n-3} as $u_{2n-1} - \alpha u_{2n-2}$ and u_{2n-2} as $u_{2n} - \alpha u_{2n-1}$ yields $(\alpha^2 + 2)u_{2n+1} + \alpha u_{2n} - 2$, and then, writing u_{2n} as $u_{2n+2} - \alpha u_{2n+1}$, yields

$$\alpha[(\alpha^2 + 2)v_{2n} - v_{2n-2} + 2\alpha] = \alpha u_{2n+2} + 2u_{2n+1} - 2 = u_{2n+3} + u_{2n+1} - 2 = \alpha v_{2n+2}$$

by Proposition 4 as required. We now prove

Proposition 7: (i) $v_{4m+2} = \alpha v_{2m+1}^2$; (ii) $v_{4m} = \alpha v_{2m}^2 + 4v_{2m}$.

To prove (i), note that, since $u_{2m}u_{2m+2} = u_{2m+1}^2 - 1$, Propositions 2 and 4 give that

$$\alpha^2 v_{2m+1}^2 = (u_{2m+2} + u_{2m})^2 = u_{2m+2}^2 + 2u_{2m+1}^2 + u_{2m}^2 - 2 = u_{4m+3} + u_{4m+1} - 2 = \alpha v_{4m+2} .$$

To prove (ii), note that, since $u_{2m+1}u_{2m-1} = u_{2m}^2 + 1$, Propositions 2 and 4 give that

$$\alpha^2 v_{2m+1}^2 + 4\alpha v_{2m} = (u_{2m+1} + u_{2m-1} - 2)^2 + 4(u_{2m+1} + u_{2m-1} - 2)$$

$$= u_{2m+1}^2 + 2u_{2m}^2 + u_{2m-1}^2 - 2$$

$$= u_{4m+1} + u_{4m-1} - 2$$

$$= \alpha v_{4m}.$$

We now start considering the sequence (w_n). First we have

Proposition 8: $w_n = w_{n-1} + u_n$.

To prove Proposition 8, let $x_n = w_n - u_n$. Since $x_n = \alpha x_{n-1} + x_{n-2} + 1$ from the definitions of (u_n) and (w_n), and since $x_1 = w_0$ and $x_2 = w_1$, we have that $x_n = w_{n-1}$ for $n \geq 1$, and hence the result. Since $w_0 = u_0$, we may immediately deduce

Corollary 9: $w_n = u_1 + u_2 + \cdots + u_n$.

Since $w_{n+1} = \alpha w_n + w_{n-1} + 1$, Corollary 7 gives that $w_{n-1} + u_n + u_{n+1} - 1 = \alpha w_n + w_{n-1}$, so that $\alpha w_n = u_n + u_{n+1} - 1$, and then

$$\alpha \begin{vmatrix} w_{n-1} & w_n \\ u_{n-1}-1 & u_n \end{vmatrix} = \alpha w_{n-1} u_n - \alpha w_n (u_{n-1} - 1) = u_n^2 - u_{n-1} u_{n+1} + u_{n-1} + u_{n+1} - 1 .$$

Using the fact that $u_n^2 - u_{n-1} u_{n+1} = (-1)^{n+1}$, Proposition 4 now gives the following result, which we used in determining $[G' : G'']$:

Proposition 10: $\begin{vmatrix} w_{n-1} & w_n \\ u_{n-1}-1 & u_n \end{vmatrix} = v_n$.

We finish with an observation. If we write $v_n(\alpha) = \sum_{i=0}^{n-1} a_{n;i} \alpha^i$ as a polynomial in α, where $a_{n;i} = 0$ if $n - i$ is even, then the coefficients $a_{n;i}$ satisfy the Fibonacci-like recurrences

$$a_{n;n-1} = 1; \qquad a_{n;0} = n; \qquad a_{n;i} = a_{n-i;i-1} + a_{n-2;i} .$$

If p is an odd prime, then p divides $a_{p;j}$, $0 \leq j \leq p - 3$, and the prime divisors of both $a_{n;i}$ and v_n have further remarkable properties, some related to the results in [6, 9, 12].

ACKNOWLEDGEMENT

The fourth author would like to thank Hilary Craig for all her help and encouragement.

REFERENCES

[1]   Campbell, C. M., Coxeter, H. S. M. and Robertson, E. F. "Some Families of Finite Groups Having Two Generators and Two Relations." *Proc. Roy. Soc. London 357A* (1977): pp. 423-438.

[2]   Campbell, C. M., Heggie, P. M., Robertson, E. F. and Thomas, R. M. One-Relator Products of Cyclic Groups and Fibonacci-Like Sequences. (Technical Report 35, Department of Computing Studies, University of Leicester, April 1990).

[3]   Campbell, C. M., Robertson, E. F. and Thomas, R. M. "Fibonacci Numbers and Groups." Applications of Fibonacci Numbers. Edited by A. F. Horadam, A. N. Philippou and G. E. Bergum. Kluwer Academic Publ. (1987): pp. 45-49.

[4]   Campbell, C. M., Robertson, E. F. and Thomas, R. M. "On Groups Related to Fibonacci Groups." Group Theory. Edited by K. N. Cheng and Y. K. Leong. Walter de Gruyter & Co. (1989): pp. 323-331.

[5]   Campbell, C. M. and Thomas, R. M. "On (2,n)-Groups Related to Fibonacci Groups." *Israel J. Math. 58* (1987): pp. 370-380.

[6]   Carmichael, R. D. "On the Numerical Factors of the Arithmetic Forms $\alpha^n \pm \beta^n$." *Ann. of Math. 15* (1913): pp. 30-70.

[7]   Doostie, H. Fibonacci-type Sequences and Classes of Groups. Ph. D. Thesis, University of St. Andrews, 1988.

[8]   Hoggatt, V. E. Jr. and Bicknell-Johnson, M. "A Primer for the Fibonacci Numbers XVII; Generalized Fibonacci Numbers Satisfying $u_{n+1}u_{n-1} - u_n^2 = (-1)^n$." *The Fibonacci Quarterly, 16* (1978): pp. 130-137.

[9]   Lehmer, D. H. "An Extended Theory of Lucas Functions." *Ann. of Math. 31* (1930): pp. 419-448.

[10]   Rabin, M. O. "Recursive Unsolvability of Group Theoretic Problems." *Ann. of Math. 67* (1958): pp. 172-194.

[11]   Thomas, R. M. "The Fibonacci Groups Revisited." Proceedings of Groups - St. Andrews (1989) Edited by C. M. Campbell and E. F. Robertson, Cambridge University Press (to appear).

[12]   Ward, M. "The Intrinsic Divisors of Lehmer Numbers." *Ann. of Math. 62* (1955): pp. 230-236.

A GENERALIZATION OF THE FIBONACCI SEARCH

Renato M. Capocelli

1. INTRODUCTION

Generalizing the Fibonacci search we define the Fibonacci search of degree r. Like the Fibonacci search, which it reduces to for $r = 2$, the Fibonacci search of degree r involves only addition and subtraction.

2. FIBONACCI SEARCH OF DEGREE r

The *Fibonacci search of degree r*, in the sequel called *r-Fibonacci search*, is a generalization of *the Fibonacci search* proposed by Ferguson [4] for searching an ordered list and extensively analyzed by Knuth in his monumental work [6]. It reduces to the Fibonacci search for $r = 2$. The important feature is that successive increments can be found by subtraction. The method makes use of *generalized Fibonacci numbers*, in the sequel called *r-Fibonacci numbers*, defined as follows

$$F_0^{(r)} = 0, \; F_1^{(r)} = 1, \; ..., \; F_j^{(r)} = 2^{j-2}, \; j = 2, 3, 4, ..., r+1;$$

$$F_j^{(r)} = F_{j-1}^{(r)} + F_{j-2}^{(r)} + F_{j-3}^{(r)} + \cdots + F_{j-r+1}^{(r)} + F_{j-r}^{(r)}, \; j \geq r+2. \tag{1}$$

The generalization is not straightforward: Indeed, in the generalized version, probes do not divide at each comparison the search interval into the same proportions.

r-Fibonacci search algorithm

Given a table of records R_1, R_2, ..., R_N whose keys are in increasing order $K_1 < K_2 < K_3 < ... < K_N$, the algorithm searches for a given argument K. For convenience in

G. E. Bergum et al. (eds.), Applications of Fibonacci Numbers Volume 4, 69–76.
© 1991 *Kluwer Academic Publishers.*

description, the algorithm assumes that $N+1$ is a perfect Fibonacci number, $F_{k+1}^{(r)}$.

$j \leftarrow 1$.

$i \leftarrow F_k^{(r)}$, $p_1 \leftarrow F_{k-1}^{(r)}$, $p_2 \leftarrow F_{k-2}^{(r)}$, ..., $p_r \leftarrow F_{k-r}^{(r)}$.

(Throughout the algorithm, p_i's will be consecutive Fibonacci numbers of degree r.)

<u>repeat</u>

<u>case</u>

$K = K_i$: SUCCESS;

$K < K_i$: If $p_2 = 0$ <u>then</u> UNSUCCESS;

$$\underline{\text{else }} i \leftarrow i - \sum_{i=2}^{r} p_i; \; j \leftarrow 1; \; (p_1, p_2, ..., p_r) \leftarrow (p_2, p_3, ...p_r, p_1 - \sum_{i=2}^{r} p_i);$$

$K > K_i$: <u>If</u> $j < r-1$ <u>then</u> <u>if</u> $p_2 = 0$ <u>then</u> UNSUCCESS;

$$\underline{\text{else }} i \leftarrow i + p_1; \; j \leftarrow j + 1;$$
$$(p_1, p_2, ..., p_r) \leftarrow (p_2, p_3, ..., p_r, p_1 - \sum_{i=2}^{r} p_i);$$
$$\underline{\text{else }} \underline{\text{if }} p_1 = 0 \quad \underline{\text{then}} \text{ UNSUCCESS};$$
$$\underline{\text{else }} i \leftarrow i + p_2; \; j \leftarrow 1;$$
$$(p_1, p_2, ..., p_r) \leftarrow (p_3, p_4, ..., p_r, p_1 - \sum_{i=2}^{r} p_i, 2p_2 - p_1);$$

<u>until</u> SUCCESS or UNSUCCESS.

The Fibonacci search of degree r never makes more than $k - 1 \approx \dfrac{log_2 N}{log_2 \Phi_1^{(r)}}$ comparisons and has performances which improve as r grows. Here $\Phi_1^{(r)}$ denotes the dominant real root of the characteristic equation $f(x) = x^r - x^{r-1} - \cdots - x - 1 = 0$ of the recurrence (1) [7].

The search can be easily interpreted by displaying its *search (decision) tree* that captures the essential properties of the search algorithm. We shall study the search algorithm through its search tree, the *Fibonacci tree of degree r*, called in the sequel *r-Fibonacci tree*. It should be remarked that r-Fibonacci trees, studied extensively in [2], constitute a beautiful class of trees with many interesting properties. In particular, tree codes associated to them lead to certain universal encodings of integers [3] that are optimal among encodings of integers that are synchronizable (i.e., that do not allow the error propagation) [3], [8].

In general, the r-Fibonacci tree of *order k*, denoted by $T_k^{(r)}$, is a balanced tree in the sense of Adelson-Velskii and Landis [1] that has $F_{k+1}^{(r)}$ terminal nodes and $F_{k+1}^{(r)} - 1$ internal nodes. By induction it can be easily seen that $T_k^{(r)}$ has height $k - 1$ [2].

For $k < 0$, the r-Fibonacci tree is the empty tree Δ. For $k = 0$ or $k = 1$, $T_k^{(r)}$ is simply 0, the root only. For $k > 1$,

$$T_k^{(r)} = \left(T_{k-1}^{(r)} \wedge \left(T_{k-2}^{(r)} \wedge \left(T_{k-3}^{(r)} \wedge \cdots \left(T_{k-r+1}^{(r)} \wedge T_{k-r}^{(r)} \right)_{k-(r-2)} \cdots \right)_{k-2} \right)_{k-1} \right)_k ;$$

where $(A \wedge B)_k$ denotes the tree having root $F_k^{(r)}$, left subtree A and right subtree B with all nodes of B increased by $F_k^{(r)}$. We call the trees $T_{k-i}^{(r)}$, $i = 1, 2, \ldots, r$, *principal subtrees* of the tree $T_k^{(r)}$. In the sequel we will find useful to consider the tree $G_k^{(r)}$ of height $r-1$ obtained from $T_k^{(r)}$ by removing all principal subtrees $T_{k-1}^{(r)}, T_{k-2}^{(r)}, \ldots, T_{k-r}^{(r)}$. We call $G_k^{(r)}$ the *skeleton (tree)* of $T_k^{(r)}$.

Figure 1 shows the Fibonacci tree of degree r and order k.

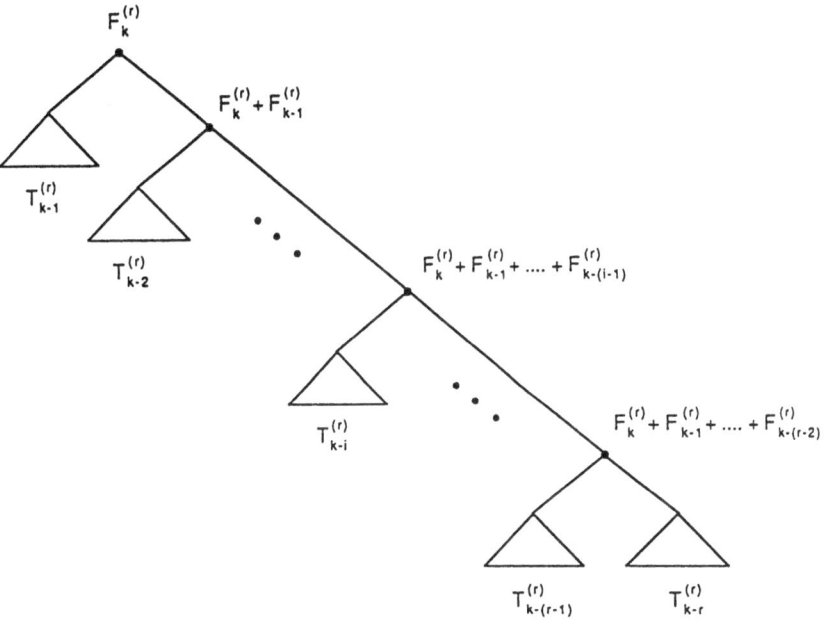

Figure 1. The r-Fibonacci tree of order k $T_k^{(r)}$. Numbers on nodes of subtrees $T_{k-i}^{(r)}$ are increased by $\sum_{j=2}^{i} F_{k-j+2}^{(r)}$.

Figure 2 shows, as an example, the Fibonacci tree of degree 3 and order 6, $T_6^{(3)}$.

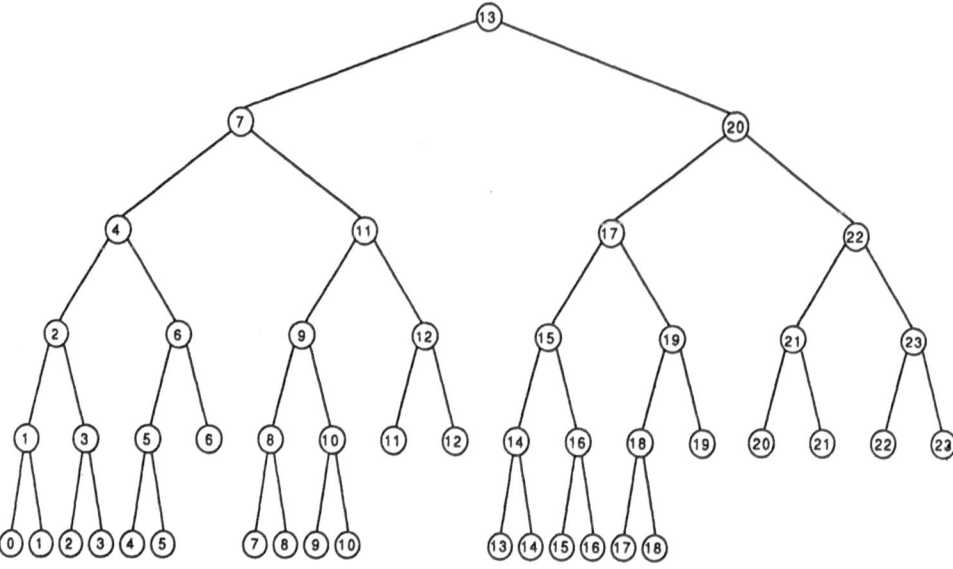

Figure 2. The 3-Fibonacci tree of order 6 $T_6^{(3)}$.

Looking at the r-Fibonacci tree it is possible to verify that, given $N = F_{k+1}^{(r)} - 1$ keys, the r-Fibonacci search starts by comparing the argument with the $F_k^{(r)}$th key. If the argument is greater than the key; at the next stage, the algorithm will compare the argument with the $(F_k^{(r)} + F_{k-1}^{(r)})$th key. The algorithm may iterate the procedure up to $r-2$ consecutive times; at which stage it compares the argument with the $(\sum_{i=0}^{r-2} F_{k-i}^{(r)})$th key. Whenever the argument is less than the current key; or else it results greater than the current key for the $(r-1)$th consecutive time, the algorithm reduces to an r-Fibonacci search on the smaller block the search has been reduced to.

Notice that probes do not divide the search interval into the same proportions, as in the standard Fibonacci search. Indeed, each probe divides the interval into two parts, with the left part about $\dfrac{1}{\Phi_1^{(r)}} \sum_{i=0}^{j} \dfrac{1}{(\Phi_1^{(r)})^i}$ times as large as the right; where j may vary from $r-2$ (when the algorithm starts) to 0 (after $r-2$ consecutive left moves). We recall that $\Phi_1^{(r)}$ denotes the dominant root of the rth degree characteristic equation.

We conclude computing the average number of comparisons required in the r-Fibonacci search for a successful search. Obviously, the maximum number of comparisons is given by the height $k - 1 \approx \dfrac{log_2 N}{log_2 \Phi_1^{(r)}}$ of the tree $T_k^{(r)}$. To establish the result we need the following

Lemma 1: In an r-Fibonacci successful search, the average number of comparisons required for searching a list of size N is asymptotically

$$C(N) = 1 + pC(pN) + q + p^2 C(P^2 N) + (p^3 + p^4 + \cdots + p^{r-1} + p^r) + p^3 C(p^3 N) \qquad (2)$$

$$+ \cdots + (p^{r-1} + p^r) + p^{r-1} C(p^{r-1} N) + p^r C(p^r N)$$

$$= A(r) + pC(pN) + p^2 C(p^2 N) + p^3 C(p^3 N) + \cdots + p^{r-1} C(p^{r-1} N) \qquad (3)$$

$$+ p^r C(p^r N).$$

where $p = 1 - q = 1 - (p^2 + p^3 + \cdots + p^r) = \dfrac{1}{\Phi_1^{(r)}}$, $A(r)$ denotes the average number of comparisons required to reach a principal subtree of $T_k^{(r)}$ (i.e., the average length of $G_k^{(r)}$) and $C(N)$ satisfies the relations $C(1) = 0$; $C(N) = 1 + pC(pN) + qC(qN), N > 1$.

Notice that the number of right moves (yes answers) encountered in the search of a given integer K provides the number of different r-Fibonacci numbers that appear in the generalized Zeckendorf expansion of that integer [2]. In particular for $r = 2$, the number of right moves provides the number of different sets containing the integer K in the game considered by R. V. Jean [5].

Proof: Roughly speaking, the Lemma says that after $i < r - 1$ comparisons with probability p^i the r-Fibonacci search has reached the principal subtree $T_{(k-i)}^{(r)}$ and reduces to the r-Fibonacci search of a list of size $(p^i N)$, in short a $(p^i N)$-element search. After $r - 1$ comparisons, i.e., when $i = r - 1$, the algorithm reduces with probability p^{r-1} to a $(p^{r-1} N)$-element r-Fibonacci search and with probability p^r to a $(p^r N)$-element r-Fibonacci search, respectively.

To prove it let us analyze carefully the r-Fibonacci search algorithm. Given a list of N keys in increasing order, the algorithm starts by comparing the argument with the (pN)th key. If $C(N)$ denotes the average number of comparisons required, in an r-Fibonacci search, to search a list of size N; $C(N)$ approximately satisfies the relations

$$C(1) = 0; \quad C(N) = 1 + pC(pN) + qC(qN), \quad N > 1. \qquad (4)$$

This happens because there is probability p (roughly) that the r-Fibonacci search reduces to a (pN)-element search, and probability q that it reduces to search a list of size (qN), after the first comparison. When N is large, we may ignore the small-order effect caused by the fact that pN, qN, etc...., are not exactly an integer.

The search for a qN-element starts by comparing the argument with the $\left((p+p^2)N\right)$th key. Then, it reduces with probability $\frac{p^2}{q}$ to a (p^2N)-element search and with probability $\frac{(p^3+p^4+\cdots+p^{r-1}+p^r)}{q}$ to the search of a list of size $\left((p^3+p^4+\ldots+p^{r-1}+p^r)N\right)$. That is,

$$C(qN) = 1 + \frac{p^2}{q}C(p^2N) + \frac{(p^3+p^4+\cdots+p^{r-1}+p^r)}{q}C\left((p^3+p^4+\cdots+p^{r-1}+p^r)N\right). \quad (5)$$

Therefore, substituting (5) into (4), one obtains

$$C(N) = 1+pC(N)+q+p^2C(p^2N)+(p^3+p^4+\cdots+p^{r-1}+p^r)C\left((p^3+p^4+\cdots+p^{r-1}+p^r)N\right);$$

where we have denoted by $C\left((p^3+p^4+\cdots+p^{r-1}+p^r)N\right)$ the average number of comparisons required to search a list of size $\left((p^3+p^4+\cdots+p^{r-1}+p^r)N\right)$.

This argument is repeated until the algorithm reduces with probability $\dfrac{p^{r-2}}{p^{r-2}+p^{r-1}+p^r}$ to a $(p^{r-2}N)$-element search and with probability $\dfrac{p^{r-1}+p^r}{p^{r-2}+p^{r-1}+p^r}$ to the search of a list of size $((p^{r-1}+p^r)N)$. Which, in turn, depending on whether the argument is less or not than the $\left(p+p^2+p^3+\cdots+p^{r-1})N\right)$th key, finally reduces with probability $\dfrac{p^{r-1}}{p^{r-1}+p^r}$ to a $(p^{r-1}N)$-element search and with probability $\dfrac{p^r}{p^{r-1}+p^r}$ to (p^rN)-element search. Equation (4) assumes thus the final form

$$C(N) = 1 + pC(pN) + q + p^2C(p^2N) + (p^3+p^4+\cdots+p^{r-1}+p^r) +$$
$$p^3C(p^3)N + \cdots + (p^{r-1}+p^r) + p^{r-1}C(p^{r-1}N) + p^rC(P^rN).$$

To prove (3) we observe that each principal subtree $T_{(k-i)}^{(r)}$ is reached with probability p^i. In addition, we observe that the length of the path from the root to $T_{(k-i)}^{(r)}$ is i if $i < r-1$, and $r-1$ if i is either $r-1$ or r. So that the average length of the skeleton $G_k^{(r)}$ is just

$$p+2p^2+3p^3+4p^4+\cdots+(r-2)p^{r-2}+(r-1)[p^{r-1}+p^r]=1+q+(p^3+p^4+\cdots+p^{r-1}+p^r)$$
$$+\cdots+(p^{r-1}+p^r). \qquad\qquad\qquad\qquad \square$$

Let us now denote by $I(r)$ the quantity $p\log_2 p^{-1} + p^2\log_2 p^{-2} + p^3\log_2 p^{-3} + p^4\log_2 p^4 \cdots$ $+ p^{r-1}\log_2 p^{-(r-1)} + p^r\log_2 p^{-r} = (p+2p^2+3p^3+4p^4+\cdots+(r-1)p^{r-1}+rp^r)\log_2 p^{-1}.$

$I(r)$ represents the average information that has to be provided to reach a principal subtree $T_{k-i}^{(r)}$, $i = 1, 2, \ldots, r$. The following theorem holds.

Theorem 1: The average number of comparisons $C(N)$ required for an r-Fibonacci successful search in a list of size N is

$$C(N) = \frac{A(r)}{I(r)} log_2 N.$$

Proof: From Lemma 1 we have

$$C(N) = A(r) + pC(pN) + p^2 C(p^2 N) + p^3 C(p^3 N) + \cdots + p^{r-1} C(p^{r-1} N) + p^r C(p^r N).$$

To determine $C(N)$ we try a solution of the form $log_x N$, obtaining

$$log_x N = A(r) + p log_x(pN) + p^2 log_x(p^2 N) + p^3 log_x(p^3 N) + \cdots + p^{r-1} log_x(p^{r-1} N) + p^r log_x(p^r N);$$

that, by moving all logarithms to the left side, can be written as

$$log_x\left(p^p p^{2p^2} p^{3p^3} \cdots p^{(r-1)p^{r-1}} p^{r p^r}\right)^{-1} = A(r).$$

From it we get

$$x = \left(p^p p^{2p^2} p^{3p^3} \cdots p^{(r-1)p^{r-1}} p^{r p^r}\right)^{\frac{-1}{A(r)}},$$

that changing to 2 the base of the logarithm gives $C(N) = \dfrac{log_2 N}{log_2 x}$ and the assertion. $\qquad\square$

In case of ordinary Fibonacci search (i.e., $r = 2$) one has that $A(r) = 1$ and $x = \left(p^p p^{2p^2}\right)^{-1}$. Therefrom one obtains that $C(N) = \dfrac{log_2 N}{(p + 2p^2) log_2 \frac{1}{p}} = \dfrac{\Phi}{\sqrt{5}} log_\Phi N$ [4].

Remark: $\dfrac{I(r)}{A(r)}$ furnishes the information provided in average in the r-Fibonacci search by a single comparison. As r becomes large, $p = \dfrac{1}{\Phi_1^{(r)}}$ goes to $\frac{1}{2}$ and the information provided in average by a single comparison reaches the maximum value 1. The average length of $G_k^{(r)}$, denoted by $A(r)$, achieves the lower bound provided by the first Shannon's Theorem, whereas $C(N)$ achieves the minimum value $log_2 N$. The r-Fibonacci search code becomes zero-redundant [2].

REFERENCES

[1]   Adelson-Velskii, G. M. and Landis, E. M. "An Algorithm for Organization of Information," *Soviet Mathematics Doklady, 3*, No. 5 (1962): pp. 1259-1263.

[2]   Capocelli, R. M. "A Generalization of Fibonacci Trees," Applications of Fibonacci
        Numbers, Vol. 3, Kluwer Academic Publishers, Boston, Massachusetts, (1990): pp. 37-56.

[3]   Capocelli, R. M. Comments and Additions to "Robust Transmission of Unbounded
        Strings Using Fibonacci Representations," *IEEE Transactions on Information Theory, 35*
        (1989): pp. 191-193.

[4]   Ferguson, D. E. "Fibonaccian Serach," *Communications to A.C.M., 21,* (1960):
        pp. 315-317.

[5]   Jean, R. V. "An Application of Zeckendorf's Theorem." In this same book.

[6]   Knuth, D. E. The Art of Computer Programming, Vol. 3. Addison-Wesley Publishing
        Company, Reading, Massachusetts (1975).

[7]   Miles, E. P. "Generalized Fibonacci Numbers and Associated Matrices," *American
        Mathematical Monthly, 67* No. 10, (1960): pp. 745-757.

[8]   Yamamoto, H. and Ochi, H. "A New Asymptotically Optimal Code for Positive
        Integers." *IEEE Transactions on Information Theory.* To appear; 1991.

PASCAL'S TRIANGLE: TOP GUN
OR JUST ONE OF THE GANG?

Daniel C. Fielder and Cecil O. Alford

INTRODUCTION

Pascal's triangle can appear as a member of classes of triangular arrays where presumably no class member should be ranked in importance over any other. Two such cases which came to mind were the multinomial triangles [6] and the Hoggatt triangles [2]. No doubt there are others. We selected the multinomial triangles. Was Pascal's triangle only a binomial triangle in a sea of trinomial, quadrinomial, pentanomial, etc., triangles, or might it exhibit a significant influence on the makeup of the other multinomial triangles? We admit a certain prejudice in our choice. Computer experimentation with partition counting, large multinomial expansions, and generating functions using computer algebra systems (muMath, Derive, Mathematica) hinted at a definite Pascal influence. A few years ago, such experimentation would have been virtually impossible.

In this note, we take advantage of features of multinomial triangles to search for influence of Pascal rows and diagonals on their counterparts in the other multinomial triangles. After having seen our results, we leave it to the reader to answer the title question.

SOME INTERESTING AND USEFUL OBSERVATIONS ON MULTINOMIAL TRIANGLES

If the coefficients of the expansions of $(1+x+x^2+x^3+\cdots+x^t)^m$ for fixed t are symmetrically arranged by rows for $m=0, 1, 2, \ldots$, the results are triangular arrays of integers popularly known as *multinomial triangles*. The degenerate triangle consisting of a single vertical

77

G. E. Bergum et al. (eds.), Applications of Fibonacci Numbers Volume 4, 77–90.
© 1991 Kluwer Academic Publishers.

line of 1's is of order $t=0$. The bi- or 2-nomial triangle of order $t=1$ is Pascal's triangle. We take the liberty of discarding the Latin or Greek prefix where convenient and use instead a numerical term, $(t+1)$-nomial. As with the Pascal triangle, all $(t+1)$-nomial triangles exhibit symmetry about a vertical centerline. The symbol $\left\langle {m \atop p} \right\rangle_t$ serves to identify uniquely the integer in position p of row m of triangle t, i.e., the $(t+1)$-nomial triangle. All m, p, and t may take on non-negative integer values 0, 1, 2, 3, ... An example row m is shown in (1).

$$\left\langle {m \atop 0} \right\rangle_t \left\langle {m \atop 1} \right\rangle_t \left\langle {m \atop 2} \right\rangle_t \cdots \left\langle {m \atop p} \right\rangle_t \cdots \left\langle {m \atop mt-2} \right\rangle_t \left\langle {m \atop mt-1} \right\rangle_t \left\langle {m \atop mt} \right\rangle_t \tag{1}$$

There are many straight paths through $(t+1)$-nomial triangles which have interesting numerical properties and which might qualify as diagonals of some type. However, there seems to be one *defacto* "diagonal" sequence designation. The defacto diagonal sequences appear as columns in the left-justified versions of multinomial triangles (See Hoggatt and Bicknell [6]). In the nomenclature of this note, diagonal d of triangle t is the series of integer coefficients

$$\left\langle {\lceil \frac{d}{t} \rceil \atop d} \right\rangle_t , \left\langle {\lceil \frac{d}{t} \rceil + 1 \atop d} \right\rangle_t , \left\langle {\lceil \frac{d}{t} \rceil + 2 \atop d} \right\rangle_t , \left\langle {\lceil \frac{d}{t} \rceil + 3 \atop d} \right\rangle_t , \left\langle {\lceil \frac{d}{t} \rceil + 4 \atop d} \right\rangle_t , \cdots . \tag{2}$$

Although the functionally equivalent left-justified version of the $(t+1)$-nomial triangle was very useful to Hoggatt and Bicknell [6] in the study of diagonal generating functions and generalized Fibonacci sequences and to Greenbury [5] for the direct display of generalized Fibonacci sequences, we use the isosceles form where it suits our purposes. Several $(t+1)$-nomial triangles through $m=5$ are shown in Figure 1. Diagonals can be traced on the triangles of Figure 1 by starting at the first integer of (2) and moving "left $t/2$, down one." Half column widths apply for odd t.

m	$t=0$			$t=1$							$t=2$							
0	1				1								1					
1	1				1		1					1		1		1		
2	1			1		2		1			1		2		3		2	1
3	1		1		3		3		1		1		3		6		7	6	3	1
4	1	1		4		6		4		1	1		4		10		16	19	16	10	4	1
5	1	1	5		10		10		5	1	1	5		15		30	45	51	45	30	15	5	1

```
m                                       t=3
0                                        1
1                            1    1    1    1
2                       1    2    3    4    3    2    1
3                  1    3    6   10   12   12   10    6    3    1
4             1    4   10   20   31   40   44   40   31   20   10    4    1
5        1    5   15   35   65  101  135  155  155  135  101   65   35   15    5    1


m                                       t=4
0                                        1
1                          1   1   1   1   1
2                      1   2   3   4   5   4   3   2   1
3                  1   3   6  10  15  18  19  18  15  10   6   3   1
4              1   4  10  20  35  52  68  80  85  80  68  52  35  20  10   4   1
5        1   5  15  35  70 121 185 255 320 365 381 365 320 255 185 121  70  35  15   5   1
```

Figure 1. Some partial (t+1)-nomial or Multinomial Triangles.

One of the features all (t+1)-triangles share in common is, except for the single 1 in row zero, that every element is the sum of the (t+1) consecutive elements centered on and placed directly above it. It is also obvious that there are $mt+1$ integers per row and that a row sum equals $(t+1)^m$. For the Pascal triangle these are often a student's first discoveries. To satisfy the construction for all $mt+1$ row elements, it may be necessary to visualize some blank positions outside the triangle filled with 0's. The construction idea is important in later manipulation of the general (t+1)-nomial triangle. In addition, the concept can be the basis of spectacular digital computer spreadsheet displays of multinomial triangles.

It can be observed that the class of isosceles form triangles is partitioned into two subclasses according to column structure. The triangles for even t have an odd number of elements in upper row sums. This leads to triangles with unbroken columns. For odd t, the result is row-staggered columns. The column difference is not obvious in left-justified triangles.

Two crucial features of multinomial triangles are:

(a) The first t coefficients, $\left\langle {m \atop 0} \right\rangle_{t-1}, \left\langle {m \atop 1} \right\rangle_{t-1}, \left\langle {m \atop 2} \right\rangle_{t-1}, \cdots \left\langle {m \atop t-1} \right\rangle_{t-1}$, of row m of triangle t-1 equal term-by-term the first t coefficients, $\left\langle {m \atop 0} \right\rangle_{t}, \left\langle {m \atop 1} \right\rangle_{t}, \left\langle {m \atop 2} \right\rangle_{t}, \cdots \left\langle {m \atop t-1} \right\rangle_{t}$, of row m of triangle t.

and

(b) The first $t+1$ coefficients, $\left\langle {m \atop 0} \right\rangle_{t}, \left\langle {m \atop 1} \right\rangle_{t}, \left\langle {m \atop 2} \right\rangle_{t}, \cdots \left\langle {m \atop t} \right\rangle_{t}$, of row m of triangle t equal term-by-term the first $t+1$ coefficients,

$$\left\langle \begin{matrix} \lceil\frac{d}{t}\rceil \\ d \end{matrix} \right\rangle, \ \left\langle \begin{matrix} \lceil\frac{d}{t}\rceil + 1 \\ d \end{matrix} \right\rangle, \ \left\langle \begin{matrix} \lceil\frac{d}{t}\rceil + 2 \\ d \end{matrix} \right\rangle, \ \dots, \ \left\langle \begin{matrix} \lceil\frac{d}{t}\rceil + t \\ d \end{matrix} \right\rangle$$

of diagonal $d=m$-1 of Pascal's triangle. It can be seen that coefficient $\left\langle \begin{matrix} m \\ t+1 \end{matrix} \right\rangle_t$ and all beyond are always less than their diagonal counterparts from Pascal's triangle.

Observation of $(t+1)$-nomial triangles reinforces the central rôle so many times accorded Pascal's triangle. Aside from being a member of other sets of triangles [5], it is a member of the class of all $(t+1)$-nomial triangles as well as a member of the subclass with row-staggered columns. Another observation indicated that sets of diagonals from Pascal's triangle are included in <u>every</u> $(t+1)$-nomial triangle. For $t \geq 2$ the diagonals 0, 1, 2, $\dots$, t of a $(t+1)$-nomial triangle are identically the same numbered diagonals of Pascal's triangle. The equality ends with diagonal t. This fact can be directly associated with the abrupt change in the form of diagonal generating functions experienced by Hoggatt and Bicknell. (See [6], bottom of p. 341 and eq. (1), p. 342.) It was the last observation, coupled with Hoggatt and Bicknells' analytic results, which challenged us to search for a layered mathematical structure for $(t+1)$-triangles where outer layers could be repeatedly "peeled" off to reveal other layers functionally related to parts of Pascal's triangle. Since this analysis of multinomial triangles appeared new, we concentrated our computer experimental efforts on it.

GENERAL DISCUSSION OF COEFFICIENT CALCULATIONS

One obvious way to find the members of row m of a $(t+1)$-nomial triangle is to expand $(1+x+x^2+\dots+x^t)^m$ and collect the (integer) coefficients of powers of x in ascending order from 0 through tm. This is an effortless task even for tremendous t and m values if muMath, Derive, Mathematica, or other computer algebra systems perform the computations. The method, however, is certainly not new.

Our approach to obtaining the coefficients of expanded $(1+x+x^2+\dots+x^t)^m$ is semi-heuristic. We found two different ways and, correspondingly, two different formulations.

In the first, we establish a tabular procedure in which column sums are the coefficients of the expansion, and the column designators are the corresponding powers of x. Through a very selective choice of row values for the examples, a pattern emerges. The pattern evokes a very safe conjecture that the coefficients of row m of triangle t are functions of row m of Pascal's triangle and rows m down through 0 of triangle $t-1$. By concentrating on how single columns are formed, we can predict general formulas for single row members, the $\left\langle \begin{matrix} m \\ p \end{matrix} \right\rangle_t$'s.

In the second, we capitalized on the fact that parts of certain diagonals of Pascal's triangle are also parts of rows of $(t+1)$-nomial triangles. In experimental examples of widely varying size, we forced diagonals of Pascal's triangles to equal, term-by-term, the row coefficients of rows of a $(t+1)$-nomial triangles. As will be detailed later, through careful (crafty?) choice of forcing terms it is shown that coefficients of $(t+1)$-triangles can be found from rows and diagonals of Pascal's triangle alone!

MULTINOMIAL TRIANGLE COEFFICIENTS. FIRST METHOD

After rephrasing, material from Parzan (see [8], page 40, formula 118) states

$$(x^0+x^1+x^2+\cdots+x^t)^m = \sum_{k_0=0}^{m} \sum_{k_1=0}^{m} \cdots \sum_{k_t=0}^{m} \binom{m}{k_0 k_1 \cdots k_t} (x^0)^{k_0} (x^1)^{k_1} \ldots (x^t)^{k_t}$$

$$k_0 + k_1 + k_t \cdots k_t = m, \tag{3}$$

where the integers

$$\binom{m}{k_0 k_1 \cdots k_t} = \left(\frac{m!}{k_0! k_1! \cdots k_t!}\right) \tag{4}$$

are universally called *multinomial coefficients*. From a combinatorics point of view, they are the numbers of permutations, or arrangements, of m objects, k_0 of one kind, k_1 of another kind, $\ldots k_t$ of the last kind. However, the coefficients found in $(d+1)$-nomial, i. e., multinomial, triangles do not, in general, equal multinomial coefficients. The reason is simple. The coefficients of multinomial triangles are coefficients of powers of x in $(x^0+x^1+x^2+\cdots+x^t)^m$. In (3) there may be several different multinomial coefficients having the same total power of x. This establishes multinomial triangle coefficients as being permutations or sums of permutations. We use restricted partitions of the $(tm+1)$ powers of x in the expansion $(x^0+x^1+x^2+\cdots+x^t)^m$ to guide us to the permutations which eventually sum to the $(d+1)$-nomial triangle coefficients.

As a first example, consider row 4 of the 3-nomial triangle. The corresponding expansion is $(x^0+x^1+x^2)^4$ for $t=2$ and $m=4$. The $mt+1 = 9$ powers of x in the expansion are 0, 1, 2, 3, 4, 5, 6, 7, 8. These are the column headings for the subsequent layers of rows. Because a power of x in the expansion is always the sum of powers of x from each of the four $(x^0+x^1+x^2)$'s, the nine powers can be partitioned into four-part partitions with 0 considered as a possible member of a partition. Before tabulating the set of four-part partitions, however, consider the same partitions with all 0's excluded except for the inevitable, single 0 for the zero power of x. Including the single zero partition, we have now the <u>restricted</u> partitions of 0, 1, 2, 3, 4, 5, 6, 7, 8 whose member size cannot exceed $t=2$ and whose number of members cannot exceed $m=4$.

Netto (see [7], page 122) suggests a very simple way of generating <u>unrestricted</u> partitions of an integer n. An algorithm based on Netto's work and adapted for computer use appears in Fielder [1] and Fielder and Alford [3]. To demonstrate the restricted partitions described in the previous paragraph, columns of unrestricted partitions of 0, 1, 2, 3, 4, 5, 6, 7, 8 are recorded as generated by the algorithm. Those partitions which cannot pass the restriction test, either because of member size or number of members or both, are crossed out. The survivors are retained in the order of their generation on their same relative rows. Table 1 verifies the choice of restricted partitions needed in our example development.

Powers of x

0	1	2	3	4	5	6	7	8
0	1	11	111	1111	~~11111~~	~~111111~~	~~1111111~~	~~11111111~~
		2	12	112	1112	~~11112~~	~~111112~~	~~1111112~~
			~~3~~	~~13~~	~~113~~	~~1113~~	~~11113~~	~~111113~~
				22	122	1122	~~11122~~	~~11122~~
				~~4~~	~~14~~	~~114~~	~~1114~~	~~11114~~
					~~23~~	~~123~~	~~1123~~	~~11123~~
					~~5~~	222	1222	~~11222~~
						~~15~~	~~115~~	1115
						~~24~~	~~124~~	1124
						~~33~~	~~133~~	1133
						~~6~~	~~223~~	~~1223~~
							~~16~~	2222
							~~25~~	~~116~~
							~~34~~	~~125~~
							~~7~~	~~134~~
								~~224~~
								~~233~~
								~~17~~
								~~26~~
								~~35~~
								~~44~~
								~~8~~

Table 1. Table of Unrestricted Partitions of Powers of x in Order Generated by Special Algorithm.
<u>Desired Restricted Partitions Shown Not Cancelled</u>

In anticipation of what is to follow, the successful partitions from Table 1, with the zero members restored, are tabulated as shown in Table 2 below.

Powers of x

0	1	2	3	4	5	6	7	8
0000	0001	0011	0111	1111				
		0002	0012	0112	1112			
				0022	1122	1122		
						0222	1222	
								2222

Table 2. Intentionally Arranged Set of $m=4$-Part Restricted Partitions of x With No Member Less Than 0 or Greater Than $t=2$

By considering the partition members as "objects", derangements of the partitions yield the permutation counts which equal the desired coefficients. This is shown in Table 3 where the "layers" are accentuated by boundary lines.

Powers of x

0	1	2	3	4	5	6	7	8
4!/4! 0!	4!/3! 1!	4!/2! 2!	4!/3! 1!	4!/4! 0!				
		4!/3! 1!	4!/2!1!1!	4!/2!1!1!	4!/3! 1!			
			4!/2!2!	4!/2!1!1!	4!/2! 2!			
						4!/3! 1!	4!/3!1!	
								4!/4!0!
1	4	10	16	19	16	10	4	1

Table 3. Permutation Counts Derived From Set of $m=4$-Part Restricted Partitions of Table 2.

The column totals are the coefficients shown at the bottom of Table 3. Now suppose that binary coefficient $\binom{4}{0}$ is factored out of each permutation of layer 0, $\binom{4}{1}$ is factored out of each coefficient of layer 1, $\binom{4}{2}$ is factored out of layer 2, $\binom{4}{3}$ is factored out of layer 3, and $\binom{4}{4}$ is factored out of layer 4.

The new tabular arrangement with the factored binomial coefficients in the left column is shown in Table 4.

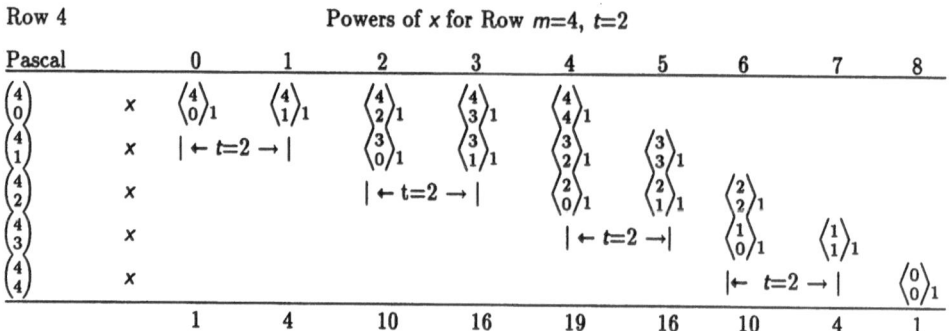

Table 4. Row $m=4$ of Pascal's Triangle Factored Out of Rows of Table 3.

The final row is the numerical version of

$$\left\langle\begin{smallmatrix}4\\0\end{smallmatrix}\right\rangle_2 \left\langle\begin{smallmatrix}4\\1\end{smallmatrix}\right\rangle_2 \left\langle\begin{smallmatrix}4\\2\end{smallmatrix}\right\rangle_2 \left\langle\begin{smallmatrix}4\\3\end{smallmatrix}\right\rangle_2 \left\langle\begin{smallmatrix}4\\4\end{smallmatrix}\right\rangle_2 \left\langle\begin{smallmatrix}4\\5\end{smallmatrix}\right\rangle_2 \left\langle\begin{smallmatrix}4\\6\end{smallmatrix}\right\rangle_2 \left\langle\begin{smallmatrix}4\\7\end{smallmatrix}\right\rangle_2 \left\langle\begin{smallmatrix}4\\8\end{smallmatrix}\right\rangle_2 \tag{5}$$

Although we chose the simplest meaningful case to illustrate our results, the orderly regularity associated with Pascal's and multinomial triangles plus many worked examples using computer algebra convinced us we could state our conclusions as "safe" conjectures. We found that the layers of partitions and their subsequent permutation counts attained multi-row heights as t increases (See the boundary lines in Table 3). Although a discussion is outside the planned scope of this note, even the partition integer member generation assumed a potentially interpretable pattern.

Special mention should be made of the $|\leftarrow t{=}2\rightarrow|$ underlap dimensions of Table 4. The consecutive underlaps of $t{=}2$ are a consequence of crucial feature (a) introduced earlier. Since the first $t{=}2$ coefficients of row 4 of triangle $t{=}1$ (Pascal's triangle in this example) already equal the first $t{=}2$ coefficients of row 4 of triangle $t{=}2$, corrective adjustment is delayed by two units. Under the assumption that the layer philosophy holds (and it appears to), the underlap progresses on consecutive rows as exemplified in Table 4.

Thus, to build row m of triangle t from a known triangle $t-1$, construct a $m+1$ by $mt+1$ table where the column headings are the powers of x of the desired row and the row headings are the coefficients $\binom{m}{0}$, $\binom{m}{1}$, $\binom{m}{2}$, $\ldots$, $\binom{m}{m}$ of row m of Pascal's triangle. Starting with the left position on the $\binom{m}{0}$ row, construct rows m, $m-1$, $m-2$, $\ldots$, 1, 0 of triangle $t-1$ with progressive row underlap equal to t for each row. Multiply the coefficients of each row, respectively, by the corresponding row heading coefficients, replacing the row coefficients with the products. The column sums are the coefficients of the powers of x for row m of the $(t+1)$-nomial triangle.

The formula for coefficient p of row m of triangle t may be deduced as

$$\left\langle {m \atop p} \right\rangle_t = \sum_{i=0}^{m} \binom{m}{i} \left\langle {m-i \atop p-it} \right\rangle_{t-1}. \tag{6}$$

Probably the most useful application of (6) is a formula for the trinomial triangle coefficients

$$\left\langle {m \atop p} \right\rangle_2 = \sum_{i=0}^{m} \binom{m}{i} \left\langle {m-i \atop p-2i} \right\rangle. \tag{7}$$

MULTINOMIAL TRIANGLE COEFFICIENTS. SECOND METHOD

The coefficient formula developed in the previous section required extensive knowledge of results of previous similar formulas. Except for (7), the formula is placed at a distinct computational disadvantage. Despite this, the development of (6) encouraged us to consider the layer concept as a vehicle to a computationally simpler result. In the second method we force a solution by periodically adjusting the integers of a Pascal diagonal $d=m-1$ in such a way that the resulting series of integers is the row m of triangle t. Influenced by the success of (6) and (7), we set up a table with column headings 0, 1, 2, . . . , mt, . . . Again, these are to be the powers of x and the generated column totals are to be the corresponding coefficients of row m of triangle t, i.e., the $(t+1)$-nomial triangle.

From the equality between the beginning $t+1$ coefficients of row m and the corresponding $t+1$ coefficients of diagonal $m-1$ of Pascal's triangle and the fact that thereafter the coefficients of the Pascal diagonal dominate, it is apparent that the first underlap equals $t+1$. Moreover, the first adjustment must be subtractive. If previous experience is to be our guide, $t+1$ appears to be a likely candidate for all overlaps.

To illustrate a meaningful example, we construct row $m=4$ of triangle $t=2$. Table 5 shows the basic construction.

```
1   4   10  20    35    56    84    120    165    220    286    364    455    560    680   (L1)
|←t+1=3→|  -4   -16  -40  -80  -140  -224  -336  -480  -660  -880  -1144  -1456  (L2)
1   4   10  16    19    16    4    -20    -59   -116   -194   -296   -425   -584   -776   (L3)
       |← t+1=3 →|   6    24     60    120    210    336    504    720    990   (L4)
1   4   10  16    19    16    10    4      1     4     16     40     79    136    214   (L5)
              |←   t+1=3   →|    -4    -16    -40    -80    -140   -224   (L6)
1   4   10  16    19    16    10    4      1     0      0      0     -1     -4    -10   (L7)
                        |←  t+1=3   →|     1      4     10   (L8)
1   4   10  16    19    16    10    4      1     0      0      0      0      0      0   (L9)
```

Table 5. Construction of Row $m=4$ of Triangle $t=2$ From Diagonal $d=3$ of Pascal's Triangle

Line (L1) is diagonal $d=3=m-1$ from Pascal's triangle. The first $t+1 = 3$ entries of (L2) are chosen to adjust the diagonal values to equal the second set of three entries of row $m=4$ in triangle $t=2$. Because the subtractive entries are equal to -4 times the first three diagonal entries, it seems like a good choice to continue the pattern for the remainder of (L2) to get (L3). Similarly, the first three entries of (L4) adjust (L3) and suggest the pattern for the remainder of (L4). The process continues until at least $mt+1=9$ values for row $m=4$ have been correctly adjusted. Table 5 carries the process to the limit so that a general procedure is evident. Note that the process emulates a generating function for the polynomial $(1+x+x^2+x^3+\cdots+x^t)^m$.

In Table 6 we summarize the above example, showing the factored out, alternating sign binomial coefficients. The process is similar to that of Table 4.

Alternating Sign Pascal Row $m=4$		Powers of x for Row $m=4$, $t=2$ Only																	
		0	1	2	3	4	5	6	7	8	–	–	–	–	–	–	–		
$\binom{4}{0}$	x	$\binom{3}{3}$	$\binom{4}{3}$	$\binom{5}{3}$	$\binom{6}{3}$	$\binom{7}{3}$	$\binom{8}{3}$	$\binom{9}{3}$	$\binom{10}{3}$	$\binom{11}{3}$	$\binom{12}{3}$	$\binom{13}{3}$	$\binom{14}{3}$	$\binom{15}{3}$	$\binom{16}{3}$	$\binom{17}{3}$			
$-\binom{4}{1}$	x	|←t+1=3→|	$\binom{3}{3}$	$\binom{4}{3}$	$\binom{5}{3}$	$\binom{6}{3}$	$\binom{7}{3}$	$\binom{8}{3}$	$\binom{9}{3}$	$\binom{10}{3}$	$\binom{11}{3}$	$\binom{12}{3}$	$\binom{13}{3}$	$\binom{14}{3}$					
$\binom{4}{2}$	x		|← t+1=3 →|	$\binom{3}{3}$	$\binom{4}{3}$	$\binom{5}{3}$	$\binom{6}{3}$	$\binom{7}{3}$	$\binom{8}{3}$	$\binom{9}{3}$	$\binom{10}{3}$	$\binom{11}{3}$							
$-\binom{4}{3}$	x			|← t+1=3 →|	$\binom{3}{3}$	$\binom{4}{3}$	$\binom{5}{3}$	$\binom{6}{3}$	$\binom{7}{3}$	$\binom{8}{3}$									
$\binom{4}{4}$	x				|← t+1=3 →|	$\binom{3}{3}$	$\binom{4}{3}$	$\binom{5}{3}$											
		1	4	10	16	19	16	10	4	1	0	0	0	0	0	0			

Table 6. Alternating Sign Row $m=4$ of Pascal's Triangle Factored Out of Rows of Table 5.

The non-zero values of the final row are the numerical versions of

$$\left\langle {4 \atop 0} \right\rangle_2 \ \left\langle {4 \atop 1} \right\rangle_2 \ \left\langle {4 \atop 2} \right\rangle_2 \ \left\langle {4 \atop 3} \right\rangle_2 \ \left\langle {4 \atop 4} \right\rangle_2 \ \left\langle {4 \atop 5} \right\rangle_2 \ \left\langle {4 \atop 6} \right\rangle_2 \ \left\langle {4 \atop 7} \right\rangle_2 \ \left\langle {4 \atop 8} \right\rangle_2 \ . \tag{8}$$

In general, to build row m of triangle t entirely from the binomial coefficients of Pascal's triangle, construct a $m+1$ by $(m+1)(t+1)$ table where the first $mt+1$ column headings are the powers of x of the desired row and the row headings are the <u>alternating sign</u> coefficients

$$\binom{m}{0}, \ -\binom{m}{1}, \ \binom{m}{2}, \ \cdots, \ (-1)^p \binom{m}{p}, \ \cdots, \ (-1)^m \binom{m}{m} \tag{9}$$

taken from row m of Pascal's triangle. Starting at the left, fill the zero row with the coefficients of Pascal diagonal $d=m-1$,

$$\binom{m\text{-}1}{m\text{-}1}, \ \binom{m}{m\text{-}1}, \ \binom{m+1}{m\text{-}1}, \ \binom{m+2}{m\text{-}1} \cdots, \ \binom{2m+mt+t\text{-}1}{m\text{-}1} . \tag{10}$$

Fill the succeeding rows with values from (10) shifted to the right by the underlap, $t+1$. Multiply the coefficients of each row, respectively, by the corresponding row heading coefficients, replacing the row coefficients with the products. The column sums are the coefficients of the powers of x for row m of the $(t+1)$-nomial triangle. Because of row symmetry only half or one more than half of the coefficients need be calculated. Thus, only the first $\lceil (mt+1)/2 \rceil$ coefficients of diagonal $m-1$ of Pascal's triangle are essential for an economical computation.

The formula for coefficient p of row m of triangle t, using only parts from Pascal's triangle is

$$\left\langle {m \atop p} \right\rangle_t = \sum_{i=0}^{m} (-1)^i \binom{m}{i} \binom{p+m\text{-}1\text{-}i(t+1)}{m\text{-}1} . \tag{11}$$

Computations can be shortened if the upper limit in (11) is $\lfloor \frac{p+m-1}{t+1} \rfloor$ instead of m.

GENERATING FUNCTIONS AND GENERAL COEFFICIENTS FOR DIAGONALS

The work of Hoggatt and Bicknell [6] further suggests that the diagonals of all multinomial triangles exhibit linear recursion of order $d+1$, independent of t. Through many experimental examples, we observed that all the recursions were "elementary." (See Fielder and Alford [4], page 80.) When the first $d+1$ forward differences of a difference triangle formed from $d+2$ consecutive coefficients of a linear recursion always terminate in zero, the recursion or difference equation is elementary of order $d+1$. The elementary property is evident in Figure 2 below for $d=5$ and $t=3$ and 4, respectively.

n	Δ^0_n	Δ^1_n	Δ^2_n	Δ^3_n	Δ^4_n	Δ^5_n	Δ^6_n
0	2						
		10					
1	12		18				
		28		15			
2	40		33		6		
		61		21		1	
3	101		54		7		0
		115		28		1	
4	216		82		8		
		197		36			
5	413		118				
		315					
6	728						

$t=3$

n	Δ^0_n	Δ^1_n	Δ^2_n	Δ^3_n	Δ^4_n	Δ^5_n	Δ^6_n
0	4						
		14					
1	18		20				
		34		15			
2	52		35		6		
		69		21		1	
3	121		56		7		0
		125		28		1	
4	246		84		8		
		209		36			
5	455		120				
		329					
6	784						

$t=4$

Figure 2. Difference Triangles for $d=5$.

It is shown in [4], that (1) the general expressions for the homogeneous difference equation, (2) the generating function in $\mathcal{Z}$-transform form, and (3) any term of the sequence of an elementary recursion of order, say, $d+1$ can be expressed as predictable, simple expressions in binary coefficients and the first $d+1$ members of the sequence. We present our general diagonal sequence results with the help of material from [4].

To simplify the notation, we use $a_0, a_1, \ldots, a_{d-1}, a_d$ as the first $d+1$ diagonal sequence terms and we let a_n be the general nth term. Through use of (2) and (11), we can calculate the $d+1$ initial values as

$$a_0 = \left\langle {\lceil\frac{d}{t}\rceil \atop d} \right\rangle_t = \sum_{i=0}^{\lceil\frac{d}{t}\rceil} (-1)^i \binom{\lceil\frac{d}{t}\rceil}{i}\binom{d+\lceil\frac{d}{t}\rceil-1-i(t+1)}{\lceil\frac{d}{t}\rceil-1},$$

$$a_1 = \left\langle {\lceil\frac{d}{t}\rceil+1 \atop d}\right\rangle_t = \sum_{i=0}^{\lceil\frac{d}{t}\rceil+1}(-1)^i\binom{\lceil\frac{d}{t}\rceil+1}{i}\binom{d+\lceil\frac{d}{t}\rceil-i(t+1)}{\lceil\frac{d}{t}\rceil},$$

$$\vdots \qquad\qquad\qquad\qquad\qquad\qquad\qquad\qquad\qquad\qquad (12)$$

$$a_{d-1} = \left\langle{\lceil\frac{d}{t}\rceil+d-1 \atop d}\right\rangle_t = \sum_{i=0}^{\lceil\frac{d}{t}\rceil+d-1}(-1)^i\binom{\lceil\frac{d}{t}\rceil+d-1}{i}\binom{d+\lceil\frac{d}{t}\rceil+d-1-i(t+1)}{\lceil\frac{d}{t}\rceil+d-2},$$

$$a_d = \left\langle{\lceil\frac{d}{t}\rceil+d \atop d}\right\rangle_t = \sum_{i=0}^{\lceil\frac{d}{t}\rceil+d}(-1)^i\binom{\lceil\frac{d}{t}\rceil+d}{i}\binom{d+\lceil\frac{d}{t}\rceil+d-i(t+1)}{\lceil\frac{d}{t}\rceil+d-1}.$$

The general homogeneous difference equation for diagonal d of triangle t becomes

$$\sum_{i-1}^{d} (-1)^i \binom{d+1}{i} a_{n+d+1-i} = 0 =$$

$$\binom{d+1}{0} a_{n+d+1} - \binom{d+1}{1} a_{n+d} + \binom{d+1}{2} a^{n+d-1} + \cdots + (-1)^{d+1} \binom{d+1}{d+1} a_n. \qquad (13)$$

For an example with $d=5$ and $t=3$ (See the left difference triangle in Figure 2.), the homogeneous difference equation is

$$\binom{6}{0} a_{n+6} - \binom{6}{1} a_{n+5} + \binom{6}{2} a_{n+4} - \binom{6}{3} a_{n+3} + \binom{6}{4} a_{n+2} - \binom{6}{5} a_{n+1} + \binom{6}{6} a_n = 0. \qquad (14)$$

For $n=0$ we have the check equation.

$$\binom{6}{0}(728) - \binom{6}{1}(413) + \binom{6}{2}(216) - \binom{6}{3}(101) + \binom{6}{4}(40) + \binom{6}{5}(12) + \binom{6}{6}(2) = 0. \qquad (15)$$

Application of the $\mathcal{Z}$-transform to (13) yields a generating function in z for diagonal d in triangle t as

$$\mathcal{Z}(a_n) = \frac{\left[\begin{array}{l} z^{d+1} a_0 + z^d (a_1 - \binom{d+1}{1} a_0) + z^{d-1} (a_2 - \binom{d+1}{1} a_1 + \binom{d+1}{2} a_0) \\ + \cdots + z\{a_{n-1} - \binom{d+1}{1} a_{n-2} + \cdots + (-1)^d \binom{d+1}{d} a_0\} \end{array} \right]}{(z-1)^{d+1}}. \qquad (16)$$

For $d=5$, $t=3$, (16) reduces to

$$\mathcal{Z}(a_n) = \frac{2z^6 - 2z^4 + z^3}{(z-1)^6} \qquad (17)$$

which checks with the left triangle fo Figure 2. (Note: Conventional generating functions in ascending powers of x are available by replacing z in (16) and (17) by $1/x$.)

Although reference [4] has a formula for the general member of elementary recursions in terms of the initial conditions of the recursion, we prefer here the simpler expression developed from (2) and (11). As the general diagonal, it is

$$a_n = \left\langle \begin{array}{c} \lceil \frac{d}{t} \rceil + n \\ d \end{array} \right\rangle_t = \sum_{i=0}^{\lceil \frac{d}{t} \rceil + n} (-1)^i \binom{\lceil \frac{d}{t} \rceil + n}{i} \binom{d + \lceil \frac{d}{t} \rceil + n - i(t+1)}{\lceil \frac{d}{t} \rceil + n - 1}. \qquad (18)$$

CONCLUDING REMARKS

By approaching the construction of multinomial triangles through partition and derangement techniques, we were able, with judicious forcing of component positions, to observe successive patterns of Pascal triangle components in the completed multinomial triangles. As a result, we were able to find finite, closed summation formulas for elements of multinomial triangles in terms of rows of the next lower order multinomial triangles and one row only of Pascal's triangle. This first formula applied to trinomial triangles uses rows of Pascal's triangle. Further work produced general formulas for elements of multiomial triangles of any size (not just trinomial) using one row and one diagonal from Pascal's triangle.

We showed that any diagonal of any multinomial triangle can be completely expressed using at most one row and one diagonal of Pascal's triangle. The diagonal properties documented were homogeneous difference equations, generating functions, and general sequence terms.

Again, we highly recommend the experimental opportunities which modern computer algebra systems offer.

REFERENCES

[1]   Fielder, D. C. "A Combinatorial-Digital Computation of a Network Parameter." *IRE Trans. on Circuit Theory*, CT-9 (1961): pp. 202-209.

[2]   Fielder, D. C. and Alford, C. O. "On a Conjecture by Hoggatt with Extensions to Hoggatt Sums and Hoggatt Triangles." *The Fibonacci Quarterly, 27* (1989): pp. 160-168.

[3]   Fielder, D. C. and Alford, C. O. "Some Computer Experiments on Selected Partitions of Numbers." Currently under review.

[4]   Fielder, D. C. and Alford, C. O. "An Investigation of Sequences Derived From Hoggatt Sums and Hoggatt Triangles." Applications of Fibonacci Numbers, Volume 3, Dordrecht: D. Reidel Publishing Company, (1990). Edited by G. E. Bergum et. al. (Pisa, Italy, July 25-29, 1988.): pp. 77-88.

[5]   Greenbury, G. J. "Tribonacci Triangle." *Mathematics Teacher, 77* (1984): pp. 14, 16.

[6]   Hoggatt, V. E., Jr. and Bicknell, M. "Diagonal Sums of Generalized Pascal Triangles." *The Fibonacci Quarterly, 7* (1969): pp. 341-358.

[7]   Netto, E. Lehrbuch der Combinatorik. (Reprint). New York: Chelsea, 19??.

[8]   Parzan, E. Modern Probability Theory and Its Applications. New York: John Wiley & Sons, 1960.

CONVERSION OF FIBONACCI IDENTITIES INTO HYPERBOLIC IDENTITIES VALID FOR AN ARBITRARY ARGUMENT

Piero Filipponi and Herta T. Freitag

1. INTRODUCTION

The representation of Fibonacci and Lucas numbers in terms of hyperbolic functions [9, p. 7 ff.] and the idea of deriving Fibonacci identities from known hyperbolic identities are not new (e.g., see [6]).

One may ask whether hyperbolic identities valid for an arbitrary argument x can be obtained from known Fibonacci identities. Such a transposition has been studied in [5], but the technique used by the author allows results which are valid only for particular values of the argument, as they depend on integral parameters only.

Using the generalized Fibonacci and Lucas numbers, defined in sec. 2 of this paper, such a restriction can be removed. In sec. 3, we show how known generalized Fibonacci identities can be converted into hyperbolic identities which are valid for an arbitrary argument. Furthermore, this conversion requires a negligible amount of manipulation. Both in sec. 3 and in sec. 4, examples of hyperbolic identities, most of which we believe to be new, are worked out.

2. GENERALITIES

Let us consider identities involving the *generalized Fibonacci numbers* $U_k(r)$ (or simply U_k, if there is no danger of confusion) and the *generalized Lucas numbers* $V_k(r)$ (or simply V_k).

91

G. E. Bergum et al. (eds.), Applications of Fibonacci Numbers Volume 4, 91–98.

Letting T stand for either U or V, these numbers are defined by the linear recurrence

$$T_k = rT_{k-1} + T_{k-2} ,\qquad (2.1)$$

where r is an *arbitrary real* quantity, together with the initial conditions

$$U_0 = 0, \ U_1 = 1, \ V_0 = 2, \ V_1 = r. \qquad (2.2)$$

For r a positive integer, these numbers have been intensively studied over the past years (e.g., see [2], [4] among others) while, for r an arbitrary quantity, some of their properties have been studied in [7].

The numbers U_k and V_k can also be given by the following closed form expressions (Binet's forms)

$$\begin{cases} U_k = (\alpha_r^k - \beta_r^k)/\Delta_r \\ V_k = \alpha_r^k + \beta_r^k \end{cases} ,\qquad (2.3)$$

where

$$\begin{cases} \Delta_r \ = \ \sqrt{r^2 + 4} \\ \alpha_r \ = \ (r + \Delta_r)/2 \\ \beta_r \ = \ (r - \Delta_r)/2 . \end{cases} \qquad (2.4)$$

Note that F_k and L_k (the Fibonacci and Lucas numbers, respectively) become special cases by letting $r = 1$ in $(2.1) - (2.4)$.

Now, denoting by $\log x$ the principal value of the logarithm of an arbitrary quantity x $(x \neq 0)$ and taking into account that $\beta_r = -1/\alpha_r$, the kth generalized Fibonacci number $U_k(r)$ can be expressed as

$$U_k(r) = (\alpha_r^k - \beta_r^k)/\Delta_r = (e^{k\log\alpha_r} - e^{k\log\beta_r})/\Delta_r$$
$$= (e^{k\ln\alpha_r} - e^{k(\ln|\beta_r|+i\pi)})/\Delta_r = (e^{k\ln\alpha_r} - e^{-k\ln\alpha_r} e^{ik\pi})/\Delta_r, \qquad (2.5)$$

where $\ln x$ denotes the natural logarithm of x and i is the imaginary unit. Since k is an integer, we have $e^{ik\pi} = \cos(k\pi) = (-1)^k$. Therefore, (2.5) can be rewritten as

$$U_k(r) = [e^{k\ln\alpha_r} - (-1)^k e^{-k\ln\alpha_r}]/\Delta_r,$$

that is

$$U_k = \begin{cases} 2\sinh(k\ln\alpha_r)/\Delta_r & (k \text{ even}) \\ 2\cosh(k\ln\alpha_r)/\Delta_r & (k \text{ odd}) \end{cases} . \qquad (2.6)$$

Analogously, it can be readily shown that the kth generalized Lucas number $V_k(r)$ can be expressed as

$$V_k = \begin{cases} 2\cosh(k\ln\alpha_r) & (k \text{ even}) \\ 2\sinh(k\ln\alpha_r) & (k \text{ odd}) \end{cases} . \tag{2.7}$$

Incidentally, it may be noted that combining (2.6) and (2.7) yields the following elegant expression for the ratio of V_k to U_k

$$\frac{V_k}{U_k} = \begin{cases} \Delta_r \coth(k\ln\alpha_r) & (k \text{ even}) \\ \Delta_r \tanh(k\ln\alpha_r) & (k \text{ odd}) \end{cases} , \tag{2.8}$$

whence the well-known limit

$$\lim_{k\to\infty} \frac{V_k}{U_k} = \Delta_r \tag{2.9}$$

results immediately.

3. OBTAINING HYPERBOLIC IDENTITIES FROM GENERALIZED FIBONACCI IDENTITIES

Identities involving numbers U_k and V_k hold obviously for any arbitrary real r so that, if we let

$$\ln\alpha_r = x, \tag{3.1}$$

in (2.6) and (2.7), it is clear that such relationships lead to identities concerning hyperbolic functions valid for any argument x. In most cases identities involving U_k and V_k involve also the quantities r and Δ_r. Therefore, suitable expressions for these quantities are needed. Let us proceed in the following way to obtain them.

By (3.1) we have

$$\alpha_r = e^x \tag{3.2}$$

and, by (2.4), we see that

$$2\alpha_r - r = \sqrt{r^2 + 4} . \tag{3.3}$$

From (3.2), and by squaring both sides of (3.3), we get

$$r = (e^{2x} - 1)/e^x = e^x - e^{-x} = 2\sinh x. \tag{3.4}$$

From (2.4), (3.4) and [1, formula (4.5.16)] it follows that

$$\Delta_r = \sqrt{4\sinh^2 x + 4} = 2\sqrt{\sinh^2 x + 1} = 2\cosh x. \tag{3.5}$$

Now, we are in a position to use generalized Fibonacci identities to obtain hyperbolic identities. Let us give some examples.

On the basis of the expressions (e.g., see [3], [4])

$$U_k = \sum_{j=0}^{[(k-1)/2]} \binom{k-1-j}{j} r^{k-1-2j} \tag{3.6}$$

$$V_k = \sum_{j=0}^{[k/2]} \frac{k}{k-j} \binom{k-j}{j} r^{k-2j}, \tag{3.7}$$

where [·] denotes the greatest integer function, substituting (3.1) in (2.6) and (2.7) and using (3.4) and (3.5), we obtain the following multiple-angle formulae expressed in terms of $\cosh x$ and powers of $\sinh x$

$$\sinh(kx) = \begin{cases} \cosh x \displaystyle\sum_{j=0}^{k/2-1} 2^{k-1-2j} \binom{k-1-j}{j} (\sinh x)^{k-1-2j} & (k \text{ even}) \\ \displaystyle\sum_{j=0}^{(k-1)/2} 2^{k-1-2j} \frac{k}{k-j} \binom{k-j}{j} (\sinh x)^{k-2j} & (k \text{ odd}) \end{cases} \tag{3.8}$$

$$\cosh(kx) = \begin{cases} \displaystyle\sum_{j=0}^{k/2} 2^{k-1-2j} \frac{k}{k-j} \binom{k-j}{j} (\sinh x)^{k-2j} & (k \text{ even}) \\ \cosh x \displaystyle\sum_{j=0}^{(k-1)/2} 2^{k-1-2j} \binom{k-1-j}{j} (\sinh x)^{k-1-2j} & (k \text{ odd}) \ . \end{cases} \tag{3.9}$$

Specifically, for $2 \le k \le 10$ in (3.8) and (3.9), we have

$$\sinh(2x) = 2\cosh x \sinh x,$$
$$\sinh(3x) = 4\sinh^3 x + 3\sinh x,$$
$$\sinh(4x) = \cosh x\,(8\sinh^3 x + 4\sinh x),$$
$$\sinh(5x) = 16\sinh^5 x + 20\sinh^3 x + 5\sinh x,$$
$$\sinh(6x) = \cosh x\,(32\sinh^5 x + 32\sinh^3 x + 6\sinh x),$$
$$\sinh(7x) = 64\sinh^7 x + 112\sinh^5 x + 56\sinh^3 x + 7\sinh x,$$
$$\sinh(8x) = \cosh x\,(128\sinh^7 x + 192\sinh^5 x + 80\sinh^3 x + 8\sinh x),$$
$$\sinh(9x) = 256\sinh^9 x + 576\sinh^7 x + 432\sinh^5 x + 120\sinh^3 x + 9\sinh x,$$
$$\sinh(10x) = \cosh x\,(512\sinh^9 x + 1024\sinh^7 x + 672\sinh^5 x + 160\sinh^3 x + 10\sinh x)$$

$$\cosh(2x) = 2\sinh^2 x + 1,$$
$$\cosh(3x) = \cosh x \ (4\sinh^2 x + 1),$$
$$\cosh(4x) = 8\sinh^4 x + 8\sinh^2 x + 1,$$
$$\cosh(5x) = \cosh x \ (16\sinh^4 x + 12\sinh^2 x + 1),$$
$$\cosh(6x) = 32\sinh^6 x + 48\sinh^4 x + 18\sinh^2 x + 1,$$
$$\cosh(7x) = \cosh x \ (64\sinh^6 x + 80\sinh^4 x + 24\sinh^2 x + 1),$$
$$\cosh(8x) = 128\sinh^8 x + 256\sinh^6 x + 160\sinh^4 x + 32\sinh^2 x + 1,$$
$$\cosh(9x) = \cosh x \ (256\sinh^8 x + 448\sinh^6 x + 240\sinh^4 x + 40\sinh^2 x + 1),$$
$$\cosh(10x) = 512\sinh^{10} x + 1280\sinh^8 x + 1120\sinh^6 x + 400\sinh^4 x + 50\sinh^2 x + 1.$$

From the identities

$$\begin{cases} U_k = rU_{k-1} + U_{k-2} = (V_{k-1} + V_{k+1})/\Delta_r^2 \\ V_k = rV_{k-1} + V_{k-2} = U_{k-1} + U_{k+1} \end{cases}$$

we readily deduce the following relations

$$\begin{aligned} \sinh(kx) &= 2\sinh x \ \cosh[(k-1)x] + \sinh[(k-2)x] \\ &= \{\sinh[(k+1)x] + \sinh[(k-1)x]\} \ / \ (2\cosh x) \end{aligned} \qquad (3.10)$$

and

$$\begin{aligned} \cosh(kx) &= 2\sinh x \ \sinh[(k-1)x] + \cosh[(k-2)x] \\ &= \{\cosh[(k+1)x] + \cosh[(k-1)x]\} \ / \ (2\cosh x) . \end{aligned} \qquad (3.11)$$

From the identities

$$\begin{cases} U_{k+1} U_{k-1} - U_k^2 = (-1)^k \quad \text{(Simson's formula)} \\ V_{k+1} V_{k-1} - V_k^2 = (-1)^{k+1} \ \Delta_r^2 \end{cases}$$

we get

$$\begin{aligned} \cosh^2 x &= \cosh[(k+1)x] \ \cosh[(k-1)x] - \sinh^2(kx) \\ &= \cosh^2(kx) - \sinh[(k+1)x] \ \sinh[(k-1)x], \end{aligned} \qquad (3.12)$$

whence $\sinh^2(kx)$ and $\cosh^2(kx)$ are immediately obtained.

Now, let us consider the identities

$$S_1 = \sum_{k=1}^{n} U_{2k} = (U_{2n+1} - 1)/r \qquad (3.13)$$

and

$$S_2 = \sum_{k=1}^{n} V_{2k-1} = (V_{2n} - 2)/r \qquad (3.14)$$

readily obtainable through (2.3) and (2.4). From (3.13) and (3.14) we have

$$S_1 + S_2 / \Delta_r = (U_{2n+1} - 1) / r + (V_{2n} - 2) / (r\Delta_r) \tag{3.15}$$

whence, by (2.6) and (2.7), we can write

$$\sum_{k=1}^{n} \sinh(2kx) + \sum_{k=1}^{n} \sinh[(2k-1)x] = \sum_{k=1}^{2n} \sinh(kx)$$

$$= \frac{\Delta_r (U_{2n+1} - 1) + V_{2n} - 2}{2r}. \tag{3.16}$$

By (3.16), (3.4) and (3.5), we obtain

$$\sum_{k=1}^{2n} \sinh(kx) = \frac{\cosh[(2n+1)x] + \cosh(2nx) - \cosh x - 1}{2\sinh x}. \tag{3.17}$$

Furthermore, considering the difference $S_2/\Delta_r - S_1$ and using an argument analogous to that used above, it is seen that

$$\sum_{k=1}^{2n} (-1)^{k-1} \sinh(kx) = \frac{-\cosh[(2n+1)x] + \cosh(2nx) + \cosh x - 1}{2\sinh x}. \tag{3.18}$$

Now consider the identity (see [7, formulae (4.6), (4.7)])

$$\sum_{k=0}^{n} (-1)^{n-k} \binom{n}{k} T_{2k+s} = r^n T_{n+s}, \tag{3.19}$$

where T stands for either U or V and s is an arbitrary integer. Using (2.6), (2.7) and (3.4) yields a set of hyperbolic identities which can be summarized as

$$\sum_{k=0}^{n} (-1)^{n-k} \binom{n}{k} \sinh[(2k+s)x] = \begin{cases} (2\sinh x)^n \sinh[(n+s)x] & (n \text{ even}) \\ (2\sinh x)^n \cosh[(n+s)x] & (n \text{ odd}) \end{cases} \tag{3.20}$$

$$\sum_{k=0}^{n} (-1)^{n-k} \binom{n}{k} \cosh[(2k+s)x] = \begin{cases} (2\sinh x)^n \cosh[(n+s)x] & (n \text{ even}) \\ (2\sinh x)^n \sinh[(n+s)x] & (n \text{ odd}). \end{cases} \tag{3.21}$$

Let us conclude this section by evaluating some infinite series involving hyperbolic functions.

In the identity (see [7, formula (5.2)])

$$\sum_{j=0}^{\infty} y^j U_{kj+s} = \frac{(-1)^{s-1} y U_{k-s} - U_s}{(-1)^{k-1} y^2 + y V_k - 1} \quad (-\alpha_r^{-k} < y < \alpha_r^{-k}), \tag{3.22}$$

we let y be a real quantity subject to the restrictions

$$-e^{-kx} < y < e^{-kx}. \tag{3.23}$$

Then, by using (2.6), (2.7), (3.2), (3.4) and (3.5), it can be shown that

$$\sum_{j=0}^{\infty} y^j \sinh[(kj+s)x] = \frac{y\sinh[(k-s)x] + \sinh(sx)}{y^2 - 2y\cosh(kx) + 1} \quad (k \text{ and } s \text{ even}) \tag{3.24}$$

and

$$\sum_{j=0}^{\infty} y^j \cosh[(kj+s)x] = \frac{y\cosh[(k-s)x] - \cosh(sx)}{2y\cosh(kx) - y^2 - 1} \quad (k \text{ even, } s \text{ odd}) . \tag{3.25}$$

Analogously, from the identity (see [7, formula (5.6)]),

$$\sum_{j=0}^{\infty} \frac{y^j U_{jk}}{j!} = [\exp(y\alpha_r^k) - \exp(y\beta_r^k)]/\Delta_r \quad (y \text{ arbitrary}) \tag{3.26}$$

we get the sum

$$\sum_{j=0}^{\infty} \frac{y^j \sinh(2jkx)}{j!} = [\exp(ye^{2kx}) - \exp(ye^{-2kx})]/2 \tag{3.27}$$

which, for given y and k, is asymptotic to $\sinh(y\,e^{2kx})$ and $\cosh(y\,e^{2kx})$ as x tends to infinity.

4. CONCLUSIONS

We have shown how hyperbolic identities, valid for an arbitrary argument, can be derived from generalized Fibonacci identities. Of course, other methods to obtain them are available, but we wish to point out that the interest of the technique presented in this paper lies in the fact that, once a Fibonacci identity is known, the amount of manipulations to obtain the corresponding hyperbolic identity is quite negligible.

We have given but a very few examples. Countless additional ones may be derived. As a minor illustration of further obtainable identities, we leave to the reader the proofs of the following identities.

Example 1: Let us consider the 2-by-2 matrix

$$R = \begin{bmatrix} y & 1 \\ 1 & 0 \end{bmatrix}, \tag{4.1}$$

where y is an arbitrary quantity. For k a nonnegative integer,

$$R^{2k} = \frac{1}{\cosh x} \begin{bmatrix} \cosh[(2k+1)x] & \sinh(2kx) \\ \sinh(2kx) & \cosh[(2k-1)x] \end{bmatrix}, \tag{4.2}$$

where $x = \ln\{[y + (y^2+4)^{1/2}] / 2\}$.

Example 2: If m and n are arbitrary natural numbers and

$$x = 2m \ln\left(\frac{\sqrt{n} + \sqrt{n+2}}{\sqrt{2}}\right) \tag{4.3}$$

(i.e., $r = \sqrt{2n}$, cf. (2.4)), then $\cosh x$ is integral. More precisely, we have (see (2.7) and (3.7))

$$\cosh x = 2^m \sum_{j=0}^{m} \frac{m}{2^j(2m-j)} \binom{2m-j}{j} n^{m-j} = \frac{V_{2m}(\sqrt{2n})}{2}. \tag{4.4}$$

Example 3: (see [8, p. 95, formulae (10') and (14')])

$$\frac{2\cosh(2x) + 1}{2\cosh(2x) - 1} = \tanh(3x)\coth x. \tag{4.5}$$

REFERENCES

[1]   Abramowitz, M. and Stegun, I. Handbook of Mathematical Functions. New York: Dover, 1972.

[2]   Bicknell, M. "A Primer on the Pell Sequence and Related Sequences." *The Fibonacci Quarterly 13.4* (1975): pp. 345-349.

[3]   Bong, N. H. "On a Class of Numbers Related to Both the Fibonacci and Pell Numbers." Fibonacci Numbers and Their Applications. Edited by A.N. Philippou, et. al. Dordrecht: D. Reidel Publ. Co., 1986, pp. 9-37.

[4]   Di Porto, A. and Filipponi, P. "A Probabilistic Primality Test Based on the Properties of Certain Generalized Lucas Numbers." *Lecture Notes in Computer Science, 330* (1988): pp. 211-223. Berlin: Springer-Verlag.

[5]   Ehrhart, E. "Associated Hyperbolic and Fibonacci Identities." *The Fibonacci Quarterly 21.2* (1983): pp. 87-96.

[6]   Hoggatt, V. E. Jr. and Bicknell, M. "Roots of Fibonacci Polynomials." *The Fibonacci Quarterly 11.3* (1973): pp. 271-274.

[7]   Horadam, A. F. and Filipponi, P. "Cholesky Algorithm Matrices of Fibonacci Type and Properties of Generalized Sequences." Fond. U. Bordoni Techn. Rept. 3T0988. *The Fibonacci Quarterly* (to appear).

[8]   Jarden, D. Recurring Sequences. 3rd ed., Jerusalem: Riveon Lematematika, 1973.

[9]   Lucas, E. The Theory of Simply Periodic Numerical Functions. Translated from French by S. Kravitz. Edited by D. Lind. Fibonacci Association, 1969.

*Work carried out in the framework of an agreement between the Fondazione U. Bordoni and the Italian PT Administration.

DERIVATIVE SEQUENCES OF FIBONACCI AND LUCAS POLYNOMIALS

Piero Filipponi and Alwyn F. Horadam

1. GENERALITIES

Let us consider the Fibonacci polynomials $U_n(x)$ and the Lucas polynomials $V_n(x)$ (or simply U_n and V_n, if there is no danger of confusion) defined as

$$U_n = xU_{n-1} + U_{n-2} \ (U_0 = 0, \ U_1 = 1) \tag{1.1}$$

and

$$V_n = xV_{n-1} + V_{n-2} \ (V_0 = 2, \ V_1 = x), \tag{1.2}$$

where x is an indeterminate. These polynomials are a natural extension of the numbers $U_n(m)$ and $V_n(m)$ considered in [1]. They have already been considered elsewhere (e.g., see [6]).

The numbers defined by (1.1) and (1.2) can also be expressed by means of the *Binet forms*

$$U_n = (\alpha^n - \beta^n)/\Delta \tag{1.3}$$

and

$$V_n = \alpha^n + \beta^n, \tag{1.4}$$

where

$$\Delta = \sqrt{x^2+4}$$
$$\alpha = (x+\Delta)/2 \tag{1.5}$$
$$\beta = (x-\Delta)/2.$$

G. E. Bergum et al. (eds.), Applications of Fibonacci Numbers Volume 4, 99–108.
© 1991 Kluwer Academic Publishers.

Observe that, if x is real (integral), then U_n and V_n are real (integral).

Further expressions for U_n and V_n (e.g., see [1], [2]) are

$$U_n = \sum_{j=0}^{\lfloor (n-1)/2 \rfloor} \binom{n-1-j}{j} x^{n-1-2j} \quad (n \geq 1),$$
(1.6)

and

$$V_n = \sum_{j=0}^{\lfloor n/2 \rfloor} \frac{n}{n-j} \binom{n-j}{j} x^{n-2j} \quad (n \geq 1),$$
(1.7)

where $\lfloor a \rfloor$ denotes the greatest integer not exceeding a.

1.1 DEFINITIONS

Let us define the polynomials U'_n and V'_n as

$$U'_n = \frac{d}{dx} U_n = \sum_{j=0}^{\lfloor (n-1)/2 \rfloor} (n-1-2j) \binom{n-1-j}{j} x^{n-2-2j} \quad (n \geq 1),$$
(1.8)

$$V'_n = \frac{d}{dx} V_n = \sum_{j=0}^{\lfloor n/2 \rfloor} \frac{n(n-2j)}{n-j} \binom{n-j}{j} x^{n-1-2j} \quad (n \geq 1),$$
(1.9)

and

$$U'_0 = V'_0 = 0.$$
(1.10)

From (1.8)-(1.10) we can write the first few elements of the sequences $\{U'_n\}_0^\infty$ and $\{V'_n\}_0^\infty$, namely,

$U'_0 = 0$	$V'_0 = 0$
$U'_1 = 0$	$V'_1 = 1$
$U'_2 = 1$	$V'_2 = 2x$
$U'_3 = 2x$	$V'_3 = 3x^2 + 3$
$U'_4 = 3x^2 + 2$	$V'_4 = 4x^3 + 8x$
$U'_5 = 4x^3 + 6x$	$V'_5 = 5x^4 + 15x^2 + 5$
$U'_6 = 5x^4 + 12x^2 + 3$	$V'_6 = 6x^5 + 24x^3 + 18x$
$U'_7 = 6x^5 + 20x^3 + 12x$	$V'_7 = 7x^6 + 35x^4 + 42x^2 + 7$
$U'_8 = 7x^6 + 30x^4 + 30x^2 + 4$	$V'_8 = 8x^7 + 48x^5 + 80x^3 + 32x.$

(1.11)

In this paper we confine ourselves to study some properties of the above sequences for the case $x=1$. Since, letting $x=1$ in (1.1)-(1.5), we have the usual Fibonacci numbers (F_n) and Lucas numbers (L_n), the sequences $\{U'_n(1)\}$ and $\{V'_n(1)\}$ will be denoted by $\{F'_n\}$ and $\{L'_n\}$ and defined as *Fibonacci* and *Lucas derivative sequences*, respectively.

From (1.11), the first few values of F'_n and L'_n are

n	0	1	2	3	4	5	6	7	8	
F'_n	0	0	1	2	5	10	20	38	71	(1.12)
L'_n	0	1	2	6	12	25	48	91	168	

2. SOME IDENTITIES INVOLVING THE POLYNOMIALS F'_n AND L'_n

Some simple properties of the polynomials F'_n and L'_n can be derived from the Binet forms (1.3) and (1.4). In fact, letting

$$\begin{cases} \alpha' = \frac{d}{dx}\,\alpha = \frac{x+\Delta}{2\Delta} = \frac{\alpha}{\Delta} \\ \beta' = \frac{d}{dx}\,\beta = \frac{\Delta-x}{2\Delta} = -\frac{\beta}{\Delta} \end{cases} \tag{2.1}$$

and

$$\begin{cases} (\alpha^n)' = \frac{d}{dx}\,\alpha^n = n\alpha^{n-1}\alpha' = n\alpha^n/\Delta \\ (\beta^n)' = \frac{d}{dx}\,\beta^n = n\beta^{n-1}\beta' = -n\beta^n/\Delta\,, \end{cases} \tag{2.2}$$

we can write

$$U'_n = \frac{d}{dx}\left(\frac{\alpha^n - \beta^n}{\Delta}\right) = \frac{[(\alpha^n)' - (\beta^n)']\Delta - (\alpha^n - \beta^n)x/\Delta}{\Delta^2}$$

$$= \frac{[(\alpha^n)' - (\beta^n)']\Delta^2 - (\alpha^n - \beta^n)x}{\Delta^3} = \frac{n(\alpha^n + \beta^n)\Delta - x(\alpha^n - \beta^n)}{\Delta^3}$$

$$= \frac{nV_n\Delta - xU_n\Delta}{\Delta^3} = \frac{nV_n - xU_n}{\Delta^2} \tag{2.3}$$

and

$$V'_n = \frac{d}{dx}(\alpha^n + \beta^n) = (\alpha^n)' + (\beta^n)' = n\frac{\alpha^n - \beta^n}{\Delta} = nU_n\,. \tag{2.4}$$

Letting $x = 1$ in (2.3) and (2.4) yields

$$F'_n = \frac{nL_n - F_n}{5} \tag{2.5}$$

and

$$L'_n = nF_n, \tag{2.6}$$

whence the expressions for the negative-subscripted elements of the Fibonacci and Lucas derivative sequences can be easily deduced

$$F'_{-n} = (-1)^{n+1} F'_n \tag{2.7}$$

and

$$L'_{-n} = (-1)^n L'_n. \tag{2.8}$$

The identities (2.5) and (2.6) allow us to state the following propositions concerning the parity of F'_n and L'_n the proofs of which are based on the well-known fact that both F_{3n} and L_{3n} are always even.

Proposition 1: F'_n is odd iff n is even and not a multiple of 3.

Proposition 2: L'_n is odd iff n is odd and not a multiple of 3.

Also the recurrence relations

$$F'_n = F'_{n-1} + F'_{n-2} + F_{n-1} \quad (F'_0 = F'_1 = 0) \tag{2.9}$$

and

$$L'_n = L'_{n-1} + L'_{n-2} + L_{n-1} \quad (L'_0 = 0, \ L'_1 = 1) \tag{2.10}$$

can be readily derived from (2.5) and (2.6), respectively.

Now let us state the following four identities.

<u>Identity 1:</u> $F'_{n+p} + (-1)^p F'_{n-p} = L_p F'_n + F_n L'_p$.

<u>Identity 2:</u> $F'_{n+p} - (-1)^p F'_{n-p} = F_p L'_n + L_n F'_p$.

<u>Identity 3:</u> $L'_{n+p} + (-1)^p L'_{n-p} = L_p L'_n + L_n L'_p$.

<u>Identity 4:</u> $L'_{n+p} - (-1)^p L'_{n-p} = nL_n F_p + pL_p F_n = 5(F_n F'_p + F_p F'_n) + 2F_n F_p$.

For the sake of brevity, we shall prove only Id. 1.

Proof of Id. 1: Using (2.5), (2.6) and identities I_{21}-I_{24} [5], we can write

$$F'_{n+p} + (-1)^p F'_{n-p} = \{n[L_{n+p} + (-1)^p L_{n-p}] + p[L_{n+p} - (-1)^p L_{n-p}] - [F_{n+p} + (-1)^p F_{n-p}]\}/5$$

$$= (nL_n L_p + 5pF_n F_p - F_n L_p)/5 = L_p(nL_n - F_n)/5 + pF_n F_p = L_p F'_n + F_n L'_p .$$

Particular cases of Id. 1-Id. 4 are

Identity 5: $(p=1$ in Id. 2$)$: $L'_n = F'_{n-1} + F'_{n+1}$.

Identity 6: $(p=1$ in Id. 4$)$: $L'_{n-1} + L'_{n+1} = 5F'_n + 2F_n$.

Identity 7: $(p=2$ in Id. 2$)$: $F'_{n+2} - F'_{n-2} = L'_n + L_n$.

Identity 8: $(n=p$ in Id. 2$)$: $F'_{2p} = F_p L'_p + L_p F'_p$.

Identity 9: $(n=p$ in Id. 3$)$: $L'_{2p} = 2L_p L'_p$.

Identity 10: $(n=2p$ in Id. 2$)$: $F'_{3p} = [L_{2p}+(-1)^p]F'_p + 2F_{2p}L'_p = [L_{2p}+(-1)^p]F'_p + F_p L'_{2p}$.

Identity 11: $(n=2p$ in Id. 3$)$: $L'_{3p} = L_p L'_{2p} + L'_p[L_{2p}-(-1)^p] = 2L_p^2 L'_p + L'_p[L_{2p}-(-1)^p]$

$$= 3L'_p [L_{2p}+(-1)^p] .$$

Further identities obtainable from (2.5) and (2.6) are

Identity 12: $(F'_n)^2 - F'_{n-p}F'_{n+p} = \dfrac{(pL_n)^2 - (-1)^{n+p}[F_p^2(5n^2-1) - 5(L'_p)^2 + L'_{2p}]}{25}$

and

Identity 13: $(L'_n)^2 - L'_{n-p}L'_{n+p} = (-1)^{n+p}[(nF_p)^2 - (L'_p)^2] + (pF_n)^2$.

For the sake of brevity we shall prove only Id. 13.

Proof of Id. 13: Using (2.6) and I_{19} [5], we can write

$$(L'_n)^2 - L'_{n-p}L'_{n+p} = (L'_n)^2 - (n^2 - p^2)[F_n^2 - (-1)^{n+p}F_p^2]$$

$$= (L'_n)^2 - (nF_n)^2 + (-1)^{n+p}(nF_p)^2 + (pF_n)^2 - (-1)^{n+p}(pF_p)^2$$

$$= (-1)^{n+p}[(nF_p)^2 - (L'_p)^2] + (pF_n)^2 .$$

Particular cases of Id. 12 and Id. 13 are the *Simson formula* analogs for $\{F'_n\}$ and $\{L'_n\}$

Identity 14: $(p=1$ in Id. 12$)$: $(F'_n)^2 - F'_{n+1}F'_{n-1} = \dfrac{L_n^2 + (-1)^n(5n^2 - 4)}{25}$

and

Identity 15: $(p=1$ in Id. 13$)$: $(L'_n)^2 - L'_{n+1}L'_{n-1} = F_n^2 - (-1)^n(n^2 - 1)$.

3. EVALUATION OF SOME SERIES INVOLVING F'_n AND L'_n

In this section several series involving the polynomials F'_n and L'_n are considered and closed form expressions for their sums are exhibited. For the sake of brevity, only a few among them are proved in detail.

The following well-known identities (e.g., see [4]), that are valid for arbitrary y, have been used throughout the proofs:

$$\sum_{i=0}^{k} iy^i = [ky^{k+2} - (k+1)y^{k+1} + y]/(y-1)^2, \tag{3.1}$$

$$\sum_{i=0}^{k} i^2y^i = [k^2y^{k+3} - (2k^2+2k-1)y^{k+2} + (k+1)^2y^{k+1} - y^2 - y]/(y-1)^3. \tag{3.2}$$

The further identities

$$\sum_{i=0}^{k} \binom{k}{i} y^i = (y+1)^k, \tag{3.3}$$

$$\sum_{i=0}^{k} \binom{k}{i} iy^i = ky(y+1)^{k-1} \tag{3.4}$$

and

$$\sum_{i=0}^{k} \binom{k}{i} i^2y^i = ky(y+1)^{k-1} + k(k-1)y^2(y+1)^{k-2} \tag{3.5}$$

have been used, the proofs of which are omitted in this context. Observe that (3.4) does not apply for $y = -1$ and $k=0$, while (3.5) does not apply for $y = -1$ and $k=0$ or 1; this rather unfortunate case occurs in the proof of (3.12).

3.1 RESULTS

The following results have been established, in which the symbols S_F and S_L refer to Fibonacci and Lucas sums of derivatives, and T_F and T_L to the corresponding sums when the derivative is multiplied by a binomial coefficient:

$$S_F(k) = \sum_{i=0}^{k} F'_i = F'_{k+2} - F_{k+3} + 1 \tag{3.6}$$

$$S_L(k) = \sum_{i=0}^{k} L'_i = L'_{k+2} - L_{k+3} + 2 \tag{3.7}$$

$$S_F^{(2)}(k) = \sum_{i=0}^{k} (F'_i)^2 = \frac{k^2L_{2k+1} - 2kF_{2k+1} + F_kF_{k+1} + 6F_{2k} - 5F'_{2k} + k(k+1)(-1)^k}{25} \tag{3.8}$$

$$S_L^{(2)}(k) = \sum_{i=0}^{k} (L'_i)^2 = \frac{k^2L_{2k+1} + 4F_{2k} - k(k+1)(-1)^k}{5} - F'_{2k} \tag{3.9}$$

$$T_F(k) = \sum_{i=0}^{k} \binom{k}{i} F'_i = \frac{F'_{2k-1}}{2} \tag{3.10}$$

$$T_L(k) = \sum_{i=0}^{k} \binom{k}{i} L'_i = kF_{2k-1} = L'_{2k-1} - (k-1)F_{2k-1} \tag{3.11}$$

$$T_F^{(2)}(k) = \sum_{i=0}^{k} \binom{k}{i} (F'_i)^2 = \begin{cases} 5^{(k-6)/2}(k^2 L_{k+2} + L_k - 5L'_k) & \text{for } k \text{ even } (k \geq 4) \\ 5^{(k-5)/2}(k^2 F_{k+2} - 5F'_k) & \text{for } k \text{ odd } (k \geq 3) \end{cases}$$

$$(T_F^{(2)}(0) = T_F^{(2)}(1) = 0, T_F^{(2)}(2) = 1) \tag{3.12}$$

(observe that $T_F^{(2)}(3) = L_4$ and $T_F^{(2)}(4) = L_8$)

$$T_L^{(2)}(k) = \sum_{i=0}^{k} \binom{k}{i} (L'_i)^2 = \begin{cases} 5^{(k-4)/2} k[(k-1)L_{k+2} + 5F_{k+1}] & \text{for } k \text{ even } (k \geq 4) \\ 5^{(k-3)/2} k[(k-1)F_{k+2} + L_{k+1}] & \text{for } k \text{ odd } (k \geq 3) \end{cases}$$

$$(T_L^{(2)}(0) = T_L^{(2)}(1) = 1, T_L^{(2)}(2) = 6) \tag{3.13}$$

$$S_F = \sum_{n=0}^{\infty} \frac{F'_n}{n!} = \frac{\alpha^2 e^\alpha - \beta^2 e^\beta}{5\sqrt{5}} \tag{3.14}$$

$$S_L = \sum_{n=0}^{\infty} \frac{L'_n}{n!} = \frac{\alpha e^\alpha - \beta e^\beta}{\sqrt{5}} \qquad \text{(cf. [3, (3.5)]) .} \tag{3.15}$$

3.2 PROOFS

The proofs of the above results are not complicated but they are rather lengthy. Therefore, as mentioned earlier, only some among them will be given in full detail.

Proof of (3.6): Using (2.5), we write

$$S_F(k) = \tfrac{1}{5} \left(\sum_{i=0}^{k} iL_i - \sum_{i=0}^{k} F_i \right)$$

whence, by (1.3), (1.4) and (3.1), we get

$$S_F(k) = [kL_{k+4} - (k+1)L_{k+3} + 4 - F_{k+2} + 1]/5 = (kL_{k+2} - F_{k+2} - L_{k+3})/5 + 1$$

$$= F'_{k+2} - (2L_{k+2} + L_{k+3})/5 + 1 = F'_{k+2} - (L_{k+5} - L_{k+1})/5 + 1$$

$$= F'_{k+2} - F_{k+3} + 1.$$

Proof of (3.9): Using (2.6), (1.3) and (3.2), we can write

$$S_L^{(2)} = \sum_{i=0}^{k} i^2 F_i^2 = \tfrac{1}{5} [\sum_{i=0}^{k} i^2 (\alpha^{2i} + \beta^{2i}) - 2 \sum_{i=0}^{k} i^2 (-1)^i]$$

$$= \tfrac{1}{5} [k^2 L_{2k+4} - (3k^2 + 2k - 1)L_{2k+2} + (3k^2 + 4k)L_{2k} - (k^2 + 2k + 1)L_{2k-2} - k(k+1)(-1)^k]$$

$$= \tfrac{1}{5} [k^2 L_{2k+1} + 4F_{2k} - k(k+1)(-1)^k] - F'_{2k}.$$

Proof of (3.11): Using (2.6), (1.3) and (3.4), we obtain

$$T_L(k) = \sum_{i=0}^{k} \binom{k}{i} i F_i = \tfrac{1}{\sqrt{5}} \left(\sum_{i=0}^{k} \binom{k}{i} i \alpha^i - \sum_{i=0}^{k} \binom{k}{i} i \beta^i \right)$$

$$= \tfrac{1}{\sqrt{5}} [k\alpha(\alpha+1)^{k-1} - k\beta(\beta+1)^{k-1}] = kF_{2k-1} \text{ or, equivalently,}$$

$$T_L(k) = L'_{2k-1} - (k-1)F_{2k-1}.$$

Proof of (3.12): Using (2.5), (1.3) and (1.4) we have

$$T_F^{(2)}(k) = \sum_{i=0}^{k} \binom{k}{i} \left(\frac{iL_i - F_i}{5} \right)^2 = \sum_{i=0}^{k} \binom{k}{i} \frac{i^2 L_i^2 + F_i^2 - 2iF_{2i}}{25}$$

$$= \tfrac{1}{25} \sum_{i=0}^{k} \binom{k}{i} i^2 [\alpha^{2i} + \beta^{2i} + 2(-1)^i] + \tfrac{1}{25} \sum_{i=0}^{k} \binom{k}{i} F_i^2 + \tfrac{2}{25\sqrt{5}} \sum_{i=0}^{k} \binom{k}{i} i(\alpha^{2i} - \beta^{2i}) .$$

Now, using (3.4) and (3.5), we can write

$$T_F^{(2)}(k) = \tfrac{1}{25} [k\alpha^2 (\alpha^2 + 1)^{k-1} + k(k-1)\alpha^4 (\alpha^2+1)^{k-2} + k\beta^2 (\beta^2+1)^{k-1}$$

$$+ k(k-1)\beta^4 (\beta^2+1)^{k-2} + 2X_k] - \tfrac{2}{25\sqrt{5}} [k\alpha^2(\alpha^2+1)^{k-1} - k\beta^2(\beta^2+1)^{k-1}]$$

$$+ \tfrac{1}{25} \sum_{i=0}^{k} \binom{k}{i} F_i^2, \tag{3.16}$$

where

$$X_k = \sum_{i=0}^{k} \binom{k}{i} i^2 (-1)^i = \begin{cases} -1 \text{ if } k = 1 \\ 2 \text{ if } k = 2 \\ 0 \text{ if } k = 0 \text{ or } k \geq 3 \text{ (cf. (3.5)).} \end{cases} \tag{3.17}$$

Taking the relations $\alpha^2 + 1 = \alpha\sqrt{5}$ and $\beta^2 + 1 = -\beta\sqrt{5}$ into account, and using identities I_{45}

and I_{47} [5] to calculate the last addend on the right-hand side of (3.16), we obtain

$$\begin{cases} T_F^{(2)}(0) = T_F^{(2)}(1) = 0 \\ T_F^{(2)}(2) = 1 \end{cases}$$

by direct calculation, and, after some simple manipulations, we can write

(i) for k ≥ 4 even

$$T_F^{(2)}(k) = k5^{(k-5)/2}(\alpha^{k+1} - \beta^{k+1}) + k(k-1)5^{(k-6)/2}(\alpha^{k+2} - \beta^{k+2})$$

$$- 2k5^{(k-6)/2}(\alpha^{k+1} + \beta^{k+1}) + 5^{(k-6)/2}L_k$$

$$= k5^{(k-4)/2}F_{k+1} + k(k-1)5^{(k-6)/2}L_{k+2} - 2k5^{(k-6)/2}L_{k+1} + 5^{(k-6)/2}L_k$$

$$= 5^{(k-6)/2}\{k[5F_{k+1} + (k-1)L_{k+2} - 2L_{k+1}] + L_k\}$$

$$= 5^{(k-6)/2}\{k[(k-1)L_{k+2} + L_{k-2}] + L_k\} = 5^{(k-6)/2}[k(kL_{k+2} - 5F_k) + L_k]$$

$$= 5^{(k-6)/2}[k^2L_{k+2} - 5kF_k + L_k] = 5^{(k-6)/2}(k^2L_{k+2} + L_k - 5L_k'),$$

(ii) for k ≥ 3 odd

$$T_F^{(2)}(k) = k5^{(k-5)/2}(\alpha^{k+1} + \beta^{k+1}) = k(k-1)5^{(k-6)/2}(\alpha^{k+2} - \beta^{k+2})$$

$$- 2k5^{(k-6)/2}(\alpha^{k+1} - \beta^{k+1}) + 5^{(k-5)/2}F_k$$

$$= 5^{(k-5)/2}[k(k-1)F_{k+2} + k(L_{k+1} - 2F_{k+1}) + F_k]$$

$$= 5^{(k-5)/2}[k(k-1)F_{k+2} + kF_{k-2} + F_k] = 5^{(k-5)/2}[k^2F_{k+2} - kL_k + F_k]$$

$$5^{(k-5)/2}(k^2F_{k+2} - 5F_k') \ .$$

Proof of (3.14): From (2.6) we have

$$S_F = \tfrac{1}{5}\left(\sum_{n=0}^{\infty} \frac{nL_n}{n!} - \sum_{n=0}^{\infty} \frac{F_n}{n!} \right) = \tfrac{1}{5}\left(\sum_{n=0}^{\infty} \frac{L_{n+1}}{n!} - \sum_{n=0}^{\infty} \frac{F_n}{n!} \right)$$

whence, by (1.3) and (1.4), one obtains

$$S_F = \tfrac{1}{5}[(\alpha - 1/\sqrt{5}) \sum_{n=0}^{\infty} \frac{\alpha^n}{n!} + (\beta + 1/\sqrt{5}) \sum_{n=0}^{\infty} \frac{\beta^n}{n!}]$$

$$= \frac{1}{5\sqrt{5}}[(\alpha\sqrt{5} - 1)e^\alpha + (\beta\sqrt{5} + 1)e^\beta] = \frac{\alpha^2 e^\alpha - \beta^2 e^\beta}{5\sqrt{5}} \ .$$

4. CONCLUDING REMARKS

In this paper we have limited the investigation to $x=1$ and to first derivatives.

A study of higher derivatives of the polynomials $U_n(x)$ and $V_n(x)$ may be subsequently undertaken. Certainly, a further examination of cases in which $x \neq 1$ is worthwhile. More particularly, when $x=2$ the polynomials give rise to the *Pell numbers* and the *Pell-Lucas numbers* which warrant still further inspection.

ACKNOWLEDGEMENT

This work was carried out in the framework of an agreement between the Italian PT Administration and the Fondazione "Ugo Bordoni".

REFERENCES

[1]   Bicknell, M. "A Primer on the Pell Sequences and Related Sequences." *The Fibonacci Quarterly 13, 4* (1975): pp. 345-349.

[2]   Di Porto, A. and Filipponi, P. "A Probabilistic Primality Test Based on the Properties of Certain Generalized Lucas Numbers." in *Lecture Notes in Computer Science, 330*, pp. 211-223. Berlin: Springer-Verlag, 1988.

[3]   Filipponi, P. and Horadam, A. F. "A Matrix Approach to Certain Identities." *The Fibonacci Quarterly 26, 2* (1988): pp. 115-126.

[4]   Gauthier, N. "Derivation of a Formula for $\sum r^k x^r$." *The Fibonacci Quarterly 27, 5* (1989): pp. 402-408.

[5]   Hoggatt, V. E. Jr., Fibonacci and Lucas Numbers. Boston: Houghton Mifflin, 1969.

[6]   Horadam, A. F. and Filipponi, P. "Cholesky Algorithm Matrices of Fibonacci Type and Properties of Generalized Sequences." Fond. U. Bordoni Rept. 3T0988. *The Fibonacci Quarterly* (to appear).

A CARRY THEOREM FOR RATIONAL BINOMIAL COEFFICIENTS

Dan Flath and Rhodes Peele

1. INTRODUCTION

Ernst Eduard Kummer proved in 1852 that for any nonnegative integers j and k and any prime p, the *exponent* of the highest power of p that divides the binomial coefficient $\binom{j+k}{k}$ equals the *number of carries* that occur when j and k are added together in the p-ary number system. This elegant theorem has been an inspiration and a point of departure for many authors. For example, it has been generalized for i) multinomial coefficients [5], [11], [2]; ii) Gaussian or q-binomial coefficients [3], [4], [6]; iii) "Fibonomial" coefficients [8], [6]; and iv) regularly divisible C-nomial coefficients [6].

In this paper we prove and use another generalization of Kummer's theorem. Our result applies to p-integral rational binomial coefficients, or more generally to p-adic integral binomial coefficients. Section 2 sets the stage, and Section 3 contains the precise statement and proof of this generalization. Section 4 develops various applications. The final section clarifies the role of the hypothesis that p is prime, and briefly discusses interconnections with Stirling numbers of the first kind and with elementary p-adic analysis.

2. PRELIMINARIES*

Let p be a fixed prime. For any nonzero rational number x, define $\mathrm{ord}_p(x)$ to be the unique integer exponent e such that $x = p^e \frac{r}{s}$ where $r, s \in \mathbf{Z}$ and r and s are nonmultiples of p.

---

*Many readers will probably be familiar with the contents of this section. It may be skipped and referred to later as the need arises.

G. E. Bergum et al. (eds.), *Applications of Fibonacci Numbers Volume 4*, 109–120.
© 1991 *Kluwer Academic Publishers.*

Define $\text{ord}_p(0) = +\infty$. Then $\mathbf{Z}_p = \{x \in \mathbf{Q} : \text{ord}_p(x) \geq 0\}$ is a subring of $\mathbf{Q}$ containing $\mathbf{Z}$, and is called the ring of *p-integral rational numbers*. For every $x \in \mathbf{Z}_p$, there is a unique sequence of integers $\mathbf{x} = (x_0, x_1, \ldots, x_i, \ldots)$ such that

$$0 \leq x_i \leq p-1 \quad \text{for all } i \geq 0; \text{ and } \text{ord}_p\left(x - \sum_{i=0}^{k} x_i p^i\right) > k \quad \text{for all } k \geq 0.$$

We often denote this sequence $\mathbf{x}$ by the formal series $\sum_{i=0}^{\infty} x_i p^i$, and refer to it as the *p-adic expansion* of x.

We denote the set of *all* integer sequences $\mathbf{x} = (x_0, x_1, \ldots, x_i, \ldots) = \sum_{i=0}^{\infty} x_i p^i$ such that $0 \leq x_i \leq p-1$ for all $i \geq 0$ by $\hat{\mathbf{Z}}_p$, and we extend the function ord_p to all of $\hat{\mathbf{Z}}_p$ by defining $\text{ord}_p(\mathbf{x}) = \min\{i : x_i > 0\}$ for $\mathbf{x} \in \hat{\mathbf{Z}}_p \backslash \mathbf{Z}$. If also $\mathbf{y} = (y_0, y_1, \ldots, y_i, \ldots) \in \hat{\mathbf{Z}}_p$, we define $\mathbf{s} = \mathbf{x} + \mathbf{y}$ (and simultaneously define an integer *carry sequence* $\mathbf{c}_+(\mathbf{x}, \mathbf{y}) = (c_0, c_1, \ldots, c_i, \ldots)$) by setting $c_0 = 0$ and requiring that

$$0 \leq s_i \leq p-1 \quad \text{and} \quad x_i + y_i + c_i = c_{i+1} p + s_i \quad \text{for all } i \geq 0.$$

Note that in fact $c_i \in \{0,1\}$ for all $i \geq 0$. We denote $|\{i : c_i = 1\}|$, which can be finite or infinite, by $\text{carry}_p(\mathbf{x}, \mathbf{y})$. Similarly we define $\mathbf{p} = \mathbf{x} \cdot \mathbf{y}$ and $\mathbf{c}_\times(\mathbf{x}, \mathbf{y})$ by setting $c_0 = 0$ and requiring that

$$0 \leq p_i \leq p-1 \quad \text{and} \quad \sum_{j=0}^{i} x_j y_{i-j} + c_i = c_{i+1} p + p_i \quad \text{for all } i \geq 0.$$

Note that in fact $c_i \in \mathbf{Z}_\geq$ for all $i \geq 0$.

The set $\hat{\mathbf{Z}}_p$ with these operations is called the *ring of p-adic integers*. The mapping $\iota : \mathbf{Z}_p \to \hat{\mathbf{Z}}_p$ defined by $x \mapsto \mathbf{x} = \sum_{i=0}^{\infty} x_i p^i$ is a ring injection, and the sequences $(x_0, x_1, \ldots, x_i, \ldots)$ in the image $\iota(\mathbf{Z}_p)$ are characterized as the *eventually periodic sequences*; i.e., those for which there exist $\mu \in \mathbf{Z}_\geq$ and $\pi \in \mathbf{Z}_>$ such that $x_i = x_{\pi+i}$ for all $i \geq \mu$. We will denote such sequences by $(x_0, \ldots, x_{\mu-1}, \overline{x_\mu, \ldots, x_{\mu+\pi-1}})$, and we will sometimes identify $\mathbf{Z}_p$ with $\iota(\mathbf{Z}_p)$.

Let $\mathbf{x} \in \mathbf{Z}_p$ (or $\hat{\mathbf{Z}}_p$). In many proofs it helps to set $t_n(\mathbf{x}) = \sum_{i=0}^{n-1} x_i p^i$ and $e_n(\mathbf{x}) = \sum_{i=n}^{\infty} x_i p^{i-n}$ for $n \in \mathbf{Z}_>$. The reason it helps is that the function $|\cdot|_p : \hat{\mathbf{Z}}_p \to \mathbf{R}_\geq$ defined by $|0|_p = 0$ and $|\mathbf{x}|_p = \dfrac{1}{p^{\text{ord}_p \mathbf{x}}}$ for $\mathbf{x} \neq 0$ is a norm, and in its induced metric topology on $\hat{\mathbf{Z}}_p$ (called the *p-adic topology*) we can think of $t_n(\mathbf{x})$ as a nonnegative integer approximation to $\mathbf{x}$ for n large, and view $p^n e_n(\mathbf{x})$ as an error term. In other words, $\mathbf{Z}_\geq$ is dense in $\hat{\mathbf{Z}}_p$ in the p-adic topology.

As a metric space, $\hat{\mathbf{Z}}_p$ is in fact the *completion* of $\mathbf{Z}_p$ with respect to the metric $|\cdot|_p$. Similarly [1], one can complete $\mathbf{Q}$ to obtain $\hat{\mathbf{Q}}_p$, the *field of p-adic numbers*, which is the field of fractions of $\hat{\mathbf{Z}}_p$.

3. GENERALIZED CARRY THEOREM

For any element q of a field of characteristic 0, and any nonnegative integer k, we may define a *generalized binomial coefficient* as follows:

$$\binom{q}{k} = \begin{cases} 1 & \text{if } k = 0; \\ \dfrac{q(q-1)(q-2)\cdots(q-k+1)}{1\cdot 2\cdot 3\cdots k} & \text{if } k > 0. \end{cases}$$

We will prove:

Theorem 3.1: Let $x \in \mathbf{Z}_p$ be a p-integral rational number and let k be a nonnegative integer. Then

a) $\left(\dfrac{x+k}{k}\right)$ is also p-integral; and

b) $\operatorname{ord}_p\left(\dfrac{x+k}{k}\right) = \operatorname{carry}_p(x,k)$.

Before proceeding, we consider some illuminating examples with $p = 3$ and $k = 5$.

Examples: Choose $x_1 = 4$, $x_2 = -1/8$, $x_3 = -4$, $x_4 = -10$.   Then $\iota(x_1) = (1,1,\bar{0})$, $\iota(x_2) = (\overline{1,0})$ $\iota(x_3) = (2,1,\bar{2})$ $\iota(x_4) = (2,2,1,\bar{2})$. We perform the additions $x_i + k$ for $i = 1,\ldots,4$, obtaining

carries	0	1	1	0	$\cdots$
$x_1 = 4$	1	1	0	0	$\cdots$
$k = 5$	2	1	0	0	$\cdots$
$x_1 + k = 9$	0	0	1	0	$\cdots$

carries	0	1	0	0	$\cdots$
$x_2 = -1/8$	1	0	1	0	$\cdots$
$k = 5$	2	1	0	0	$\cdots$
$x_2 + k = 39/8$	0	2	1	0	$\cdots$

carries	0	1	1	1	$\cdots$
$x_3 = -4$	2	1	2	2	$\cdots$
$k = 5$	2	1	0	0	$\cdots$
$x_3 + k = 1$	1	0	0	0	$\cdots$

carries	0	1	1	0	$\cdots$
$x_4 = -10$	2	2	1	2	$\cdots$
$k = 5$	2	1	0	0	$\cdots$
$x_4 + k = -5$	1	1	2	2	$\cdots$

The theorem predicts that

$$\operatorname{ord}_3\binom{9}{5} = 2, \quad \operatorname{ord}_3\binom{39/8}{5} = 1, \quad \operatorname{ord}_3\binom{1}{5} = +\infty, \quad \operatorname{ord}_3\binom{-5}{5} = 2$$

and one verifies by direct calculation that these values are correct. The first and last examples are linked by the reciprocity identity $(-1)^k \binom{-q}{k} = \binom{q+k-1}{k}$.

Our proof of Theorem 3.1 depends on the classical Kummer carry theorem of Section 1 [7], [11] and the two lemmas that follow. (See Section 2 for notation.)

Lemma 3.2 Let $f(X) \in \mathbf{Q}[X]$ be a polynomial with rational coefficients such that $f(n) \in \mathbf{Z}$ for all nonnegative integers n. Then

      a) $f(x) \in \mathbf{Z}_p$ for all $x \in \mathbf{Z}_p$

      b) $\lim\limits_{k \to \infty} \operatorname{ord}_p(f(t_k(x))) = \operatorname{ord}_p(f(x))$ for all $x \in \mathbf{Z}_p$

Proof: Let $f(X) = \sum_{n=0}^{N} a_n X^n$ where $a_n \in \mathbf{Q}$. Let $x \in \mathbf{Z}_p$. Compute:

$$a_n x^n = a_n[t_k(x) + p^k e_k(x)]^n$$

$$= a_n t_k(x)^n + a_n p^k \left[\sum_{j=1}^{n} \binom{n}{j} p^{k(j-1)} t_k(x)^{n-j} e_k(x)^j \right]$$

$$= a_n t_k(x)^n + a_n p^k w_{n,k}$$

where $w_{n,k} \in \mathbf{Z}_p$. Summing over n, we get

$$f(x) = f(t_k(x)) + y_k \qquad\qquad (*)$$

where

$$y_k = \sum_{n=0}^{N} a_n p^k w_{n,k}.$$

For every positive integer L there is an integer K such that $a_n p^k \in \mathbf{Z}_p$ and $\operatorname{ord}_p(a_n p^k) > L$ for all n and all $k \geq K$. It follows that

$$y_k \in \mathbf{Z}_p \text{ for all sufficiently large } k$$

and that

$$\lim_{k \to \infty} \operatorname{ord}_p(y_k) = \infty. \qquad\qquad (**)$$

We prove a). By hypothesis, $f(t_k(x)) \in \mathbf{Z} \subset \mathbf{Z}_p$. Therefore, for sufficiently large k, equation $(*)$ expresses $f(x)$ as a sum of two elements of $\mathbf{Z}_p$. It follows that $f(x) \in \mathbf{Z}_p$.

We prove b). There are two cases.

Case 1: $f(x) \neq 0$.

There exists K such that $\operatorname{ord}_p(y_k) > \operatorname{ord}_p(f(x))$ for all $k \geq K$. Hence, for $k \geq K$, $\operatorname{ord}_p(f(t_k(x))) = \operatorname{ord}_p(f(x) - y_k) = \operatorname{ord}_p(f(x))$.

Case 2: $f(x) = 0$.

Since $f(t_k(x)) = -y_k$, we have by (**) that

$$\lim_{k\to\infty} \operatorname{ord}_p f(t_k(x)) = \lim_{k\to\infty} \operatorname{ord}_p(y_k) = \infty = \operatorname{ord}_p(0) = \operatorname{ord}_p(f(x)). \qquad \text{QED}$$

Lemma 3.3: Let a be a nonnegative integer, and let $b \in \mathbf{Z}_p$. Then

$$\lim_{k\to\infty} \operatorname{carry}_p(a, t_k(b)) = \operatorname{carry}_p(a, b).$$

Proof: Again there are two cases.

Case 1: $\operatorname{carry}_p(a, b)$ is finite.

Let K be a positive integer so large that $a_i = 0$ for all $i \geq K$ and there is no carry to any column beyond column $K - 1^*$ in the sum of the p-adic expansions of a and b. It is evident that $\operatorname{carry}_p(a, t_k(b)) = \operatorname{carry}_p(a, b)$ for all $k \geq K$.

Case 2: $\operatorname{carry}_p(a, b)$ is infinite.

Let K be such that $a_i = 0$ for all $i \geq K$. Then $b_i = p - 1$ for all $i \geq K$, and a 1 is carried to column i for all $i \geq K$. If $k \geq K$, then the sum of a and $t_k(b)$ must have carries to all columns numbered $K, K + 1, \ldots, k$. Thus $\operatorname{carry}_p(a, t_k(b)) > k - K$. Hence

$$\lim_{k\to\infty} \operatorname{carry}(a, t_k(b)) = \infty. \qquad \text{QED}$$

Proof of Theorem 3.1: To prove a), we just apply Lemma 3.2a to the polynomial $f(X) = \binom{X+k}{k}$. To prove b), we apply the classical carry theorem to deduce that for all nonnegative integers n,

$$\operatorname{ord}_p \binom{t_n(x) + k}{k} = \operatorname{carry}_p(t_n(x), k),$$

let $n \to \infty$, and use Lemma 3.2b and 3.3. QED

To conclude this section, we will sketch a (fancier) proof of a generalization of Theorem 3.1.

Theorem $\widehat{3.1}$: Let $\mathbf{x} \in \hat{\mathbf{Z}}_p$ be a p-adic integer, and let k be a nonnegative integer. Then

a) $\binom{\mathbf{x}+k}{k}$ is a p-adic integer:

b) $\operatorname{ord}_p \binom{\mathbf{x}+k}{k} = \operatorname{carry}_p(\mathbf{x}, k)$.

Proof: Proceed as above, but replace Lemma 3.2 by an analog for $\hat{\mathbf{Z}}_p$ as follows:

---

*Column numbers refer to subscripts so that the leftmost column is column zero.

Lemma 3.2: Let $f(X) \in \mathbb{Q}[X]$ be a polynomial with rational coefficients such that $f(n) \in \mathbb{Z}$ for all nonnegative integers n. Then

a) $f(x) \in \hat{\mathbb{Z}}_p$ for all $x \in \hat{\mathbb{Z}}_p$.

b) $\lim_{k \to \infty} \operatorname{ord}_p(f(t_k(x))) = \operatorname{ord}_p(f(x))$ for all $x \in \hat{\mathbb{Z}}_p$.

Proof: a) $f : \hat{\mathbb{Z}}_p \to \hat{\mathbb{Q}}_p$ is a continuous function, the nonnegative integers $\mathbb{Z}_{\geq}$ are dense in $\hat{\mathbb{Z}}_p$, $f(\mathbb{Z}_{\geq}) \subset \hat{\mathbb{Z}}_p$, and $\hat{\mathbb{Z}}_p$ is closed.

b) $\operatorname{ord}_p(f(\,\cdot\,)) : \hat{\mathbb{Z}}_p \to \{0, 1, \ldots, +\infty\}$ is a continuous function. (Neighborhoods of $+\infty$ are the sets containing $+\infty$ and all sufficiently large integers.) QED

4. APPLICATIONS

Let p prime, $q \in \mathbb{Z}_p$ or $\hat{\mathbb{Z}}_p$, and $m \in \mathbb{Z}_{>} \cup \{+\infty\}$ all be fixed, and define an infinite *p-exponent matrix* $A(p; q; m)$ as follows:

$$
a_{i,j} = \begin{cases} \operatorname{ord}_p \dbinom{q+i}{j} & \text{if } \operatorname{ord}_p \dbinom{q+i}{j} < m; \\ + & \text{otherwise.} \end{cases} \qquad (i \in \mathbb{Z};\ j \in \mathbb{Z}_{\geq})
$$

It is suggestive to represent such a matrix as a multicolored pattern in which each symbol s in the alphabet $\{0, 1, \ldots, m-1, +\}$ is assigned a different color, and all "pixels" (i, j) for which $a_{i,j} = s$ are colored the color assigned to s. All of the applications that we develop in this section can be interpreted as proofs of certain attributes of these patterns. They fall into three broad categories : i) periodicity properties; ii) replication properties; iii) surjectivity properties.

4i. Periodicity: For $m \geq 1$ define the funciton

$$
\phi_m : \{0, 1, \ldots, +\infty\} \to \{0, 1, \ldots, m-1, +\}
$$

by the formula

$$
\phi_m(x) = \begin{cases} x & \text{if } x < m; \\ + & \text{if } x \geq m. \end{cases}
$$

COLUMNS.

Proposition 4i.1: Let $1 \leq m < +\infty$, let $r \geq 1$, and let $b \in \hat{\mathbb{Z}}_p$. Then

$$
\phi_m \operatorname{ord}_p \dbinom{b + p^{r+m-1}}{j} = \phi_m \operatorname{ord}_p \dbinom{b}{j}
$$

for all j such that $0 \le j < p^r$. Hence the sequence $c_i = \phi_m \mathrm{ord}_p \binom{b+i}{j}$ is periodic for every b and j.

Proof: Let $a = b - j = (a_0, a_1, \ldots)$. By the generalized carry theorem we must prove that

$$\phi_m(\mathrm{carry}_p(j, a + p^{r+m-1})) = \phi_m(\mathrm{carry}_p(j, a)).$$

There are two cases.

Case 1: In the sum of j and a there is no carry added to column $r+m-1$; i.e., $(j+a)_{r+m-1} = a_{r+m-1}$. In this case,

$$\mathrm{carry}_p(j, a + p^{r+m-1}) = \mathrm{carry}_p(j, a),$$

so the result is clear.

Case 2: In the sum of j and a there is a carry added to column $r+m-1$. Then there must in fact be carries added to all columns from column r to column $r+m-1$ inclusive, from which it follows that

$$\phi_m(\mathrm{carry}_p(j, a + p^{r+m-1})) = \phi_m(\mathrm{carry}_p(j, a)) = + . \qquad \text{QED}$$

ROWS.

Proposition 4i.2: Let $b \in \hat{Z}_p$, and $1 \le m < +\infty$. Then the sequence $r_j = \phi_m \mathrm{ord}_p \binom{b}{j}$ is eventually periodic if and only if $b \in Z$.

Proof: Suppose $b \in Z$. If $b \ge 0$, then (r_j) is eventually periodic since $r_j = +$ for $j > b$. If $b < 0$, choose $k \ge 0$ so that $b_i = p-1$ for $i \ge k$. Let $j \in Z_\ge$, and consider the p-adic summation

carries	$c_0\, c_1 \cdots c_k$	$\cdots c_{k+m-1}\, c_{k+m}$	$\cdots$
j	$j_0\, j_1 \cdots j_k$	$\cdots j_{k+m-1}\, j_{k+m}$	$\cdots$
$b-j$	$*\; * \cdots\; *$	$\cdots\quad *$	$*\; \cdots$
b	$b_0\, b_1 \cdots p-1 \cdots p-1$	$p-1\; \cdots$

The carries c_i, $0 \le i \le k+m-1$ are the same as the carries to the corresponding columns in the sum of $t_{k+m-1}(j)$ and $b - t_{k+m-1}(j)$, because changing digits in the summands in columns $k+m-1$ and beyond can only affect carries to columns $k+m$ and beyond. If $c_{k+m-1} = 0$, then $c_i = 0$ for $i \ge k+m-1$, and so $\mathrm{carry}_p(j, b-j) = \mathrm{carry}_p(t_{k+m-1}(j), b-t_{k+m-1}(j))$. If $c_{k+m-1} = 1$, then working backwards, we see that $c_{k+m-2} = c_{k+m-3} = \cdots = c_k = 1$, so that $\mathrm{carry}_p(j, b-j) \ge m$ and

$\mathrm{carry}_p(t_{k+m-1}(j), b - t_{k+m-1}(j)) \geq m$. In all cases, $r_j = r_{t_{k+m-1}(j)}$. Hence the value of r_j is determined by the first $k+m-1$ p-ary digits of j. So (r_j) is periodic with period p^{k+m-1}.

It suffices to prove the converse for the case $m = 1$. Suppose that $b \in \hat{Z}_p \backslash Z$. We will show that (r_j) is not eventually periodic by showing that it contains strings of consecutive $+$'s of arbitrary finite length, but it does not contain a string of consecutive $+$'s of infinite length.

Let $(b_0, b_1, b_2, \ldots)$ be the p-adic expansion of b. Since b is not a *negative* integer, there are arbitrarily large subscripts k such that $b_k < p - 1$. Then for all the p^k consecutive integers $j = \sum_{i=0}^{k} j_i p^i$ with $j_k = p - 1$, we have $\mathrm{carry}_p(j, b - j) \geq 1$ since the p-ary addition of j and $b - j$ necessarily entails a carry to column $k+1$. Hence $r_j = \,+$. On the other hand, since b is not a *nonnegative* integer, there are arbitrarily large l such that $0 < b_l$. Letting $j = 1 \cdot p^l$, we see that $\mathrm{carry}_p(j, b - j) = 0$ since the one nonzero p-ary digit of j does not produce a carry when added to the corresponding digit of $b - j$. Hence $r_j = 0$, and the two claims of the previous paragraph are proved. QED

4ii. Replication: Imagine that a finite portion of one of the color patterns described at the beginning of this section is displayed on a color monitor. The following result essentially says that if q is nonintegral, then the displayed portion of the pattern is exactly replicated infinitely often in both the vertical and horizontal directions, and that in the vertical direction this replication is periodic.

Proposition 4ii: Let p be prime. Let $b \in \hat{Z}_p \backslash Z$. Let A, B, C be three integers such that $A \leq B$ and $0 \leq C$.

a) There exists a positive integer m such that

$$\mathrm{ord}_p \binom{b+i+kp^m}{j} = \mathrm{ord}_p \binom{b+i}{j}$$

for all integers i, j, k such that $A \leq i \leq B$, $0 \leq j \leq C$, and $-\infty < k < +\infty$.

b) There exist arbitrarily large integers $n > 0$ such that

$$\mathrm{ord}_p \binom{b+i}{j+p^n} = \mathrm{ord}_p \binom{b+i}{j}$$

for all integers i and j such that $A \leq i \leq B$ and $0 \leq j \leq C$.

Proof: Since $b \notin \mathbf{Z}$, we have that $b + i - j \notin \mathbf{Z}$ for integers i and j. Two things follow.

In the first place, $\mathrm{carry}_p(j, b + i - j)$ is finite for all integers i and j. This means that there exists an integer m such that $C < p^m$ and all carries in the sums of the p-adic expansions of j and $b + i - j$ where $A \leq i \leq B$ and $0 \leq j \leq C$ are added to columns 0 through $m - 1$. Since $b + i - j + kp^m$ can differ from $b + i - j$ only in columns m and beyond and j is 0 from column m on out, we get

$$\mathrm{carry}_p(j, b + i - j + kp^m) = \mathrm{carry}_p(j, b + i - j)$$

for $A \leq i \leq B$ and $0 \leq j \leq C$. Part a) follows directly from Theorem 3.1. *

In the second place, there are arbitrarily large positive integers n such that $(b + i - j)_n > 0$ for all i and j such that $A \leq i \leq B$ and $0 \leq j \leq C$. For such n, $(b + i - j - p^n)_n < p - 1$. If moreover, $p^n > C \geq j$ and all carries in the sum of j and $b + i - j$ are added to the columns 0 through $n - 1$, then the sum of $j + p^n$ and $b + i - j - p^n$ will have no carry from column n, which is the only column where it differs from the sum of j and $b + i - j$. Hence

$$\mathrm{carry}_p(j + p^n, \, b + i - j - p^n) = \mathrm{carry}_p(j, b + i - j)$$

which is equivalent to b) by Theorem 3.1.

4iii. Surjectivity: Suppose one chooses $m = +\infty$, so that the color pattern has (potentially) infinitely many colors. The next proposition says that, indeed, all ∞ colors except possibly for the infinite color corresponding to $s = +\infty$ occur in *all* columns of the pattern except the zeroth column, and that if q is nonintegral, then all finite colors occur in *all* rows of the pattern.

Proposition 4iii: Let p be prime, and let $b \in \hat{\mathbf{Z}}_p$.

a) If $j > 0$, then for every nonnegative integer n there exists $i \in \mathbf{Z}$ such that
$$\mathrm{ord}_p \binom{b + i}{j} = n.$$

b) If $b \notin \mathbf{Z}$, then for every nonnegative integer n there exists $j \geq 0$ such that
$$\mathrm{ord}_p \binom{b}{j} = n.$$

---

*Part a) can also be easily deduced from Proposition 4i.1.

Proof: a) Let n be a nonnegative integer. Let $j = \sum_{k=0}^{K} j_k p^k = (j_0, j_1, ..., j_K, \bar{0})$ be the p-adic expansion of j, where $j_K \neq 0$. Let i be an integer such that the p-adic expansion of $a = b + i - j$ satisfies

$$a_k = 0 \text{ for } 0 \leq k < K;$$
$$a_k = p - 1 \text{ for } K \leq k < K + n;$$
$$a_{K+n} = 0;$$
$$a_k \text{ unrestricted for } k > K + n.$$

Then it is evident that $\operatorname{ord}_p \binom{b+i}{j} = \operatorname{carry}_p(j, b + i - j) = n$.

b) Let $b = (b_0, b_1, ...) \in \hat{\mathbb{Z}}_p \backslash \mathbb{Z}$. Because b is not a negative integer there are arbitrarily large k such that $b_k < p - 1$. Suppose that there were a K such that $b_k < p - 1$ implied $b_{k+1} = 0$ for $k > K$. Then choosing $k > K$ such that $b_k < p - 1$, we would get $b_l = 0$ for all $l > k$, which contradicts the fact that b is not a nonnegative integer. Thus there is no such K. Hence there are infinitely many nonnegative integers k such that $b_k < p - 1$ and $b_{k+1} > 0$.

Let $k_1, k_2, ...$ be a sequence of subscripts such that $b_{k_i} < p - 1$, $b_{k_i+1} > 0$, and $k_i + 1 < k_{i+1}$ for all i. Let n be a nonnegative integer, and let $j = \sum_{i=1}^{n}(b_{k_i} + 1)p^{k_i}$. Note that j is a nonnegative integer. It is easy to compute

$$\operatorname{ord}_p \binom{b}{j} = \operatorname{carry}_p(j, b - j)$$

$$= \sum_{i=1}^{n} \operatorname{carry}_p((b_{k_i} + 1)p^{k_i}, [(p-1) + (b_{k_i+1} - 1)p]p^{k_i})$$

$$= \sum_{i=1}^{n} 1 = n.$$

QED

5. CONCLUDING REMARKS

Note that the fact that p was prime was used only through the carry theorems of Section 3. In other words, one can perfectly well study the matrix $\operatorname{carry}_n(j, b + i - j)$ where $b = (b_0, b_1, ...)$ is an n-adic sequence for any positive integer $n \geq 2$, prime or composite. The only thing lacking if n is not prime is the interpretation of the matrix in terms of binomial coefficients - apart from this, the results of Section 4 (and their proofs) remain valid. In fact, one could even replace the place-value sequence $1 < p < p^2 < \cdots$ by any sequence

$1 < n_1 < n_2 < \cdots$ of place values, for example $n_1 = 2$, $n_2 = 3$, $\ldots, n_k = k+1, \ldots$ and study the resulting carry matrix. This is potentially interesting even for $b = (0, 0, 0, \ldots)$.

Note that the proof of Theorem 3.1 does not use Lemma 3.2 in full generality since it sets $f(X) = \binom{X+k}{k}$, a specific sequence of polynomials. Let $\binom{X+k}{k} = \sum_{j=0}^{k} a_j X^j$. It is well known [12] that $k! a_j = s(k+1, j+1)$, where the *Stirling number of the first kind* $s(k, j)$ is the number of permutations of a k-element set with exactly j disjoint cycles. Using Legendre's lemma $\operatorname{ord}_p k! = \dfrac{k - \sum_{i=0}^{\infty} k_i}{p-1}$ for $k = \sum_{i=0}^{\infty} k_i p^i$, we obtain $\operatorname{ord}_p a_j = \operatorname{ord}_p s(k+1, j+1) + \dfrac{\sum_{i=0}^{\infty} k_i - k}{p-1}$. Therefore the question "Which numbers K work in the proof of Lemma 3.2?" is linked with the apparently open problem of determining $\operatorname{ord}_p s(k, j)$. The *p-divisibility pattern* of the matrix $S = (s(k, j))$ (this is just the p-exponent pattern with $m = 1$) is studied in [9] and [10].

Not only do the generalized binomial coefficients that were defined at the beginning of Section 3 make sense for $q \in \hat{Z}_p$, but also one can show (in analogy with Newton's binomial series theorem) that the series $\sum_{k=0}^{\infty} \binom{q}{k} x^k$ converges in $\hat{Q}_p$ for $\operatorname{ord}_p(x) > \dfrac{1}{p-1}$. Moreover, if we set $E(x) = \sum_{k=0}^{\infty} \dfrac{x^k}{k!}$ (which also converges in $\hat{Q}_p$ for $\operatorname{ord}_p(x) > \dfrac{1}{p-1}$) and $L(1+x) = \sum_{k=1}^{\infty} \dfrac{(-1)^{k-1}}{k} x^k$ (which converges in $\hat{Q}_p$ for $\operatorname{ord}_p(x) > 0$), we can (again in analogy with the field R) *define* $(1+x)^q$ to be $E(qL(1+x))$. Finally, it can be shown [1] that

$$(1+x)^q = \sum_{k=0}^{\infty} \binom{q}{k} x^k. \qquad (q \in \hat{Z}_p; \operatorname{ord}_p x > \tfrac{1}{p-1}).$$

The name *generalized binomial coefficient* is therefore very appropriate for $\binom{q}{k}$.

REFERENCES

[1]   Bachman, G., Introduction to p-adic Numbers and Valuation Theory, Academic Press (1964).

[2]   Dodd, F. and Peele, R., "Some Counting Problems Involving the Multinomial Expansion." *Math. Mag.* (to appear).

[3]   Fray, R., "Congruence Properties of Ordinary and q-Binomial Coefficients." *Duke Math. J. 34* (1967) pp. 467-480.

[4]   Howard, F. T., "Prime Divisors of q-Binomial Coefficients." *Rend. Sem. Mat. Univ. Padova* (1973) pp. 181-188.

[5]   Howard, F. T., "The Number of Multinomial Coefficients Divisible by a Fixed Power of a Prime." *Pacific J. Math. 50* (1974) pp. 99-108.

[6]   Knuth, D. and Wilf, H., "The Power of a Prime that Divides a Generalized Binomial Coefficient." *J. Reine Angew. Math. 396* (1989) pp. 212-219.

[7]   Kummer, E. E., "Über die Ergänzungßetze zu den Allgemeinen Reciprocitätsgesetzen." *J. für Math. 44* (1852) pp. 115-116.

[8]   Lucas, E. "Théorie des Fonctions Numériques Simplement Périodiques." *Amer. J. Math. 1* (1878) pp. 184-240.

[9]   Peele, R., "Divisibility Patterns for Some Combinatorial Sequences." Proceedings of Combinatorics '88 (Ravello, May 1988) (to appear).

[10]   Peele, R. and Wilf, H., "Congruence Problems Involving Stirling Numbers of the First Kind." (submitted).

[11]   Singmaster, D., "Divisibility of Binomial and Multinomial Coefficients by Primes and Prime Powers." *18th Anniversary Volume of the Fibonacci Association* (1980) pp. 98-113.

[12]   Stanley, R., <u>Enumerative Combinatorics</u>, Wadsworth & Brooks/Cole (1986).

ON CO-RELATED SEQUENCES INVOLVING
GENERALIZED FIBONACCI NUMBERS

H. T. Freitag and G. M. Phillips

1. INTRODUCTION

We shall consider the general sequence satisfying the recurrence relation which generates the Fibonacci and Lucas sequences $\{F_n\}$ and $\{L_n\}$:

$$A_{n+1} = A_n + A_{n-1}, \quad n \in \mathbb{Z} \tag{1}$$

with A_0, $A_1 \in \mathbb{R}$ given. (See Walton and Horadam [5] for a lengthy list of references on this sequence.) There is a nearly symmetrical relation between $\{F_n\}$ and $\{L_n\}$ exhibited by the well known identities

$$F_{n+1} + F_{n-1} = L_n \tag{2}$$

and

$$L_{n+1} + L_{n-1} = 5F_n. \tag{3}$$

On surveying the very large number of identities involving $\{F_n\}$ and $\{L_n\}$ (see, for example, Hoggatt [1], Long [2], Vajda [3], Vorob'ev [4]) we have observed that some are equivalent with respect to the identities (2) and (3). We will illustrate this in §2. This encourages us to seek other pairs of sequences satisfying (1) which are also inter-related in a symmetrical way, as in (2) and (3) for $\{F_n\}$ and $\{L_n\}$.

Given any $\{A_n\}$ which satisfies (1), let us consider a sequence $\{B_n\}$ defined by

$$B_n = \lambda(A_{n+1} + A_{n-1}), \tag{4}$$

121

G. E. Bergum et al. (eds.), Applications of Fibonacci Numbers Volume 4, 121–125.
© 1991 Kluwer Academic Publishers.

where $\lambda \neq 0 \in \mathbf{R}$. On using (4) and (1) we obtain

$$A_n = (B_{n+1} + B_{n-1})/5\lambda. \tag{5}$$

While noting that the constants on the right sides of (4) and (5) differ unless $5\lambda^2 = 1$, there is a symmetry in the relationship between $\{A_n\}$ and $\{B_n\}$ and we will say that such sequences are *co-related*. Thus $\{F_n\}$ and $\{L_n\}$ are co-related.

Note that the sequence $\{B_n\}$ which is co-related with $\{A_n\}$, as in (4), also satisfies (1) with $B_0 = \lambda(-A_0 + 2A_1)$ and $B_1 = \lambda(2A_0 + A_1)$. In particular, with $\lambda = 1$ in (4), we obtain $\{A'_n\}$, a sequence co-related with $\{A_n\}$, given by

$$A'_n = A_{n+1} + A_{n-1}. \tag{6}$$

With this notation we have $F'_n = L_n$ and $L'_n = 5F_n$. Also, we have the canonical form

$$A_n = \tfrac{1}{2} (A_0 F'_n + A'_0 F_n) \tag{7}$$

which we find a useful alternative to the familiar

$$A_n = A_0 F_{n-1} + A_1 F_n. \tag{8}$$

We note in passing that (8) leads immediately to

$$A'_n = A_0 L_{n-1} + A_1 L_n. \tag{9}$$

Since $F_{-n} = (-1)^{n-1} F_n$ and $L_{-n} = (-1)^n L_n$ we obtain from (7)

$$A_{-n} = (-1)^n \tfrac{1}{2}(A_0 F'_n - A'_0 F_n). \tag{10}$$

We also define $\{A''_n\}$ from

$$A''_n = A'_{n+1} + A'_{n-1} \tag{11}$$

and note from (6), (4) and (5) that $A''_n = 5A_n$.

We will use the concept of co-related sequences first to show in §2 how some well known identities involving $\{F_n\}$ and $\{L_n\}$ are related and second to derive some new identities in §3.

2. APPLICATIONS OF CO-RELATION

We consider the well known identity

$$F_{m+n} + (-1)^n F_{m-n} = F_m L_n \tag{12}$$

and use co-related sequences to derive other well known identities from it. Keeping n fixed and using co-relation (as in (6)) on both sides of (12) with respect to m, we obtain

$$L_{m+n} + (-1)^n L_{m-n} = L_m L_n. \tag{13}$$

Keeping m fixed and using co-relation on n we first obtain

$$((-1)^n L_{m-n})' = (-1)^{n+1} L_{m-(n+1)} + (-1)^{n-1} L_{m-(n-1)}$$
$$= (-1)^{n-1} 5 F_{m-n}$$

and thus derive from (13)

$$F_{m+n} - (-1)^n F_{m-n} = L_m F_n \tag{14}$$

on cancelling the common factor 5. Finally, we keep n fixed in (14) and use co-relation on m to give

$$L_{m+n} - (-1)^n L_{m-n} = 5 F_m F_n. \tag{15}$$

Thus the identities (12)-(15) are all equivalent via co-related sequences.

We leave the reader to experiment with this process, for example to transform the identity

$$L_n F_{m+d} - L_{m+d} F_n = 2(-1)^n F_{m-n+d}, \tag{16}$$

given in Long [2], into another identity cited in [2].

3. IDENTITIES INVOLVING CO-RELATED SEQUENCES

Let us multiply (12) by $\frac{1}{2} A_0'$, (13) by $\frac{1}{2} A_0$ and add to deduce, using (7), that

$$A_{m+n} + (-1)^n A_{m-n} = A_m L_n. \tag{17}$$

We note that (17) generalizes (12) and (13). Similarly we may generalize (14) and (15) to give

$$A_{m+n} - (-1)^n A_{m-n} = A_m' F_n. \tag{18}$$

Combining (17) and (18) gives

$$A_{m+n} = \frac{1}{2}(A_m F_n' + A_m' F_n) \tag{19}$$

and

$$A_{m-n} = (-1)^n \frac{1}{2}(A_m F_n' - A_m' F_n) \tag{20}$$

which generalize (7) and (10) respectively.

Let $\{A_n\}$ and $\{B_n\}$ denote any two sequences satisfying (1). Then, using (7), we obtain

$$A_m B_n - A_n B_m = \tfrac{1}{4}(A_0' B_0 - A_0 B_0')(F_m F_n' - F_n F_m'). \qquad (21)$$

In (21) we will make use of (20) with $\{A_n\} = \{F_n\}$ and also write

$$\tfrac{1}{2}(A_0' B_0 - A_0 B_0') = A_1 B_0 - A_0 B_1$$

to give

$$A_m B_n - A_n B_m = (-1)^n (A_1 B_0 - A_0 B_1) F_{m-n}. \qquad (22)$$

In particular, with $\{B_n\} = \{A_n'\}$, (22) yields

$$A_m A_n' - A_n A_m' = (-1)^n (A_1 A_0' - A_0 A_1') F_{m-n}. \qquad (23)$$

We now use co-related sequences to generalize two other standard identities:

$$F_n L_n = F_{2n} \qquad (24)$$

$$L_n^2 + 5F_n^2 = 2L_{2n}. \qquad (25)$$

For any two sequences $\{A_n\}$ and $\{B_n\}$ satisfying (1), we apply (7) to obtain

$$A_m B_n' + A_m' B_n = \tfrac{1}{2}(A_0 B_0' + A_0' B_0) L_{m+n}$$

$$+ \tfrac{1}{2}(A_0' B_0' + 5\, A_0 B_0) F_{m+n}, \qquad (26)$$

also making use of the identities

$$L_m L_n + 5F_m F_n = 2L_{m+n}$$

and

$$F_m L_n + F_n L_m = 2F_{m+n}$$

The right side of (26) can be rewritten, using (7), to give

$$A_m B_n' + A_m' B_n = A_{m+n} B_0' + A_{m+n}' B_0 \qquad (27)$$

and we note that the right side of (27) can be expressed alternatively as $A_0 B_{m+n}' + A_0' B_{m+n}$. Indeed, we observe from (27) that, for fixed m and n, the expression $A_{m+k} B_{n-k}' + A_{m+k}' B_{n-k}$ is constant for all $k \in \mathbf{Z}$. As a special case of (27), putting $m = n$ and taking $\{B_n\} = \{A_n\}$ we obtain

$$2A_n A_n' = A_0' A_{2n} + A_0 A_{2n}', \qquad (28)$$

which can be seen as a generalization of (24). Again let us take $m = n$ in (27), but this time choose $\{B_n\} = \{A_n'\}$ to give

$$A_n A_n'' + (A_n')^2 = A_{2n} A_0'' + A_{2n}' A_0',$$ (29)

which generalizes (25).

REFERENCES

[1]   Hoggatt, Verner E. Jr. Fibonacci and Lucas Numbers. Houghton Mifflin, Boston, 1969.

[2]   Long, Calvin T., "Discovering Fibonacci Identities," *The Fibonacci Quarterly 24* (1986): pp. 160-167.

[3]   Vajda, S. Fibonacci & Lucas Numbers, and the Golden Section. John Wiley & Sons, New York, 1989.

[4]   Vorob'ev, N. N. Fibonacci Numbers. Pergamon, 1961.

[5]   Walton, J. E. and Horadam, A. F. "Some Aspects of Generalized Fibonacci Numbers," *The Fibonacci Quarterly 12* (1974): pp. 241-250.

FIBONACCI AND B-ADIC TREES IN MOSAIC GRAPHS

Heiko Harborth and Sabine Jäger

A (p,q)-mosaic graph $G_{p,q}$ is a plane graph with all vertices of degree q and only p-gons as its faces, $p, q \geq 3$. These graphs are infinite if $(p-2)(q-2) \geq 4$ is assumed which only excludes the five platonic solid graphs. Starting with any p-gon C_0 we can construct $G_{p,q}$ by adding coronas C_n which consist of all p-gons having one vertex or one edge in common with a p-gon of C_{n-1} (see Figure 1 with C_0 to C_2 of $G_{7,3}$ which is one of the two Fibonacci mosaic graphs in [3]).

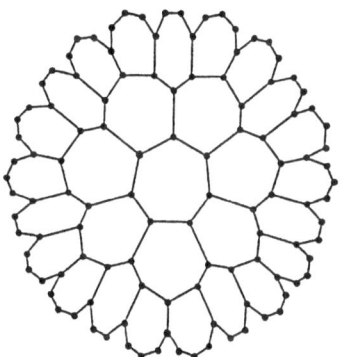

Figure 1.

A b-adic tree $T_b(n)$, $b \geq 2$, is a plane tree with one root vertex of degree b, with b^n end vertices, and with all other vertices of degree $b+1$. A plane Fibonacci tree $T_F(n)$ has vertices of

127

G. E. Bergum et al. (eds.), Applications of Fibonacci Numbers Volume 4, 127–132.

two different types, A and B, on levels 2 to n. Starting with a root vertex A on level 2, each vertex of type A on level h is joined by two edges to vertices on level $h+1$, right to a vertex of type A, and left to a vertex of type B, and each vertex of type B on level h is joined by an edge to a vertex of type A on level $h+1$ (see [2]). Thus $T_F(n)$ is a subtree of $T_2(n-2)$. The numbers of vertices on levels h are b^h for $T_b(n)$, $0 \leq h \leq n$, and F_h for $T_F(n)$, $2 \leq h \leq n$, with $F_0 = 0$, $F_1 = 1$, $F_{n+1} = F_n + F_{n-1}$.

We now consider realizations of trees T in $G_{p,q}$ where p-gons of $G_{p,q}$ correspond to the vertices of T, p-gons of adjacent vertices of T have an edge in common, and no other common points of these p-gons occur, excluded those vertices of $G_{p,q}$ which are the common vertex of a consecutive sequence of neighbored of these p-gons.

Which trees $T_b(n)$ and $T_F(n)$ do allow a realization in (p,q)-mosaic graphs $G_{p,q}$? An answer is given if the maximum numbers $N_b(p,q)$ and $N_F(p,q)$ of levels of realizable trees $T_b(n)$ and $T_F(n)$ are determined. We write $N_b(p,q) = \infty$ and $N_F(p,q) = \infty$ if realizations exist for all levels n.

Since $N_F(p,q) \geq N_2(p,q) + 2$ let us first discuss $N_b(p,q)$. Trivially $N_b(p,q) = 0$ for $b > p$ so that $b \leq p$ can be assumed in the following. We proceed in decreasing bases b. The values $N_p(p,3) = 0$ and $N_p(p,q) = V_1 \geq 4$ are easily figured out.

Theorem 1: $N_{p-1}(p,q) = \lfloor \frac{q-2}{2} \rfloor$.

Proof: In C_1 there exist two p-gons of level 1 which are neighbored to C_0 by two successive edges of C_0. Between these two p-gons there are exactly $q-3$ consecutive p-gons of $G_{p,q}$ in C_1. Two of them are used for each further level of $T_b(n)$ since $b = p-1$. At least one of them cannot be used so that the unique realization ends at level $1 + \lfloor (q-3-1)/2 \rfloor = \lfloor (q-2)/2 \rfloor$.

Theorem 2: $N_{p-2}(p,3) = 0$, $N_2(4,4) = 3$, $N_3(5,4) = N_4(6,4) = 2$,

$$N_{p-2}(p,4) = 1 \text{ for } p \geq 7 \text{ and } N_{p-2}(p,q) = \infty \text{ in all other cases.}$$

Proof: It is immediate that $N_{p-2}(p,3) = 0$. Realizations for $b = p-2$ and $(p,q) = (4,4)$, $(5,4)$, and $(6,4)$ with largest possible level 3, 2, and 2, respectively, are easily figured out (see Figures 2 to 4).

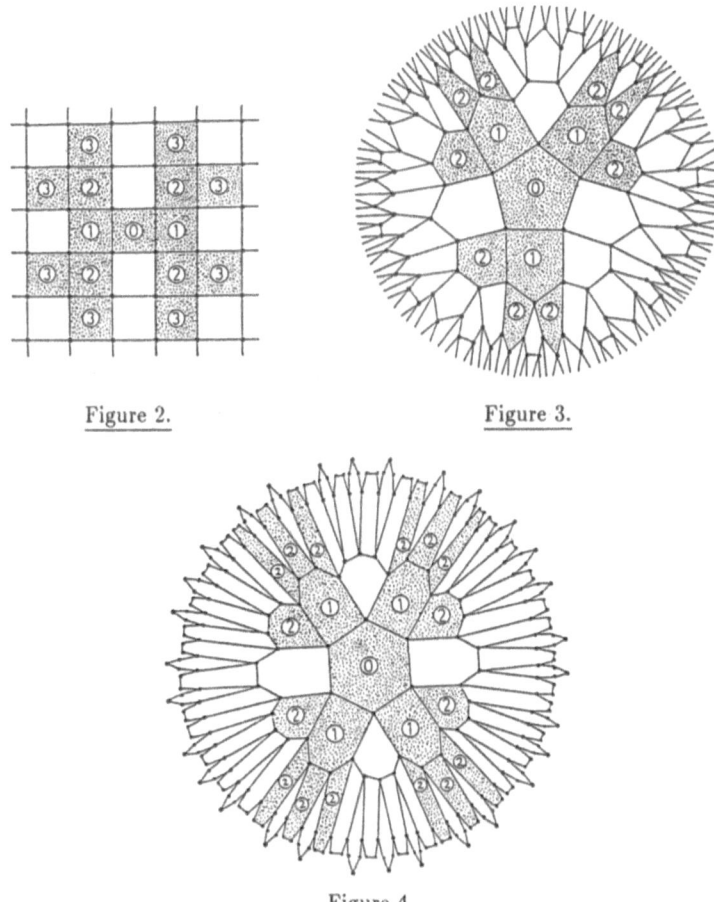

Figure 2.   Figure 3.

Figure 4.

For $p \geq 7$ the pigeonhole principle guaranties that at least three of $p-2$ neighbored p-gons of C_0 are neighbors of three consecutive edges of C_0. Because of $q = 4$ there is at least one of them which can have at most $p-3$ further neighbors, and thus $N_{p-2}(p,4) = 1$ for $p \geq 7$.

It remains to prove $N_{p-2}(p,q) = \infty$ for $q \geq 5$, $p \geq 4$, which follows from the following inductive construction for the realization of the infinite T_{p-2}. In the corona C_r, $r \geq 1$, we first choose all p-gons which have an edge in common with a chosen p-gon of C_{r-1}. Further we choose in C_r the left or right neighbor to every such p-gon if its neighbored chosen p-gon of C_{r-1} has in C_{r-1} a neighbor to right or to left, respectively. Then all chosen pairs of p-gons are

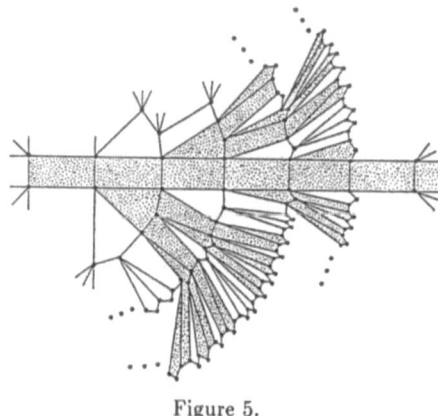

Figure 5.

separated in C_r by at least one other p-gon since $q \geq 5$ (see Figure 5 for $(p,q) = (5,4)$). Thus every chosen p-gon besides C_0 has exactly one edge without a neighbor, that is, the desired degree $b + 1 = p - 1$ in the realization of T_{p-2}.

Theorem 3: $N_{p-j}(p,3) = 0$ for $3 \leq j \leq \lfloor \frac{p-1}{2} \rfloor$,

$\qquad N_{\lfloor p/2 \rfloor}(p,3) = 1$, $N_2(6,3) = 3$,

$\qquad N_{(p-2)/2}(p,3) = \frac{p-2}{2}$ for $p \geq 8$, $p \equiv 0 \pmod 2$,

$\qquad$ and $N_{p-j}(p,q) = \infty$ for $j \geq 3$ in all other cases.

Proof: For $q = 3$, at most $\lfloor \frac{p}{2} \rfloor$ neighbors can be added to the root p-gon. Thus $b = p - j \geq \lfloor \frac{p}{2} \rfloor + 1$ implies the first equation of Theorem 3. Any p-gon of level 1 has to have $b + 1$ neighbor p-gons of levels 0 and 2. This is impossible for $j = \lfloor \frac{p+1}{2} \rfloor$, since $b = p - j = \lfloor \frac{p}{2} \rfloor$ and $q = 3$ which proves $N_{\lfloor p/2 \rfloor}(p,3) = 1$.

For even p the next j to discuss is $j = \lfloor \frac{p+3}{2} \rfloor = 1 + \frac{p}{2}$. First $N_2(6,3)$ is easily figured out. For $p \geq 8$ and $q = 3$ there exists at least one triple of consecutive p-gons in C_1 such that only the central one does not belong to level 1 of the tree. Since $b + 1 = \frac{p}{2}$ and $q = 3$, exactly two p-gons of each further level are neighbors of this central p-gon. Therefore the realization ends at level $N_{(p-2)/2}(p,3) = \frac{p-2}{2}$.

The remaining cases are $j \geq \lfloor \frac{p}{2} \rfloor + 2$ for $q = 3$, and $j \geq 3$ for $q \geq 4$. For $q = 3$ any p-gon of corona C_r, $r \geq 1$, has one or two edges in common with C_{r-1}, and thus at least $p - 4$ edges in common with C_{r+1}. Thus for any chosen p-gon in C_r its b neighbors corresponding to the next level of the tree can be chosen as every second neighbor in C_{r+1} if $b = p - j \leq \lfloor \frac{p-4+1}{2} \rfloor$, that is,

$j \geq \lfloor \frac{p}{2} \rfloor + 2$, and so the infinite T_b can be realized. For $q \geq 4$ and $2 \leq b = p - j \leq p - 3$ a realization of T_b always exists where all b subsequent neighbor p-gons of any realized p-gon can be chosen in the subsequent corona of $G_{p,q}$.

Summarizing, it follows from Theorem 1 to 3 that many of the (p,q)-mosaic graphs allow realizations of the infinite trees T_b, $b < p$. What about the special case of Fibonacci trees $T_F(n)$?

Theorem 4: $N_F(3,6) = 5$, $N_F(4,4) = N_F(6,3) = 6$,

$N_F(3,q) = q$, for $q \geq 7$, and $N_F(p,q) = \infty$ otherwise.

Proof: The first three values are determined in [1,2]. Since Fibonacci trees are subtrees of the binary trees Theorems 2 and 3 prove Theorem 4 in all cases besides $N_F(3,q) = q$, for $q \geq 7$.

To realize $T_F(n)$ in $G_{3,q}$ we choose in corona 1 one neighbor triangle of the root triangle (level 2 of the tree) for the A-vertex of level 3, and counter-clockwise the next neighbor triangle for the B-vertex of level 3. All further triangles of the sequence of consecutive A-vertices have to be chosen clockwise as consecutive triangles of corona 1, and thus $n \leq N_F(3,q) \leq q$. To realize $T_F(q)$ we further choose in corona 1 counter-clockwise the next three triangles of the A-vertex triangle of level 3, and the next two triangles of the B-vertex triangle of level 3 which correspond to vertices of type B of level 4, A of level 5, B of level 6, and A of level 4, B of level 5, respectively. If possible, the remaining triangles always are chosen in the next following corona. Then between any pair of triangles of C_r, $r \geq 2$, which are neighbors of two consecutive edges of the boundary of C_{r-1}, there remain at most $q - 5$ vertices of $T_F(q)$ to be chosen so that a realization is guaranteed.

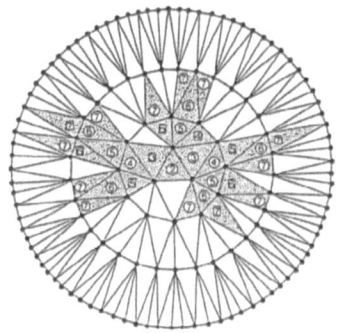

Figure 6.

Figure 6 shows the example for $q = 7$ (circles and rectangles correspond to vertices of types A and B respectively).

Summarizing, all (p,q)-mosaic graphs $G_{p,q}$ are determined which allow a realization of $T_b(n)$ or $T_F(n)$. It may be a subsequent problem to ask for the mosaic numbers $M_b(p,q;n)$ or $M_F(p,q;n)$ which are sums of $b^{n+1} - 1$ or $F_{n+2} - 2$, and of the smallest numbers of additional vertices of degree 2 inserted into the edges of $T_b(n)$ or $T_F(n)$, respectively, so that the resulting trees are realizable in $G_{p,q}$. In [1] for $(p,q) = (3,6)$, $(4,4)$, and $(6,3)$ the mosaic numbers $M_2(p,q;n)$ and $M_F(p,q;n)$ were determined to have orders of magnitude 2^n and F_n, respectively. In all other cases where $N_b(p,q)$ or $N_F(p,q)$ exists, and where $b \leq \lfloor \frac{p}{2} \rfloor$ for $q = 3$, and $b \leq p - 1$ for $q \geq 4$, that is (see Theorems 1 to 4), for $b = \lfloor \frac{p-1}{2} \rfloor$ and $b = \lfloor \frac{p}{2} \rfloor$ if $q = 3$, for $b = p - 2$ and $b = p - 1$ if $q = 4$, for $b = p - 1$ if $q \geq 5$, and for Fibonacci trees in $(3,q)$-mosaic graphs, $q \geq 7$, the corresponding mosaic numbers still have to be determined or estimated.

REFERENCES

[1]   Lohmann, S. Mosaikzahlen von Bäumen in regulären Parkettierungen. Dissertation, TU Braunschweig 1989.

[2]   Harborth, H. and Lohmann, S. "Mosaic Numbers of Fibonacci Trees." G. E. Bergum, et. al. (eds.), Applications of Fibonacci Numbers, Vol. 3, Kluwer Acad. Publ. 1990, pp. 133-138.

[3]   Harborth, H. "Concentric Cycles in Mosaic Graphs." G. E. Bergum, et. al. (eds.), Applications of Fibonacci Numbers, Vol. 3, Kluwer Acad. Publ. 1990, pp. 123-128.

FIBONACCI REPRESENTATIONS OF GRAPHS

Heiko Harborth and Arnfried Kemnitz

An integral representation $D(G)$ of a graph G in the Euclidean space E^d ($d \geq 2$) is a mapping of the vertices of G into distinct points of E^d, also called vertices, and a mapping of the edges of G into straight line segments of integer length, also called edges, which connect the corresponding vertices in E^d such that two edges have at most one point in common, either a vertex or a crossing. If only Fibonacci numbers F_n occur as edge lengths in $D(G)$ then we call $D(G)$ a Fibonacci representation $F(G)$. The Fibonacci numbers are defined by $F_0 = 0$, $F_1 = 1$, $F_{n+1} = F_n + F_{n-1}$. To decide for any graph G whether a Fibonacci representation $F(G)$ exists in E^d seems to be hopeless in general. Here some first results are given, mainly for $d = 2$.

The triangle inequality implies that Fibonacci representations of the complete graph K_3 are possible only as isosceles triangles with base F_n and legs F_k for any $k \geq n - 1$. If in addition the area of these triangles is asked to be an integer then the existence of such so-called Fibonacci triangles is discussed in [4].

Next we ask for a Fibonacci representation $F(K_4)$ of the complete graph K_4.

Theorem 1: A Fibonacci representation $F(K_4)$ does not exist.

Proof: Let F_n denote the length of a largest edge in $F(K_4)$. This largest edge is a common edge of two of three possible types of triangles in one of the twelve cases of Figure 1. The assumption that the length x of the dotted edge also is a Fibonacci number leads to contradictions in all cases as follows. Since all edges cannot be of length 1 we can assume $n \geq 3$.

G. E. Bergum et al. (eds.), Applications of Fibonacci Numbers Volume 4, 133–138.
© 1991 *Kluwer Academic Publishers.*

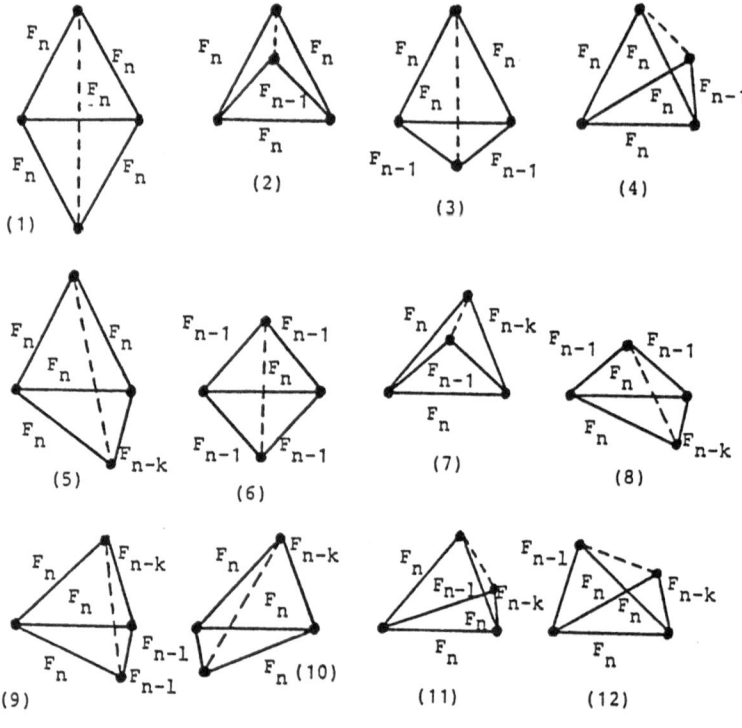

Figure 1

(1) $x = F_n\sqrt{3} \neq F_m$.

(2)   The triangle inequality yields $x = F_{n-1}$. Then $F_{n-1} = \frac{1}{3}\sqrt{3}\,F_n$ would follow which is impossible.

(3)   The only possibility is $x = F_n$. Then

$$\left(\tfrac{1}{2}F_n\right)^2 + (F_n - \tfrac{1}{2}\sqrt{3}\,F_n)^2 = F_{n-1}^2,$$

which is equivalent to $2 - (F_{n-1}/F_n)^2 = \sqrt{3}$, a contradiction.

(4)   Since $x = F_{n-k}$, and then $k = 1$, we have case (3).

(5)   It must hold $x = F_n$ which is impossible.

(6)   Assume $x = F_j$ for $j \leq n$. Then $F_j^2 + F_n^2 = 4F_{n-1}^2$. This yields the contradictions $F_n/F_{n-1} = \sqrt{2}$ for $j = n$, and $F_n/F_{n-1} = \sqrt{3}$ for $j = n-1$. For $j \leq n-2$ we have (see [7])

$$F_j^2 + F_n^2 \leq F_{n-2}^2 + F_n^2 = F_{n-1}^2 + 2F_{n-2}F_n =$$
$$= F_{n-1}^2 + 2(F_{n-1}^2 + (-1)^{n-1}) = 3F_{n-1}^2 + 2(-1)^{n-1} <$$
$$< 4F_{n-1}^2.$$

(7)   Only $x = F_{n-1}$ could be possible. Then

$$\left(F_{n-1} + \sqrt{F_{n-1}^2 - \tfrac{1}{4}F_{n-k}^2}\,\right)^2 + \tfrac{1}{4}F_{n-k}^2 = F_n^2$$

implies

$$F_n^4 = F_{n-1}^2(4F_n^2 - F_{n-k}^2)$$

which is impossible because of $(F_n, F_{n-1}) = 1$ (see [7]).

(8)   If $x = F_{n-1}$ we have case (7). It remains $x = F_n$ and thus $k = 1$. Then there is a trapezoid, and it holds

$$F_{n-1}^2 - (\tfrac{1}{2}F_n - \tfrac{1}{2}F_{n-1})^2 = F_n^2 - (\tfrac{1}{2}F_n + \tfrac{1}{2}F_{n-1})^2$$

which implies

$$F_{n-1}^2 = F_n^2 - F_nF_{n-1} = F_nF_{n-2}.$$

The last equality is impossible since $(F_{n-1}, F_n) = (F_{n-1}, F_{n-2}) = 1$.

(9)   For $k \neq 1$ we have $x = F_{n-k}$ or $x = F_{n-1}$ which both are impossible. Thus $k = 1$, and $x = F_{n-k+1}$. If $k = 1$ we have case (3), and for $k \geq 2$ it holds

$$(F_n - \sqrt{F_n^2 - \tfrac{1}{4}F_{n-k+1}^2})^2 + \tfrac{1}{4}F_{n-k+1}^2 = F_{n-k}^2$$

which implies

$$F_n^2(4F_{n-k}^2 - F_{n-k+1}^2) = F_{n-k}^4.$$

However, from $5F_{n-k} \geq 3F_{n-k+1}$ we obtain the contradiction

$$\frac{F_n^2}{F_{n-k}^2} = \frac{F_{n-k}^2}{4F_{n-k}^2 - F_{n-k+1}^2} < 1.$$

(10)   This is impossible since $x \leq F_n$ was assumed.

(11)   Discussing $\ell = k$ and $\ell \neq k$ we obtain cases corresponding to those of case (9).

(12)   Since $x < F_n$ we conclude $k = \ell = 1$, that is, $x = F_{n-1}$, and only that trapezoid of case (8) occurs.

It obviously follows from Theorem 1 that no Fibonacci representation $F(K_n)$ exists for $n \geq 4$. Thus one may ask for the maximum number $J(n,d)$ of edges a graph G with n vertices can have such that $F(G)$ exists in E^d. For the plane trivially $J(3,2) = 3$, and Theorem 1 implies $J(4,2) = 5$.

Theorem 2: $J(5,2) = 8$.

Proof: From Figures 2 and 3 where two Fibonacci triangles are turned round a common vertex to get the vertex of degree 4 follows $J(5,2) \geq 8$. Since 5 points with 9 Fibonacci distances would imply an $F(K_4)$ Theorem 1 proves $J(5,2) \leq 8$.

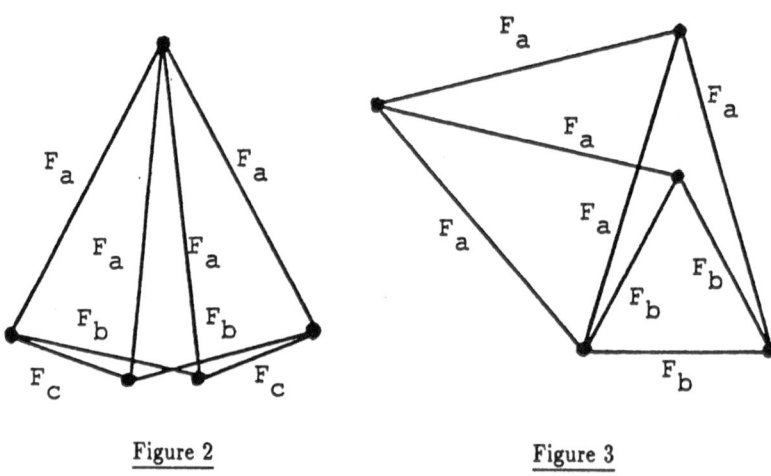

Figure 2   Figure 3

By reflection of Figure 3 we obtain Figure 4 which proves $J(6,2) \geq 11$. Theorem 1 implies $J(6,2) \leq 12$. We conjecture $J(6,2) = 11$. One has to prove that no Fibonacci representation of the octahedron graph exists, and this could be concluded if there are no other examples of 5 points and 8 edges of Fibonacci length than those of Figures 2 and 3.

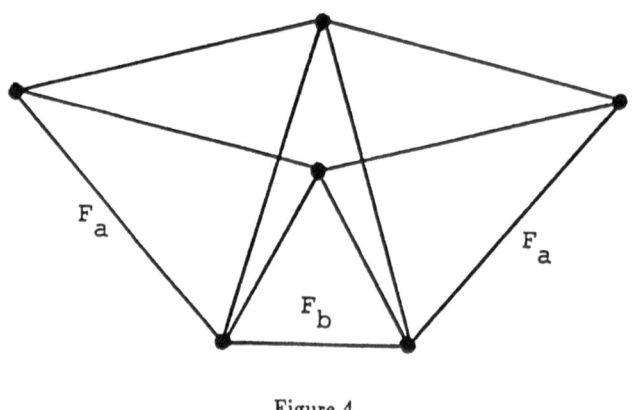

Figure 4

First poor general estimations of $J(n,2)$ are as follows.

Theorem 3: $\frac{8}{3}(n-1) - 3 \le J(n,2) \le \frac{1}{3}n^2$.

Proof: The upper bound corresponds to the Turán number of K_4 which is the smallest number of edges a graph with n vertices can have without containing a subgraph K_4 (see [2], p. 18).

For the lower bound we choose $\frac{1}{3}(n-3)$ isosceles triangles of Fibonacci length which have a common base, furthermore we choose the equilateral triangle of this base, and to right so as to left $\frac{1}{3}(n-3)$ points each having 3 Fibonacci distances to the chosen triangles (see Figure 4 for n = 6). We then have n = 3t points with $\frac{1}{3}(8n - 15)$ Fibonacci distances. Deleting one or two points of degree 3 yields the lower bound for all $n \ge 3$.

We can remark that a better lower bound is obtained for $n = 2^t + 1$ if Figure 2 is turned round the vertex of degree 4 such that corresponding points on the circle have another Fibonacci distance less than the maximum F_a. If this procedure is repeated then the number j(t) of Fibonacci edges is deduced from $j(1) = 3$, and $j(t) = 2j(t-1) + 2^{t-1}$ to be $j(t) = (t+2)2^{t-1}$. Thus $J(2^t + 1, 2) \ge (t+2)2^{t-1}$, and with $n = 2^t + 1$ this corresponds to $J(n,2) \ge cn \log n$ for a constant c. However, Figure 5 shows $J(9,2) \ge 21$ which already is better by one edge. An asymptotically even better bound,

$$J(n,2) \ge n^{1 + c/\log\log n},$$

c being a constant, is the lower bound of P. Erdös for the maximum number of unit distances in a plane point set of n points (see [1]). Nevertheless, the gap remains large.

Only some few examples are known of Fibonacci representations in dimensions $d \ge 3$ (see [3,5,6]). Clearly, the unit tetrahedron in E^d proves $J(n,n-1) = \binom{n}{2}$. For an equilateral triangle of side length 3 both points of E^3 which have distance 2 to all vertices of that triangle have distance 2 from one another. This proves $J(5,3) = \binom{5}{2} = 10$, and implies

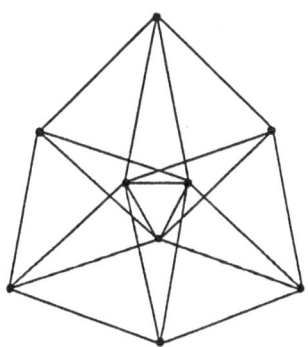

Figure 5

$J(n,3) \geq 3n - 5$. Examples of 8 points in E^6, and 10 points in E^8 with distances 1, 2, and 3 only (see [5]) determine $J(8,6) = 28$, and $J(10,8) = 45$. These examples suggest to ask in general for the smallest dimension $D(n)$ for which a Fibonacci representation $F(K_n)$ exists. The above results correspond to $D(3) = 2$, $D(4) = 3$, $D(5) = 3$, $D(8) \leq 6$, $D(10) \leq 8$. Furthermore, one can ask for those Fibonacci representations of smallest diameter.

REFERENCES

[1]   Erdös, P. Hickerson, D. and Pach, J. "A Problem of Leo Moser about Repeated Distances in Space". *American Math. Monthly 96* (1989), pp. 569-575.

[2]   Harary, F. Graph Theory. Addison-Wesley, 1969.

[3]   Harborth, H. and Kemnitz, A. "Diameters of Integral Point Sets". Colloquia Math. Soc. János Bolyai 48. Intuitive Geometry, Siófok (Hungary) 1985. North-Holland, Amsterdam 1987, pp. 255-266.

[4]   Harborth, H. and Kemnitz, A. "Fibonacci Triangles". Applications of Fibonacci Numbers, Vol. 3, edited by G. E. Bergum, et. al. Kluwer Acad. Publ. 1990, pp. 129-132.

[5]   Harborth, H. and Piepmeyer, L. Point Sets with Small Integral Distances. Festschrift Victor Klee 65, to appear.

[6]   Harborth, H. and Kemnitz, A. Integral Representations of Graphs. Contemporary Methods in Graph Theory, edited by R. Bodendiek. Bibl. Inst. Mannheim 1990, pp. 359-367.

[7]   Hoggatt, V. E. Jr. Fibonacci and Lucas Numbers. Houghton Mifflin Co., Boston, 1969.

ON THE SIZES OF ELEMENTS IN THE COMPLEMENT
OF A SUBMONOID OF INTEGERS

Chung-wu Ho, James L. Parish, and Jau-shyong Shiue*

1. THE MAIN RESULT

Let N be the set of nonnegative integers. If we select an arbitrary subset A of k elements from the set N in such a way that the set $A' = N - A$ is closed under addition, how large can the elements of A be? In the following, we will show that $\sum_{a \in A} a$ is at most k^2, and that is the best possible bound. In fact, we will establish in this paper an n-dimensional version of this result.

In the following, we shall use the term *monoid* to mean a nonempty set together with an associative binary operation and an identity element. Suppose M is a monoid. A nonempty subset A of M is called a *submonoid* of M if A is closed under the binary operation inherited from M. Let N^n be the set of all n-tuples of nonnegative integers. N^n is a monoid under the pointwise addition. Define the norm $\| a \|$ for each $(a_1, a_2, ..., a_n)$ in N^n by $\| a \| = \sum_{i=1}^{n} a_i$

Theorem: Let A be a finite subset of k elements of N^n such that the set $A' = N^n - A$ is a submonoid of N^n. Then, $\sum_{a \in A} \| a \| \leq k^2$. The equality holds if and only if

$$A = \{(0,...,1,...,0), \ (0,...,3,...,0), \ ..., \ (0,...,2k-1,...,0)\} .$$

*The research of the third author was partially supported by the University Research Council of the University of Nevada, Las Vegas.

G. E. Bergum et al. (eds.), Applications of Fibonacci Numbers Volume 4, 139–144.
© 1991 Kluwer Academic Publishers.

The n-dimensional case may be a little surprising in the sense that adding more components does not increase the bound for the sizes of the elements of the set A. The one-dimensional version of the problem arises in algebraic geometry, in connection with the theory of Weierstrass points. Given a point P on a compact Riemann Surface C, one may consider those natural numbers n for which there is a function on C holomorphic away from P and having a pole of order exactly n at P. It is easy to see that these numbers form a monoid; it is less obvious that the complement of the monoid is finite. There are many interesting and useful relations between the monoids associated with various points P and the geometry of C; some of these are explored in Exercise Set E of Chapter 1 of [1].

In the following, we shall collect some necessary preliminaries in Section 2. The theorem will then be proved in Section 3. The authors would like to thank Professor Paul Erdös for his suggestions and advice.

2. PRELIMINARIES

Let a, b, and c be elements of N^n such that $a + b = c$. We shall call the (unordered) pair {a, b} a summand pair of the element c.

Lemma 1: Each element $c = (c_1, c_2, ..., c_n)$ of N^n has exactly $[\frac{1}{2}((c_1 + 1)(c_2 + 1) \cdots (c_n + 1) + 1)]$ distinct summand pairs, where $[\ \cdot\ ]$ is the greatest integer function.

Proof: Consider an arbitrary element $c = (c_1, c_2, ..., c_n)$ of N^n. We first count the number of all ordered pairs (a, b) of elements a and b in N^n such that $a + b = c$. Since, letting $a = (a_1, a_2 ..., a_n)$, the component a_i can vary from 0 to c_i for each i, there are clearly $(c_1 + 1)(c_2 + 1) \cdots (c_n + 1)$ such ordered pairs.

Now, for the unordered pairs, we define an equivalence relation R among the set of all such ordered pairs such that (a, b) R (c, d) if and only if (a, b) = (c, d) or (a, b) = (d, c). We need then to count the number of equivalence classes of R.

Case 1: If at least one of the c_i's is odd, then each equivalence class of R consists of exactly two elements. Thus, the number of distinct summand pairs of c

$$= \tfrac{1}{2} (c_1 + 1) (c_2 + 1) \cdots (c_n + 1)$$

$$= [\, \tfrac{1}{2} ((c_1 + 1) (c_2 + 1) \cdots (c_n + 1) + 1) \,]$$

Case 2: If all the c_i's are even, then there is exactly one equivalence class of R consisting of a single element, the class $\{(a, b)\}$, where

$$a = b = \left(\tfrac{c_1}{2}, \tfrac{c_2}{2}, \cdots, \tfrac{c_n}{2} \right).$$

The other equivalence classes of R all consists of two elements. Thus, the number of distinct summand pairs of c

$$= \tfrac{1}{2} ((c_1 + 1) (c_2 + 1) \cdots (c_n + 1) - 1) + 1$$

$$= [\, \tfrac{1}{2} ((c_1 + 1) (c_2 + 1) \cdots (c_n + 1) + 1) \,]$$

This finishes the proof.

Corollary: Let c be an element of N^n with $\| c \| = m$ for some nonnegative integer m. Then c has at least $[\, \tfrac{m}{2} \,] + 1$ distinct summand pairs in N^n.

Proof: It is easy to prove that for any $c = (c_1, c_2, \ldots, c_n)$ in N^n, $1 + \sum_{i=1}^{n} c_i \leq \prod_{i=1}^{n} (c_i + 1)$. The corollary then follows from the lemma by means of this inequality.

Definition: A subset $A \subset N^n$ will be called an _admissible set_ if the set $A' = N^n - A$ is a submonoid of N^n. An admissible set A of k elements will be called a _maximal_ admissible set if for any other admissible set B of k elements,

$$\sum_{b \in B} \| b \| \leq \sum_{a \in A} \| a \|.$$

Lemma 2: Let A be an admissible set of k elements. No element of A can have a norm greater than $2k - 1$.

Proof: Suppose A has an element c with norm m > 2k − 1; then c has at least $[\frac{m}{2}] + 1 > k - 1 + 1 = k$ distinct summand pairs. Note that since A is admissible, A must contain at least one element in each summand pair of c. Thus, A contains more than k elements, contradiction.

3. PROOF OF THE THEOREM

We first show that for each k, the set

$$A = \{(0,\ldots,1,\ldots,0),\ (0,\ldots,3,\ldots,0),\ \ldots,\ (0,\ldots,2k-1,\ldots,0)\}$$

is a maximal admissible set of k elements. First observe that a set of such form is indeed admissible. We now prove the maximality of such sets by induction on k. The case for k=1 follows immediately from Lemma 2. Now assume the case for k and consider a set of k + 1 elements

$$A = \{(0,\ldots,1,\ldots,0),\ (0,\ldots,3,\ldots,0),\ \ldots,\ (0,\ldots,2k+1,\ldots,0)\}\ .$$

Let B be any admissible set of k+1 elements. Choose any $b_0 \in B$ with maximal norm. By Lemma 2, $\| b_0 \| \leq 2\,(k + 1) - 1 = 2k + 1$. Note that the set $B_1 = B - \{b_0\}$ is still an admissible set. By the induction hypothesis

$$\sum_{b \in B_1} \| b \| \leq \sum_{a \in A_1} \| a \|,$$

where $A_1 = A - \{(0,\ldots,2k + 1,\ldots,0)\}$. Hence,

$$\sum_{b \in B} \| b \| = \| b_0 \| + \sum_{b \in B_1} \| b \| \leq (2k + 1) + \sum_{a \in A_1} \| a \| = \sum_{a \in A} \| a \|.$$

This finishes the inductive step.

Conversely, we claim that any maximal admissible set of k elements must of of the form

$$A = \{(0,\ldots,1,\ldots,0),\ (0,\ldots,3,\ldots,0),\ \ldots,\ (0,\ldots,2k-1,\ldots,0)\}\ .$$

We prove this again by induction. The case for $k = 1$ again follows from Lemma 2. Now assume the assertion for k (≥ 1) and consider a maximal admissible subset B of $k + 1$ elements of N^n. Fix an element $b_0 \in B$ with maximal norm. By Lemma 2, $\| b_0 \| \leq 2(k+1) - 1 = 2k + 1$. Since B is a maximal admissible set

$$\sum_{b \in B} \| b \| \geq \sum_{a \in A_{k+1}} \| a \| ,$$

where $A_{k+1} = \{(1,0,\dots,0),\ (3,0,\dots,0),\ \dots,\ (2k+1,0,\dots,0)\}$. Hence,

$$\sum_{b \in B - \{b_0\}} \| b \| \geq \sum_{a \in A_k} \| a \| ,$$

where $A_{k+1} = \{(1,0,\dots,0),\ (3,0,\dots,0),\ \dots,\ (2k-1,0,\dots,0)\}$,

but from the first part of the proof, A_k is a maximal admissible set. Thus, the set $B - \{b_0\}$ must also be a maximal admissible set of k elements. By the induction hypothesis, $B - \{b_0\}$ must be a set of the form

$$\{(0,\dots,1,\dots,0),\ (0,\dots,3,\dots,0),\ \dots,\ (0,\dots,2k-1,\dots,0)\} ,$$

where the odd integers occur, say at the rth component of the vectors. Furthermore, $\| b_0 \|$ must be $2k + 1$; otherwise B would have a sum of norms less than that of A_{k+1}. Let $b_0 = (c_1, c_2, \dots, c_n)$. We first show that $c_i = 0$ for each i $\neq$ r. Suppose that there exists a $c_i > 0$ for some i $\neq$ r. Since the set B is admissible, B must contain at least one element of the summand pair $\{(0,\dots,1,\dots,0)$ and $(c_1,\dots,c_i-1,\dots,c_n)\}$, where the element 1 is at the ith component. Note that $(0,\dots,1,\dots,0)$ does not belong to $B - \{b_0\}$ for each element of $B - \{b_0\}$ has a non-zero component only at the rth position. Furthermore, $(0,\dots,1,\dots,0)$ does not equal b_0, for $\| b_0 \| = 2k + 1$, where $k \geq 1$. Thus, B must contain the element $(c_1,\dots,c_i-1,\dots,c_n)$. But $(c_1,\dots,c_i-1,\dots,c_n) \neq (c_1,\dots,c_i,\dots,c_n) = b_0$. Hence, $(c_1,\dots,c_i-1,\dots,c_n) \in B - \{b_0\}$. Consequently, $c_j = 0$ for each j $\neq$ i or r, and $c_i = 1$, $c_r = 2k$. But this leads to a contradiction for the rth component of all the elements of $B - \{b_0\}$ are odd integers.

We may therefore conclude that $c_i = 0$ for all $i \neq r$. But $\| b_0 \| = 2k + 1$. We conclude that $b_0 = (0,\ldots,2k+1,\ldots,0)$ where $2k+1$ is at the rth component and

$$B = \{(0,\ldots,1,\ldots,0), \ (0,\ldots,3,\ldots,0), \ \ldots, \ (0,\ldots,2k+1,\ldots,0)\} \ .$$

This finishes the inductive step.

We have now proved that A is a maximal admissible set of k elements if and only if A is of the form

$$\{(0,\ldots,1,\ldots,0), \ (0,\ldots,3,\ldots,0), \ \ldots, \ (0,\ldots,2k-1,\ldots,0)\} \ .$$

Since $1 + 3 + \cdots + (2k-1) = k^2$, our main theorem follows immediately.

REFERENCE

[1]   Arbarello, E., Cornalba, M., Griffiths, P. A. and Harris, J. Geometry of Algebraic Curves, Vol. I, Springer, Berlin, Heidelberg, New York, 1985.

GENOCCHI POLYNOMIALS

A. F. Horadam

1. INTRODUCTION

The purpose of this investigation is to exhibit some of the fundamental properties of $G_n^{(k)}(x)$, the *generalized Genocchi polynomials of order k*.

When k = 1, we have the *Genocchi polynomials* $G_n^{(1)}(x) \equiv G_n(x)$.

If x = 0, we obtain the *generalized Genocchi numbers of order k*, $G_n^{(k)}(o) \equiv G_n^{(k)}$. When k = 1, the *Genocchi numbers* $G_n(o) \equiv G_n$ result.

Information on generalized Bernoulli numbers $B_n^{(k)}(x)$ and generalized Euler polynomials $E_n^{(k)}(x)$, to which $G_n^{(k)}(x)$ may be related, is to be found in [3], [4], [11], [13], [15], and [16]. Nörlund in particular has extensively investigated these polynomials. Several references to Genocchi's work occur in [10]. Also see, for example, Genocchi [5].

While many of the properties of Genocchi polynomials bear a close resemblance to the corresponding properties of Bernoulli and Euler polynomials, some properties are rather different. Indeed, Genocchi polynomials are worthy of an investigation *per se*. To the best of my knowledge, the material on Genocchi polynomials presented here has not previously appeared. Some of the data relating to Genocchi numbers may also be fresh.

Genocchi Numbers

Genocchi numbers may be defined - cf. [2] - by

$$\sum_{n=0}^{\infty} G_n \frac{t^n}{n!} = \frac{2t}{e^t+1} = t\left(1 - \tanh \frac{t}{2}\right) \qquad (1.1)$$

G. E. Bergum et al. (eds.), Applications of Fibonacci Numbers Volume 4, 145–166.

so that

$$G_0 = 0. \tag{1.1}'$$

Since $\tanh \frac{t}{2}$ is an odd function, $t \tanh \frac{t}{2}$ is even in t. Hence we deduce that

$$\left.\begin{array}{l} G_1 \quad = 1 \\ G_{2n+1} = 0 \end{array} \quad (n = 1, 2, 3, \ldots) \right\} \tag{1.2}$$

The first few Genocchi numbers are [2]:

n	0	1	2	4	6	8	10	12	14	16	18	20
G_n	0	1	-1	$1-3$	17	-155	2073	-38227	929567	-28820619	1109652905	

$$\tag{1.3}$$

Designating Bernoulli numbers by B_n ($\equiv B_n^{(1)}(0)$) we know [2], [10] that

Genocchi's Theorem: $\qquad\qquad G_{2m} = 2(1 - 2^{2m}) B_{2m} \tag{1.4}$

Note that Euler's numbers E_n are given [2] by $E_n = 2^n E_n(\frac{1}{2})$ and not by $E_n(0)$ ($\equiv E_n^{(1)}(0)$). In fact,

$$G_{2m} = 2m\, E_{2m-1}(0) \tag{1.5}$$

A recent appearance of Genocchi numbers occurs in [6] where the author examines integers related to the Bessel function $J_1(z)$, and where Genocchi numbers are part of the value of $\sigma_{2n}(-\frac{1}{2})$, $\sigma_{2n}(v)$ being the Rayleigh function.

2. GENERALIZED GENOCCHI POLYNOMIALS

Definition and Basic Properties

Generalized Genocchi polynomials of order k, $G_n^{(k)}(x)$, are defined by

$$\sum_{n=0}^{\infty} G_n^{(k)}(x) \frac{t^n}{n!} = \left(\frac{2t}{e^t + 1}\right)^k e^{tx} \quad (k = 0, 1, 2, 3, \ldots) \tag{2.1}$$

whence

$$G_n^{(k)}(x) = 0 \qquad (n < k) \tag{2.1}'$$

and

$$G_n^{(o)}(x) = x^n \tag{2.2}$$

The first few generalized Genocchi polynomials are:

$$G_k^{(k)}(x) = k!$$

$$G_{k+1}^{(k)}(x) = (k+1)! \left\{ x - \frac{k}{2} \right\}$$

$$G_{k+2}^{(k)}(x) = \frac{(k+2)!}{2!} \left\{ x^2 - kx + \frac{k(k-1)}{4} \right\}$$

$$G_{k+3}^{(k)}(x) = \frac{(k+3)!}{3!} \left\{ x^3 - \frac{3k}{2} x^2 + \frac{3k(k-1)}{4} x - \frac{k^2(k-3)}{8} \right\} \tag{2.3}$$

$$G_{k+4}^{(k)}(x) = \frac{(k+4)!}{4!} \left\{ x^4 - \frac{4k}{2} x^3 + \frac{6k(k-1)}{4} x^2 - \frac{4k^2(k-3)}{8} x \right.$$
$$\left. + \frac{k(k-1)(k^2-5k-2)}{16} \right\}$$

$$G_{k+5}^{(k)}(x) = \frac{(k+5)!}{5!} \left\{ x^5 - \frac{5k}{2} x^4 + \frac{10k(k-1)}{4} x^3 - \frac{10k^2(k-3)}{8} x^2 \right.$$
$$\left. + \frac{5k(k-1)(k^2-5k-2)}{16} x - k^2 \frac{k^3 - 10k^2 + 15k + 10}{32} \right\}$$

In particular, when $k = 1$, so that we omit the superscript,

$$G_0(x) = 0$$

$$G_1(x) = 1$$

$$G_2(x) = 2x - 1 \qquad\qquad = 2(x - \tfrac{1}{2})$$

$$G_3(x) = 3x^2 - 3x \qquad\qquad = 3x(x-1) \tag{2.4}$$

$$G_4(x) = 4x^3 - 6x^2 + 1 \qquad = 4(x - \tfrac{1}{2})(x^2 - x - \tfrac{1}{2})$$

$$G_5(x) = 5x^4 - 10x^3 + 5x \qquad = 5x(x-1)(x^2 - x - 1)$$

$$G_6(x) = 6x^5 - 15x^4 + 15x^2 - 3 \qquad = 6(x - \tfrac{1}{2})(x^4 - 2x^3 - x^2 + 2x + 1)$$

$$G_7(x) = 7x^6 - 21x^5 + 35x^3 - 21x \qquad = 7x(x-1)(x^4 - 2x^3 - 2x^2 + 3x + 3)$$

Putting $x = 0$ in (2.4) we obtain the simplest values (1.2) for G_n. It is worthwhile writing out the values of some Genocchi numbers of order k by putting $x = 0$ in (2.3).

Comparing (2.1) with the definition of $E_n^{(k)}(x)$ in, say, [13] we readily see that

$$G_n^{(k)}(x) = n(n-1)\ldots(n-k+1) E_{n-k}^{(k)}(x) \quad n \geq k \geq 0 \tag{2.5}$$

Because of this simple relationship, many results for $G_n^{(k)}(x)$ will parallel those of $E_n^{(k)}(x)$, and, to a lesser extent, those of $B_n^{(k)}(x)$. However, some results are unique to the theory of Genocchi polynomials.

Standard techniques, which are sometimes suppressed in this presentation to conserve space, allow us to produce a variety of results for $G_n^{(k)}(x)$. It is useful to put $k = 1$ in each case to see what happens for the Genocchi polynomials $G_n(x)$. Some theorems valid only when $k = 1$, i.e., for which no generalization exists, are considered in a separate section of this paper.

Differentiating both sides of (2.1) w.r.t. x and simplifying gives the "Appell property"

$$\frac{d}{dx} G_n^{(k)}(x) = n\, G_{n-1}^{(k)}(x) \quad (n-1 \geq 0) \tag{2.6}$$

whence

$$\frac{d^p}{dx^p} G_n^{(k)}(x) = n(n-1)(n-p+1)\, G_{n-p}^{(k)}(x) \quad (n-p \geq 0) \tag{2.7}$$

so that

$$\frac{d^n}{dx^n} G_n^{(k)}(x) = n! \tag{2.8}$$

Integration of (2.6) yields $G_n^{(k)}$ as the constant of integration. Polynomials $A_n(x)$ are called *Appell polynomials* [4] if they have the property

$$\frac{d\, A_n(x)}{dx} = n\, A_{n-1}(x) \tag{2.9}$$

Thus, Bernoulli, Euler and Genocchi polynomials are Appell polynomials.

If in (2.1) we write

$$\Phi = \left(\frac{2t}{e^t + 1}\right)^k e^{tx} \tag{2.10}$$

then

$$\frac{\partial \Phi}{\partial x} = t\Phi \tag{2.11}$$

(which may be used to establish (2.6)) and

$$t\frac{\partial \Phi}{\partial t} - \left\{\frac{k + tx}{t} - \frac{ke^t}{e^t + 1}\right\}\frac{\partial \Phi}{\partial x} = 0 \tag{2.12}$$

From (2.6), we have, on anticipating (2.22),

$$\int_0^1 G_n^{(k)}(x)\, dx = \frac{G_{n+1}^{(k)}(1) - G_{n+1}^{(k)}}{n+1}$$

$$= \begin{cases} 0 & \text{if n is even} \\ -\dfrac{2\, G_{n+1}^{(k)}}{n+1} & \text{if n is odd} \end{cases} \quad \text{by (1.2)} \tag{2.13}$$

More generally,

$$\int_x^{x+1} G_n^{(k)}(x)\, dx = \frac{G_{n+1}^{(k)}(1+x) - G_{n+1}^{(k)}(x)}{n+1} \tag{2.14}$$

Also from (2.6)

$$G_n^{(k)}(x) - G_n^{(k)} = n \int_0^x G_{n-1}^{(k)}(x)\, dx \tag{2.15}$$

If we know the $G_n^{(k)}$, then (2.15) used in conjunction with (2.3) provides us with a means of successively calculating the $G_n^{(k)}(x)$, though this becomes complicated when n is large.

Substitution of $k = 1$ in (2.13) − (2.15) yields important special cases.

Recurrence Relations

Using Cauchy's rule for the multiplication of power series, we derive from (2.1) the *summation formula*

$$G_n^{(k)}(x+y) = \sum_{j=0}^n \binom{n}{j} G_j^{(k)}(x)\, y^{n-j} \tag{2.16}$$

Special cases of (2.16) include, when $x = 0$ and y is replaced by x, the *explicit form*

$$G_n^{(k)}(x) = \sum_{j=0}^n \binom{n}{j} G_j^{(k)} x^{n-j} \tag{2.17}$$

and, when $y = k$, the *difference, or recurrence, relation*

$$G_n^{(k)}(k+x) - G_n^{(k)}(x) = \sum_{j=0}^{n-1} \binom{n}{j} G_j^{(k)}(x)\, k^{n-j} \tag{2.18}$$

In (2.18), $k + x$ may be replaced by $1 + x$ without loss of generality, in which case the factor k^{n-j} becomes 1, so that

$$G_n^{(k)}(1+x) - G_n^{(k)}(x) = \sum_{j=0}^{n-1} \binom{n}{j} G_j^{(k)}(x) \tag{2.19}$$

that is

$$G_n^{(k)}(1+x) = \sum_{j=0}^n \binom{n}{j} G_j^{(k)}(x) \tag{2.19}'$$

Equivalence of (2.14) and (2.19) gives

$$\int_x^{x+1} G_{n-1}^{(k)}(x)\, dx = \frac{1}{n} \sum_{j=0}^{n-1} \binom{n}{j} G_j^{(k)}(x) \tag{2.14}'$$

From (2.17) or (2.19)

$$G_n^{(k)}(1) = \sum_{j=0}^n \binom{n}{j} G_j^{(k)} \tag{2.17}'$$

We now prove

Theorem 1: $G_n^{(k)}(1+x) + G_n^{(k)}(x) = 2n\, G_{n-1}^{(k-1)}(x)$ $(n \geq 1)$ (2.20)

Proof: $\sum_{n=1}^{\infty} \left[G_n^{(k)}(1+x) + G_n^{(k)}(x) \right] \frac{t^n}{n!}$

$$= \left(\frac{2t}{e^t+1}\right)^k e^{t(1+x)} + \left(\frac{2t}{e^t+1}\right)^k e^{tx} \quad \text{by (2.1)}$$

$$= 2t \left(\frac{2t}{e^t+1}\right)^{k-1} e^{tx}$$

$$= 2n \sum_{n=1}^{\infty} G_{n-1}^{(k-1)}(x) \frac{t^n}{n!} \qquad\qquad \text{by (2.1)}$$

whence the result follows on equating coefficients of $\frac{t^n}{n!}$.

Corollary (k = 1): $G_n(1+x) + G_n(x) = 2n\, x^{n-1}$ $(n \geq 1)$ (2.21)

by (2.20) and (2.2), whence

$$G_n(1) = -G_n \quad (n>1) \qquad\qquad (2.22)$$

Taking $(2.17)'$, with $k = 1$, in conjunction with (2.22) we deduce the *recurrence relation* for Genocchi numbers

$$G_n = -\tfrac{1}{2} \sum_{j=0}^{n-1} \binom{n}{j} G_j \quad (n > 1) \qquad\qquad (2.23)$$

Next, from (2.16), on simplification and use of (1.2), we observe that

$$G_n(x) + (-1)^{n-1} G_n(-x) = 2n\, x^{n-1} \qquad\qquad (2.24)$$

and

$$G_n(x) + (-1)^n G_n(-x) = 2 \sum_{i=1}^{\left[\frac{N}{2}\right]} \binom{n}{2i} G_{2i}\, x^{n-2i} \qquad\qquad (2.25)$$

where $N = n$ or $n-1$ according as n is even or odd.

For n even in (2.24), we deduce that

$$G_n(x) - G_n(-x) = 2n\, x^{n-1} \qquad\qquad (2.24)'$$

whence

$$G_n(1) - G_n(-1) = 2n \qquad\qquad (2.24)''$$

Combining (2.21) and $(2.24)'$ by subtraction yields

$$G_n(1+x) + G_n(-x) = 0 \qquad\qquad (2.26)$$

In evaluating the left-hand side of (2.24), the determining factor is $1-(-1)^j$, while in the case of (2.25) it is $1+(-1)^j$.

Finally, if we replace x by $x-1$ in (2.21), or use (2.1), it follows that

$$G_n(x) + G_n(x-1) = 2n(x-1)^{n-1} \tag{2.27}$$

(Cf. Nielsen [14] for $E_n(x)$.)

The Operators Δ, ∇

The symbols Δ and ∇ ("nabla") represent the operations of obtaining the *difference*, and taking the *mean*, of $f(1+x)$ and $f(x)$, thus:

$$\Delta\, f(x) = f(1+x) - f(x) \tag{2.28}$$

and

$$\nabla\, f(x) = \frac{f(1+x) + f(x)}{2} \tag{2.29}$$

Hence, by (2.21),

$$\nabla\, G_n(x) = nx^{n-1} \tag{2.29'}$$

(so $\nabla\, G_n(0) = 0$ as in (2.22), $\nabla\, G_n(1) = n$, $\nabla\, G_n(-1) = (-1)^{n-1}n$) while from (2.19)

$$\Delta\, G_n(x) = \sum_{j=0}^{n-1} \binom{n}{j} G_j(x) \tag{2.19''}$$

More generally, from (2.20)

$$\nabla\, G_n^{(k)}(x) = n\, G_{n-1}^{(k-1)}(x) \qquad (n \geq 1) \tag{2.20'}$$

i.e., the operator ∇ depresses the order k by one and the degree n by one.

Repeated application of ∇ on $G_n^{(k)}(x)$ $n-k$ times leads by (2.2) to

$$(\nabla)^{n-k} G_n^{(k)}(x) = n(n-1)\ldots(n-k+1)\, x^{n-k} = \frac{n!}{(n-k)!}\, x^{n-k} \tag{2.30}$$

Finally,

$$\sum_{n=k}^{\infty} \frac{t^n}{n!}\, (\nabla)^{n-k} G_n^{(k)}(x) = t^k\, e^{tx} \tag{2.31}$$

Use of Δ and ∇ will recur subsequently in this article.

Complementary Argument

The functions x and $k-x$ are called complementary arguments for the Genocchi polynomials. We now prove the "complementary argument theorem", which has a symmetry property.

Theorem 2: $G_n^{(k)} (k-x) = (-1)^{k+n} G_n^{(k)} (x)$ (2.32)

Proof: $\sum_{n=0}^{\infty} G_n^{(k)} (k-x) \frac{t^n}{n!} = \left(\frac{2t}{1+e^t}\right)^k e^{t(k-x)}$ by (2.1)

$$= (-1)^k \left(\frac{2(-t)}{e^{-t}+1}\right)^k e^{-tx}$$

$$= (-1)^k \sum_{n=0}^{\infty} G_n^{(k)} (x) \frac{(-t)^n}{n!} \quad \text{by (2.1)}$$

$$= \sum_{n=0}^{\infty} (-1)^{k+n} G_n^k (x) \frac{t^n}{n!}$$

Equating coefficients of $\frac{t^n}{n!}$ we deduce the theorem.

Corollary 1: Putting $k = 1$ in (2.32) we derive

$$\left.\begin{array}{ll} G_n (1-x) = G_n (x) & \text{if n is odd} \\ G_n (1-x) = -G_n (x) & \text{if n is even} \end{array}\right\}$$ (2.33)

Corollary 2: Replace x by $x + \frac{k}{2}$ in (2.32). It follows that

$$\left.\begin{array}{ll} G_n^{(k)} \left(\frac{k}{2}+x\right) = G_n^{(k)} \left(\frac{k}{2}-x\right) & \text{if n+k is even} \\ = -G_n^{(k)} \left(\frac{k}{2}-x\right) & \text{if n+k is odd} \end{array}\right\}$$ (2.34)

so that

$$\left.\begin{array}{ll} G_n \left(\frac{1}{2}+x\right) = G_n \left(\frac{1}{2}-x\right) & \text{n odd} \\ = -G_n \left(\frac{1}{2}-x\right) & \text{n even} \end{array}\right\}$$ (2.35)

whence, with $x = \frac{1}{2}$, the results in (2.22), and (2.40) and (2.41) follow.

Two Summation Formulas

Theorem 3: $\sum_{s=1}^{m} (-1)^s s^n = \frac{(-1)^m G_{n+1}(1+m) - G_{n+1}(1)}{2(n+1)}$ (2.36)

Proof:
$$\frac{1}{2}\frac{\left[(-1)^m G_{n+1}(1+m) - G_{n+1}(1)\right]}{n+1}$$

$$= \frac{1}{2} \sum_{s=1}^{m} (-1)^s \frac{\left[G_{n+1}(1+s) + G_{n+1}(s)\right]}{n+1}$$

$$= \sum_{s=1}^{m} (-1)^s \frac{\nabla G_{n+1}(s)}{n+1} \qquad \text{by (2.29)}$$

$$= \sum_{s=1}^{m} (-1)^s s^n \qquad \text{by (2.20)', (2.2)}$$

Miline-Thomson [13] remarks that this method can clearly be applied if the sth term in a finite series is a polynomial in s. In fact, by (2.20), we may generalize the result to read

$$\sum_{s=1}^{m} (-1)^s \, G_n^{(k-1)}(x) = \frac{(-1)^m \, G_{n+1}^{(k)}(1+m) - G_{n+1}^{(k)}(1)}{2(n+1)} \qquad (2.36)'$$

Theorem 4: $G_n^{(k)}(x) = \sum_{r=0}^{n} \binom{n}{r} G_r^{(k)} \left(\tfrac{1}{2}\right) \left(x - \tfrac{1}{2}\right)^{n-r}$ \hfill (2.37)

Proof: $\left(\dfrac{2t}{e^t+1}\right)^k e^{tx} = \left(\dfrac{2t}{e^t+1}\right)^k e^{\frac{t}{2}} \, e^{(x-\frac{1}{2})t}$

i.e., $\displaystyle\sum_{n=0}^{\infty} G_n^{(k)}(x) \frac{t^n}{n!} = \sum_{r=0}^{\infty} G_r^{(k)} \left(\tfrac{1}{2}\right) \frac{t^r}{r!} \sum_{m=0}^{\infty} \left(x - \tfrac{1}{2}\right)^m \frac{t^m}{m!}$ by (2.1)

Cauchy's rule for multiplication of power series produces the result.

Some Numerical Results

Let $x = 0$, $n = 2m$ in (2.32). Then

$$G_{2m}^{(k)}(k) = (-1)^k \, G_{2m}^{(k)} \qquad (2.38)$$

while

$$G_{2m+1}^{(k)}(k) = (-1)^{k+1} \, G_{2m+1}^{(k)} \qquad (2.38)'$$

Furthermore, $x = \frac{k}{2}$, $n = 2m$ in (2.32) yield

$$G_{2m}^{(k)} \left(\tfrac{1}{2} k\right) = (-1)^k \, G_{2m}^{(k)} \left(\tfrac{1}{2} k\right) \qquad (2.39)$$

Now, let $k = 1$. Results (2.33) − (2.35) then lead to the following numerical values for $n > 1$.

n even:
$$\left. \begin{array}{l} G_n\left(\tfrac{1}{2}\right) = 0 \\[4pt] G_n(1) = -\, G_n \\[4pt] G_n\left(\tfrac{1}{4}\right) = G_n\left(\tfrac{3}{4}\right), \; G_n\left(\tfrac{1}{3}\right) = -G_n\left(\tfrac{2}{3}\right) \end{array} \right\} \qquad (2.40)$$

n odd:
$$\left. \begin{array}{l} G_n(1) = G_n = 0 \\[4pt] G_n\left(\tfrac{1}{4}\right) = G_n\left(\tfrac{3}{4}\right), \; G_n\left(\tfrac{1}{3}\right) = G_n\left(\tfrac{2}{3}\right) \end{array} \right\} \qquad (2.41)$$

From the above, we conclude that $G_{2n}(x)$ has a factor $x - \frac{1}{2}$ whereas $G_{2n+1}(x)$ has a factor $x\,(x-1)$. See (2.4). Next,

$$\int_0^{\frac{1}{2}} G_n\,(x)\,dx = \frac{G_{n+1}\left(\frac{1}{2}\right) - G_{n+1}}{n+1} \qquad \text{from (2.6)}$$

$$\left.\begin{aligned} &= -\frac{G_{n+1}}{n+1} && \text{n even by (2.40)}\\[2mm] &= \frac{G_{n+1}\left(\frac{1}{2}\right)}{n+1} && \text{n odd by (1.2)} \end{aligned}\right\} \qquad (2.42)$$

Relationships of use in computation include

$$G_n\left(\tfrac{1}{2}\right) = \tfrac{n}{2^{n-1}}\,E_{n-1} = n\,E_{n-1}\left(\tfrac{1}{2}\right) \qquad (2.43)$$

and

$$G_{n+1} = \tfrac{n+1}{2^n}\,C_n = (n+1)\,E_n(0) = (-1)^n(n+1)\,E_n(1) \qquad (2.44)$$

in which $C_n = 2^n E_n(0)$ are the *voisins* ("cousins") of E_n in Nörlund's phraseology.

Applying these, we may substitute appropriately in various formulas in [3], [11] and [16] to obtain, for example,

$$\sum_{m=0}^{n} \binom{2n}{2m} \frac{2^m}{2m+1}\,G_{2m+1}\left(\tfrac{1}{2}\right) = 0 \qquad (2.45)$$

$$G_{n+1}^{(k)}(x) = (n+1) \sum_{m=k-1}^{n} \binom{n}{m} \frac{G_{m+1}^{(k)}}{m+1}\,x^{n-m} \qquad (2.46)$$

$$\sum_{m=0}^{n} \binom{n}{m} \frac{G_{m+1}}{m+1} + \frac{G_{n+1}}{n+1} = 0 \quad \text{by (2.40), (2.46)} \qquad (2.46)'$$

$$G_{2m} = 2^{4m} \frac{2^{2m}-1}{2^{2m-1}-1}\,B_{2m}\left(\tfrac{1}{4}\right). \qquad (2.47)$$

Integral formulas adaptable form [16] include

$$G_{n+1}(2x) = 2^{n+1}(n+1) \int_x^{x+\frac{1}{2}} B_n(x)\,dx \qquad (2.48)$$

whence

$$G_{n+1} = 2^{n+1}(n+1) \int_0^{\frac{1}{2}} B_n\,dx \quad (x = 0 \text{ in (2.48)}) \qquad (2.48)'$$

and

$$G_{n+1}\left(\tfrac{1}{2}\right) = 2^{n+1}(n+1) \int_{\frac{1}{4}}^{\frac{3}{4}} B_n\left(\tfrac{1}{4}\right)\,dx \quad (x = \tfrac{1}{4} \text{ in (2.48)}) \qquad (2.48)''$$

Relationships between G_n and E_n are, e.g.,

$$E_{n-1} = \frac{1}{2n} \sum_{j=0}^{n} \binom{n}{j} 2^j G_j \qquad \text{by (2.16), (2.43)} \qquad (2.49)$$

and

$$G_n = \frac{2n}{2^n} \sum_{j=0}^{n-1} (-1)^j \binom{n-1}{j} E_{n-1-j} \qquad \text{by [16], (2.44)} \qquad (2.50)$$

while another connection between G_n and $E_n (0)$ is

$$G_{n+1} = -(n+1) \sum_{s=0}^{n} \binom{n}{s} E_{n-s} (0). \qquad (2.51)$$

Graphs of $G_n (x)$ in the Interval $(0,1)$

As we have seen in (2.40) and (2.41), $G_{2n} (x)$ has a zero at $x = \frac{1}{2}$ whereas $G_{2n+1} (x)$ has zeros at $0,1$. Following a routine argument for $B_n (x)$ (cf. [7], [13]), we may establish that these are the only zeros of $G_n (x)$ in $(0,1)$.

Because of the connection (2.5) between $G_n (x)$ and $E_n (x)$, the graphs for $G_n (x)$ will be similar to those for $E_n (x)$ and are therefore not reproduced here. For the graphs of $- E_1 (x)$, $- E_2 (x)$, $(-1)^n E_{2n-1} (x)$ and $(-1)^n E_{2n} (x)$, we refer the reader to [16]. One may compare these graphs with those of $B_n (x)$ given in [7].

Recurrence Relation between Polynomials of Successive Orders

Theorem 5: $G_n^{(k+1)} (x) = \frac{2n(k-x)}{k} G_{n-1}^{(k)}(x) + \frac{2}{k} (n-k) G_n^{(k)}(x)$ $\qquad (2.52)$

Proof: Differentiate both sides of (2.1) w.r.t. t, and then multiply by t. We have

$$\sum_{n=0}^{\infty} G_n^{(k)} (x) \frac{n t^{n-1}}{n!} \cdot t$$

$$= \left(\frac{2t}{e^t+1}\right)^k x e^{tx} \cdot t + k t^{k-1} \left(\frac{2}{e^t+1}\right)^k e^{tx} \cdot t - \frac{k}{2} \left(\frac{2}{e^t+1}\right)^{k+1} \cdot t^k e^{t(1+x)} \cdot t$$

$$= x \sum_{n=0}^{\infty} G_n^{(k)} (x) \frac{t^{n+1}}{n!} + k \sum_{n=0}^{\infty} G_n^{(k)} (x) \frac{t^n}{n!} - \frac{k}{2} \sum_{n=0}^{\infty} G_n^{(k+1)} (1+x) \frac{t^n}{n!}$$

Equate coefficients of t^{n+1}. Then

$$G_{n+1}^{(k)} (x) = x\, G_n^{(k)} (x) + k \frac{G_{n+1}^{(k)}}{n+1} (x) - \frac{k}{2} \frac{G_{n+1}^{(k+1)} (1+x)}{n+1} \quad \ldots \ldots (\alpha)$$

Now, by (2.20),

$$G_{n+1}^{(k+1)} (1+x) = 2(n+1) G_n^{(k)} (x) - G_{n+1}^{(k+1)} (x) \qquad \ldots \ldots (\beta)$$

Hence, by (α) and (β),

$$G_{n+1}^{(k)}(x) = x\, G_n^{(k)}(x) + \frac{k}{n+1}\left\{ G_{n+1}^{(k)}(x) - \tfrac{1}{2}\big(2(n+1)\,G_n^{(k)}(x) - G_{n+1}^{\{k+1\}}(x)\big)\right\}$$

On tidying up these expressions, we obtain

$$G_{n+1}^{(k+1)}(x) = \frac{2(n+1)}{k}(k-x)\,G_n^{(k)}(x) + \frac{2}{k}\left\{n+1-k\right\}G_{n+1}^{(k)}(x) \quad \dots \dots (\gamma)$$

Replacing $n+1$ by n, we get

$$G_n^{(k+1)}(x) = \frac{2n(k-x)}{k}\,G_{n-1}^{(k)}(x) + \frac{2}{k}(n-k)\,G_n^{(k)}(x) \qquad \dots \dots (\delta)$$

This corresponds to the recurrence relations for k^{th} order Euler polynomials and Bernoulli polynomials which are given in [11], [13] and [16]. Observe the existence of an extra term on the right-hand side in the Genocchi case. This is due to the necessity to differentiate the extra factor t^k in (2.1) which is absent in the Euler and Bernoulli cases.

When $x = 0$ in (δ), we have

$$G_n^{(k+1)} = 2n\, G_{n-1}^{(k)} + \frac{2}{k}(n-k)\, G_n^{(k)} \qquad\qquad (2.52)'$$

Connection with Bernoulli Polynomials

Theorem 6: $G_n^{(k)}(x) = 2^n \sum\limits_{r=0}^{k} (-1)^{r+k}\binom{k}{r} B_n^{(k)}\left(\frac{x+r}{2}\right)$ $\qquad\qquad$ (2.53)

Proof: $\sum\limits_{n=0}^{\infty} G_n^{(k)}(x)\,\frac{t^n}{n!} = 2^k \left(\dfrac{\frac{t}{2}}{e^{\frac{t}{2}}+1}\right)^k e^{\frac{t}{2}x}$ $\qquad$ by (2.1)

$$= \frac{t^k\, e^{\frac{t}{2}x}\,(e^{\frac{t}{2}}-1)^k}{(e^t-1)^k} \qquad\qquad (A)$$

$$= (-1)^k \sum_{r=0}^{k} \frac{(-1)^r \binom{k}{r} e^{\frac{(x+r)t}{2}}}{(e^t-1)^k}\, t^k$$

$$= 2^n \sum_{n=0}^{\infty}\sum_{r=0}^{k} (-1)^{r+k}\binom{k}{r} B_n^{(k)}\left(\frac{x+r}{2}\right)\frac{t^n}{n!}$$

on using the definition of $B_n^{(k)}(x)$ [16]. Equating the coefficients of $\frac{t^n}{n!}$ in the summation, we obtain the formula (2.53).

When $k = 1$, we may proceed quickly to the formula from (A). Alternatively, substitute $k = 1$ in the formula. In either case, we have

$$G_n(x) = 2^n\left[B_n\left(\frac{x+1}{2}\right) - B_n\left(\frac{x}{2}\right)\right] \qquad\qquad (2.53)'$$

There does not appear to be a corresponding result in the literature for $E_n^{(k)}(x)$, though the case $k = 1$ is given in Erdelyi [3].

Result $(2.53)'$ also follows from (2.5) and $[4, p.41]$.

Equivalently, $(2.53)'$ may be expresses as

$$G_n(x) = 2 \left[2^n B_n \left(\frac{x+1}{2} \right) - B_n(x) \right] \qquad (2.53)''$$

Lastly, we mention that the method of [12] for reciprocals of the kth power of *Pell-Lucas numbers* Q_n allows us to determine that

$$\frac{1}{Q_n^k} = \frac{1}{2^k \, m^k \, (\delta^x \, \gamma^{1-x})^n} \sum_{r=0}^{\infty} G_r^{(k)}(x) \log \left(\frac{\delta}{\gamma} \right)^r \cdot \frac{n^r}{r!} \qquad (2.54)$$

where $m = n \log (\delta/\gamma)$ and $\gamma = 1 + \sqrt{2}$, $\delta = 1 - \sqrt{2}$. Specialization involving Genocchi numbers occurs in (2.54) when $k = 1$, $x = 0$.

3. GENOCCHI POLYNOMIALS OF THE FIRST ORDER

While many results for Genocchi polynomials $G_n^{(1)}(x)$ can be generalized, some cannot. An important instance of this is the *multiplication theorem* for $G_n(mx)$ for which we must consider two cases, m even and n odd. In the case m even, Bernoulli polynomials occur.

Multiplication Theorem

Theorem 7 (Multiplication Theorem):

$$\left. \begin{array}{ll} G_n(mx) = m^{n-1} \sum_{s=0}^{m-1} (-1)^s G_n \left(x + \frac{s}{m} \right) & \text{m odd} \\ \\ G_n(mx) = -2 \, m^{n-1} \sum_{s=0}^{m-1} (-1)^s B_n \left(x + \frac{s}{m} \right) & \text{m even} \end{array} \right\} \qquad (3.1)$$

Proof: m odd

$$\sum_{n=0}^{\infty} \frac{t^n}{n!} \sum_{s=0}^{m-1} (-1)^s G_n \left(x + \frac{s}{m} \right) = \sum_{s=0}^{m-1} \frac{2t}{e^t + 1} \cdot (-1)^s e^{tx} \, e^{\frac{s}{m}t} \qquad \text{by (2.1)}$$

$$= \frac{2t}{e^t + 1} \cdot e^{tx} \left\{ 1 - e^{\frac{t}{m}} + e^{\frac{2t}{m}} - \ldots + (-1)^{m-1} e^{\frac{m-1}{m}t} \right\}$$

$$= \frac{2t}{e^t + 1} e^{tx} \frac{\{ 1 - (-1)^m e^t \}}{1 + e^{\frac{t}{m}}} \qquad \text{by the geometric progression formula}$$

$$= \frac{2t \, e^{tx}}{1 + e^{\frac{t}{m}}} \qquad \text{since m is odd}$$

$$= \frac{2m \cdot \frac{t}{m} e^{mx\frac{t}{m}}}{1 + e^{\frac{t}{m}}}$$

$$= \sum_{n=0}^{\infty} 2m \frac{t^n}{m^n \cdot n!} G_n (mx) \qquad \text{by (2.1)}$$

Hence, $G_n (mx) = m^{n-1} \sum_{s=0}^{m-1} (-1)^s G_n \left(x + \frac{s}{m}\right)$

m even

$$\sum_{n=0}^{\infty} \frac{t^n}{n!} \sum_{s=0}^{m-1} (-1)^s B_n \left(x + \frac{s}{m}\right) = \sum_{s=0}^{m-1} \frac{t}{e^t - 1} (-1)^s e^{(x + \frac{s}{m})t} = \frac{t\, e^{tx}}{e^t - 1} \cdot \frac{1 - (-1)^m e^t}{1 + e^{\frac{t}{m}}}$$

as in the case m odd

$$= - \frac{t\, e^{tx}}{1 + e^{\frac{t}{m}}} \qquad\qquad\qquad \text{for m even}$$

$$= - \frac{2m}{2} \cdot \frac{t}{m} \frac{e^{mx \cdot \frac{t}{m}}}{1 + e^{\frac{t}{m}}}$$

$$= - \sum_{n=0}^{\infty} \frac{1}{2} m \frac{t^n}{m^n \cdot n!} G_n (mx) \qquad \text{by (2.1)}$$

Hence, $G_n (mx) = - 2m^{n-1} \sum_{s=0}^{m-1} (-1)^s B_n \left(x + \frac{s}{m}\right) \qquad$ m even

For example, with m = 3, x = 0,

$$G_n = 3^{n-1} \left(G_n - G_n \left(\tfrac{1}{3}\right) + G_n \left(\tfrac{2}{3}\right) \right)$$

Our multiplication theorem corresponds to *Raabe's theorem* for multiplication for Bernoulli polynomials. Setting m = 2 in (3.1), we revisit (2.53)′ for $G_n (2x)$.

Orthogonality - Type Relation

Theorem 8: $\int_0^1 G_u (x) G_v (x) \, dx = \begin{cases} 2 (-1)^u \dfrac{u! \, v!}{(u+v)!} G_{u+v} & \text{if } u + v \text{ is even} \\ 0 & \quad\ldots\ldots\ldots \text{ odd} \end{cases}$ (3.2)

Proof:

$$\int_0^1 G_u (x) G_v (x) \, dx = \frac{1}{v+1} \left[G_u(x) G_{v+1}(x)\right]_0^1 - \frac{1}{v+1} \int_0^1 G_{v+1}(x) \cdot u \, G_{u-1} \, dx \qquad \text{by (2.6)}$$

$$= - \frac{u}{v+1} \int_0^1 G_{u-1} (x) G_{v+1} (x) \, dx \qquad\qquad \text{by (2.40), (2.41)} \quad \ldots \text{(i)}$$

$$= (-1)^{u-1} \frac{u! \, v!}{(u+v)!} \left[G_{u+v} (x)\right]_0^1 \qquad \text{by repeated application of (i)}$$

$$= (-1)^{u-1} \frac{u! \, v!}{(u+v)!} \left(G_{u+v} (1) - G_{u+v} \right)$$

$$= 2(-1)^u \frac{u! \, v!}{(u+v)!} G_{u+v} \quad \text{if } u + v \text{ is even}$$

$$= 0 \qquad\qquad \dots\dots\dots \text{odd} \qquad \Big\} \quad \text{by (2.40), (2.41)}$$

Some interesting special cases arise when $u = v = 2w$; $u = v = 2w + 1$; $u = 2$, $v = 2w$; $u = 1$, $v = 2w - 1$ (the last one leading to the value $-\frac{G_{2w}}{w}$ for the integral). See Jordan [7] for corresponding results for the Euler polynomials.

Interchanging u and v in (3.2) we have $(-1)^u \, G_{u+v} = (-1)^v \, G_{u+v}$. Next, if $u = v + 1$ we then have $G_{2u+1} = -G_{2u+1}$, that is $G_{2u+1} = 0$ as we know (1.2).

Theorem 8 may be generalized for $G_u^{(k)}(x)$ and $G_v^{(k)} (x)$, the limits of integration then being k and 0). Recall - cf. (2.38), (2.38)′ − that generally $G_{u+v}^{(k)} \neq 0$ when $u + v$ is of odd parity.

Fourier Expansions in the Interval (0,1)

Two cases for the Fourier development of Genocchi polynomials over the interval $(0,1)$ need to be considered, namely, n even and n odd.

Following standard methods ([4], [7]), we eventually obtain

$$G_{2n} (x) = (-1)^n \, 4 \, (2n)! \sum_{m=0}^{\infty} \frac{\cos (2m+1)\pi x}{[(2m+1)\pi]^{2n}} \qquad (3.3)$$

and

$$G_{2n+1} (x) = (-1)^n \, 4 \, (2n+1)! \sum_{m=0}^{\infty} \frac{\sin (2m+1)\pi x}{[(2m+1)\pi]^{2n+1}} \qquad (3.4)$$

Fourier expansions (3.3) and (3.4) are in agreement with the corresponding Fourier developments for Euler polynomials given in, say, [9], [11] and [16], which Jordan [7] attributes to Lindelöf [9]. See also Jordan [7] for Fourier expansions for $E_n(x)$ different from those in the sources quoted.

Using (1.4) and [8, p. 238] with $x = 0$, we deduce that

$$\frac{(-1)^m \, \pi^{2m} \, G_{2m}}{4(2m)!} = 1 + \frac{1}{3^{2m}} + \frac{1}{5^{2m}} + \frac{1}{7^{2m}} + \cdots \qquad (3.5)$$

(giving known series for $\frac{\pi^2}{8}$, $\frac{\pi^4}{96}$, and $\frac{\pi^6}{980}$ when $m = 1, 2, 3$, respectively).

Furthermore, from (1.4) and the "remarkable relation" [4, p. 37], we derive

$$\frac{(-1)^{m-1} (2\pi)^{2m} G_{2m}}{4 (2m)! (1-2^{2m})} = \zeta (2m) \tag{3.6}$$

where $\zeta (2m)$ is the *Riemann ζ - function* $\sum\limits_{n=1}^{\infty} \frac{1}{n^{2m}}$. With m = 1, 2, 3 in turn there appear the known infinite series summations for $\frac{\pi^2}{6}, \frac{\pi^4}{90},$ and $\frac{\pi^6}{945}$ respectively.

Use of Boole's Theorem for Polynomials

A polynomial P(t) may be expanded in terms of Euler polynomials by Boole's Theorem thus ([13], see [1] also):

Boole's Theorem: $P(x+y) = \nabla P(x) + E_1(y) \nabla P'(x) + \frac{1}{2!} E_2(y) \nabla P''(x) + \dots \dots$ $\tag{3.7}$

(a) Genocchi Polynomials in terms of Euler's $E_n(0)$

Put y = 0 in (3.7) and take $P(x) \equiv G_n(x)$.

Then Boole's expansion for $G_n(x)$ is

$$G_n(x) = \nabla G_n(x) + E_1(0) \nabla G'_n(x) + \frac{1}{2!} E_2(0) \nabla G''_n(x) + \dots \tag{3.8}$$

The series on the right-hand side terminates after a finite number of terms.

Using $(2.29)'$, (1.3) and (1.5) we verify, for example, that $G_4(x) = 4x^3 - 6x^2 + 1$, as we know from (2.4).

(b) Bernoulli Polynomials in terms of Genocchi Polynomials

In (3.7), put x = 0 and replace y by x and P by B_n. Then

$$\left.\begin{aligned}
B_n(x) &= \nabla B_n(0) + E_1(x) \nabla B'_n(0) + \frac{1}{2!} E_2(x) \nabla B''_n(x) + \dots \\
&= \nabla B_n(0) + \frac{1}{2} G_2(x) \nabla B'_n(0) + \frac{1}{3!} G_3(x) \nabla B''_n(x) + \dots
\end{aligned}\right\} \tag{3.9}$$

by (2.5).

For example, $\nabla B_5(x) = x^5 + \frac{5}{3} x^3 - \frac{1}{6} x$ on calculation, whence

$B_5(x) = \frac{1}{6} G_6(x) + \frac{5}{12} G_4(x) - \frac{1}{12} G_2(x) = x^5 - \frac{5}{2} x^4 + \frac{5}{3} x^3 - \frac{1}{6} x$ on calculation

Genocchi Polynomials in terms of Bernoulli Polynomials

The derivative of a polynomial may be expanded in terms of Bernoulli polynomials by the Euler-Maclaurin Theorem thus [13]:

Euler-Maclaurin Theorem: $P'(x+y) = \Delta P(x) + B_1(y) \Delta P'(x) + \frac{1}{2!} B_2(y) \Delta P''(x) + \dots$ $\tag{3.10}$

Putting $x = 0$ and replacing y by x and P' by G_n in (3.10), we derive the Euler-Maclaurin expansion for Genocchi polynomials

$$G'_n(x) = \triangle G_n(0) + B_1(x) \triangle G'_n(0) + \frac{1}{2!} B_2(x) \triangle G''_n(0) + \frac{1}{3!} B_3(x) \triangle G'''_n(0) + \dots \quad (3.11)$$

where $G'_n(x) = n\, G_{n-1}(x)$ by (2.6).

Accordingly, with $n = 5$ (for example), $\triangle G_5(x) = 20x^3 - 10x$ on calculation, so $G_4(x) = 4\, B_3(x) - 2\, B_1(x)$.

Inverse Mean ∇^{-1} of the Genocchi Polynomial

Defining the inverse mean function ∇^{-1} symbolically by $\nabla\, \nabla^{-1} = 1$, we proceed to give some brief information about $\nabla^{-1} G_n(x)$. [Refer to (2.29)' for $\nabla G_n(x)$.]

Following the method used by Jordan [7] we obtain, after several steps in the manipulation, two equivalent forms

$$\left.\begin{aligned} \nabla^{-1} G_n(x) &= 2\left[(1 - x\, G_n(x) + \frac{G_{n+1}(x)}{n+1}\right] \\ &= \sum_{j=0}^{n} \binom{n}{j} \frac{G_j\, G_{n+1-j}(x)}{n+1-j} \qquad \text{by (2.17), (2.29)'} \end{aligned}\right\} \qquad (3.12)$$

Combining these two forms when $x = 0$, we establish that

$$G_n + \frac{n}{n+1} G_{n+1} = \sum_{j=0}^{n} \binom{n}{j} \frac{G_j\, G_{n+1-j}}{2(n+1-j)} \qquad (3.13)$$

This illustration shows how ∇^{-1} may be used to discover relationships among the $G_n(x)$, and the G_n.

Calculation in (3.13) when $n = 7,\ 8$ (say) leads to the common values $29\frac{3}{4}$, 34 respectively of each side of the equation.

4. DIFFERENTIAL EQUATIONS ASSOCIATED WITH $G_n(X)$

Earlier in (2.12), a partial differential equation was seen to appear naturally in the theory of $G_n^{(k)}(x)$. We now establish some differential equations which arise in a more special way when $k = 1$.

Descending Diagonal Functions

Imagine the information in (2.4) is slant-wise re-tabulated so that the following functions (*descending diagonal functions*) are emphasized $\left(|x| < 1\right)$:

$$g_1(x) = G_1(1+2x+3x^2+4x^3+5x^4+\ldots\ldots\ldots) = G_1(1-x)^{-2}$$

$$g_2(x) = G_2(1+3x+6x^2+10x^3+15x^4+\ldots\ldots) = G_2(1-x)^{-3}$$

$$g_3(x) = G_3(1+4x+10x^2+20x^3+35x^4+\ldots\ldots) = G_3(1-x)^{-4} = 0 \qquad (4.1)$$

$$g_4(x) = G_4(1+5x+15x^2+35x^3+70x^4+\ldots\ldots) = G_4(1-x)^{-5}$$

$$\ldots\ldots\ldots\ldots\ldots\ldots\ldots\ldots\ldots\ldots\ldots\ldots\ldots\ldots$$

$$g_n(x) = G_n(1+\binom{n+1}{1}x+\binom{n+2}{2}x^2+\binom{n+3}{3}x^3+\ldots) = G_n(1-x)^{-(n+1)}$$

Clearly

$$g_{2n+1}(x) = 0 \qquad (n = 1, 2, 3, \ldots) \quad \text{by (1.2)} \qquad (4.1)'$$

To obtain (4.1), the values of G_n in (1.3) are utilized along with the combinatorial formula (2.17) for $k = 1$. Observe that

$$g_n(x) = G_n \sum_{j=0}^{\infty} \binom{n+j}{j} x^j \qquad (4.2)$$

In particular

$$g_n(0) = G_n$$

$$g_n\left(\frac{1}{2}\right) = 2^{n+1} G_n \left(= 2^{n+1} g_n(0)\right)$$

$$g_n\left(\frac{3}{4}\right) = 2^{n+1} g_n\left(\frac{1}{4}\right) \qquad (4.3)$$

$$g_n(1) \text{ is not defined.}$$

Write

$$D \equiv D(x,y) = \sum_{n=1}^{\infty} g_n(x) y^{n-1} = \sum_{n=1}^{\infty} G_n (1-x)^{-(n+1)} y^{n-1} \qquad (4.4)$$

Then we can derive the partial differential equation

$$(n+1)y \frac{\partial D}{\partial y} - (n-1)(1-x) \frac{\partial D}{\partial x} = 0 \qquad (4.5)$$

From (4.1), it follows that

$$(1-x) \frac{d\, g_n(x)}{dx} = (n+1) g_n(x) \qquad (4.6)$$

whence, on integrating by parts,

$$\int g_n(x)\,dx = \frac{1-x}{n}\,g_n(x) + c \tag{4.7}$$

Extending the preceding theory to general k, we find that

$$g_n^{(k)}x = G_{n+k-1}^{(k)}\,(1-x)^{-(n+k)} \tag{4.1a}$$

leading to a differential equation which is identical with (4.4) except that $n+1$ is replaced by $n+k$, but with D now standing for $\sum\limits_{n=1}^{\infty} g_n^{(k)}(x)\,y^{n-1}$.

Rising Diagonal Functions

Suppose next that in the re-tabulation of (2.4) we concentrate on the *rising diagonal functions* $h_n(x)$:

$$
\left.
\begin{aligned}
h_1(x) &= G_1 \\
h_2(x) &= G_2 + x\,G_0 \\
h_3(x) &= G_3 + 2x\,G_1 \\
h_4(x) &= G_4 + 3x\,G_2 + x^2\,G_0 \\
h_5(x) &= G_5 + 4x\,G_3 + 3x^2\,G_1 \\
h_6(x) &= G_6 + 5x\,G_4 + 6x^2\,G_2 + x^3\,G_0 \\
h_7(x) &= G_7 + 6x\,G_5 + 10x^2\,G_3 + 4x^3\,G_1
\end{aligned}
\right\} \tag{4.8}
$$

. . . .

Generally

$$h_n(x) = \sum_{j=0}^{[n/2]} \binom{n-j}{j} G_{n-2j}\,x^j \tag{4.9}$$

Obviously

$$
\left.
\begin{aligned}
h_n(0) &= G_n = g_n(0) && \text{by (4.3)} \\
&= 0 && \text{for n odd, } >1
\end{aligned}
\right\} \tag{4.10}
$$

Now let

$$
\left.
\begin{aligned}
R \equiv R(x,y) &= \sum_{n=1}^{\infty} h_n(x)\,y^{n-1} \\
&= (1-xy^2)^{-2}\,G_1 + y\,(1-xy^2)^{-3}\,G_2 + y^2\,(1-xy^2)^{-4}\,G_3 + \dots
\end{aligned}
\right\} \tag{4.11}
$$

from (4.8).

Then

$$\frac{\partial R}{\partial x} = y^2 \, \psi \tag{4.12}$$

and

$$\frac{\partial R}{\partial y} = 2xy \, \psi + \phi \tag{4.13}$$

where

$$\psi \equiv 2 \, (1 - xy^2)^{-3} \, G_1 + 3y \, (1 - xy^2)^{-4} \, G_2 + 4y^2 \, (1 - xy^2)^{-5} \, G_3 + \dots \tag{4.14}$$

and

$$\phi \equiv (1 - xy^2)^{-3} \, G_2 + 2y \, (1 - xy^2)^{-4} \, G_3 + 3y^2 \, (1 - xy^2)^{-5} \, G_4 + \dots \tag{4.15}$$

Using (4.12) and (4.13), we are led to

$$\frac{\partial^2 R}{\partial y \partial x} = 2y \, \psi + y^2 \, \frac{\partial \psi}{\partial y} \tag{4.16}$$

and

$$\frac{\partial^2 R}{\partial x \partial y} = 2y \, \psi + 2xy \, \frac{\partial \psi}{\partial x} + \frac{\partial \phi}{\partial x} \tag{4.17}$$

Applying Bernoulli's Theorem, we derive

$$\frac{\partial \phi}{\partial x} = y^2 \, \frac{\partial \psi}{\partial y} \; - \; 2xy \, \frac{\partial \psi}{\partial x} \tag{4.18}$$

For general k,

$$h_n^{(k)}(x) = \sum_{j=0}^{[n/2]} \binom{n+k-1-j}{j} G_{n+k-1-2j}^{(k)} \, x^j \tag{4.19}$$

and

$$R^{(k)}(x,y) = (1 - xy^2)^{-(k+1)} \, G_1 + y \, (1 - xy^2)^{-(k+2)} \, G_2 + y^2 \, \cdot$$
$$(1 - xy^2)^{-(k+3)} \, G_3 + \dots \tag{4.20}$$

The theory delineated above then produces (4.18) again, though ϕ and ψ are now fairly obvious extensions of the expressions in (4.15) and (4.14). Notice that in general $G_m^{(k)} \neq 0$ $(m = k, \, k+1, \, \dots)$.

An alternative approach for the case $k = 1$ is to treat n even and n odd separately whence differential equations are obtained for each parity of n.

5. CONCLUDING REMARKS

Other developments germane to the material in this paper are suppressed here in order to conserve space: e.g., the theory of *Genocchi polynomials of negative order* $G_n^{(-k)}(x)$ a treatment of which will be presented in a separate submission.

Finally, it would be nice to know something more about the life and mathematical works of Angelo Genocchi whose researches have prompted these investigations. Various snippets of mathematical information are available, e.g., in [10] and in L.E. Dickson's "History of the Theory of Numbers", Chelsea (1952). Furthermore, one finds in an 1885 *Encyclopaedia Britannica* article by Moritz Cantor an adverse comment on Genocchi's belief that Fibonacci's sexagesimal solution in his book *Flos* of the John of Palermo problem (namely: solve $x^3 + 2x^2 + 10x = 20$) involved knowledge of a method associated with Cardan in the sixteenth century.

Any new information of a biographical nature about Genocchi would be appreciated.

ACKNOWLEDGEMENT

The author wishes to thank the referee, particularly for drawing references 17-19 to his attention. While the material in these references is applicable to Genocchi polynomials, it is felt that use of the techniques employed in them could be treated in a separate article. Otherwise, this paper would become excessively lengthy.

REFERENCES

[1]   Boole, G. A Treatise on Differential Equations. Macmillan (1872).

[2]   Comtet, L. Advanced Combinatorics. Reidel (1974).

[3]   Erdelyi, A. (ed). Higher Transcendental Functions (3 Vols). McGraw-Hill (1953).

[4]   Fort, T. Finite Differences. Oxford University Press (1948).

[5]   Genocchi, A. "Intorno all'espressione generale de'Numeri Bernulliani. Nota." *Annali di Scienze Matematiche e Fisiche 3* (1852): pp. 395-405.

[6]   Howard, F. T. "Integers Related to the Bessel Function $J_1(z)$", *The Fibonacci Quarterly 23*, No. 3 (1985): pp. 249-257.

[7]   Jordan, C. Calculus of Finite Differences. Chelsea (1950).

[8]   Knopp, K. Theory and Application of Infinite Series. Blackie (1948).

[9]   Lindelöf, E. Le Calcul des Résidus. Chelsea (1947).

[10]   Lucas, E. Théorie des Nombres. Blanchard (1950).

[11]   Magnus, W., Oberhettinger, F. and Soni, R. P. Formulas and Theorems for the Special Functions of Mathematical Physics. Springer-Verlag (1966).

[12]   Mahon, J. M. and Horadam, A. F. "Infinite Series Summation Involving Reciprocals of Pell Polynomials." Fibonacci Numbers and Their Applications. Edited by A. N. Philippou, G. E. Bergum and A. F. Horadam. Reidel Publishing (1986).

[13] Milne-Thomson, L. M. The Calculus of Finite Differences. Macmillan (1933).

[14] Nielsen, N. Traité Elémentaire des Nombres de Bernoulli. Gauthier-Villars (1923).

[15] Nörlund, N. E. "Mémoire sur les polynomes de Bernoulli." *Acta Math.* XLIII (1920): pp. 121-196.

[16] Nörlund, N. E. Vorlesungen über Differenzenrechnung. Chelsea (1954).

[17] Subramanian, P. R. "A Short Note on the Bernoulli Polynomial of the First Kind." *Math. Student* 42 (1974): pp. 57-59.

[18] Subramanian, P. R. and Devanathan, V. "Recurrence relations for angular momentum traces." *J. Phys. A: Math. Gen. 13* (1980): pp. 2689-2693.

[19] Subramanian, P. R. and Devanathan, V. "Generation of angular momentum traces and Euler polynomials by recurrence relations." *J. Phys. A: Math. Gen. 18* (1985): pp. 2909-2915.

AN APPLICATION OF ZECKENDORF'S THEOREM

Roger V. Jean

INTRODUCTION

The author is involved in research in biomathematics, more precisely in the application of mathematics to plant biology. From the very beginning he was confronted with Fibonacci numbers. Indeed these numbers arise in what is considered to be the bugbear of botany. Their overwhelming presence in the secondary spirals on plants has puzzled many research workers. Many theories and models have been elaborated but the problem is still unsolved. The author produced dozens of articles and the book [2] mentioned in the references on the subject. This book has been reviewed in many places especially in [1]. Another book by the same author on the same subject is in preparation.

As tools for teaching, the author produced articles that may be particularly attractive for Fibonacci's disciples. Some of them are listed in the references ([3, 4, 5]).

Fibonacci numbers are not only involved in advanced mathematics and research subjects, but also in recreational amusements and tricks. These tricks often apply important mathematical results to which the attention is happily attracted. The paper presents such a trick, based on an application of Zeckendorf's theorem on the completeness of the Fibonacci sequence.

A MATHEMATICAL MYSTIFICATION WITH CARDS

Consider the nine sets of integers from 1 to 75, given at the end of the paper. You will notice that any given number can appear in several of the sets. Each of the sets are on separate

167

G. E. Bergum et al. (eds.), Applications of Fibonacci Numbers Volume 4, 167–170.
© 1991 *Kluwer Academic Publishers.*

cards which can be cut out and shuffled before making the trick, to make it more mysterious. The idea is to ask someone to choose a number between 1 and 75, and to guess that number when he or she finished telling you on which cards is the number he or she chose. The only prerequisite to make the trick a convincing demonstration of mental power and mental reading is to know how to add integers. The theorem (see below) at the basis of the trick is not so simple however, but its knowledge is not required for the performance.

Let us take an example. Suppose that 19 was chosen. You are then told that this number is on cards 1, 4, and 6. You announce immediately that the number is precisely 19 by summing up the first numbers in each set, here $1 + 5 + 13$. The numbers in the sum are all Fibonacci numbers of course, as is the first number on every card. Given the theme of the Conference you have certainly noticed that peculiarity of the cards, but as we move away from Fibonacci's circles the trick will certainly appear to be more fantastic. Try again: you are told that the number chosen is on cards 3, 5, and 8. The sum to perform is $3 + 8 + 34$, and the number is 45.

DISSOLVING THE MYSTERY - THE CONSTRUCTION RULE

How can this trick be explained? Zeckendorf's theorem states that every positive integer can be decomposed in a unique way as the sum of distinct Fibonacci numbers, such that no two such numbers are consecutive in the Fibonacci sequence. Because of Zeckendorf's theorem there is a unique way to decompose an integer and thus to put it on the cards, and we say that the Fibonacci sequence is complete. For each Fibonacci number there is a card containing integers, and each card starts with a different Fibonacci number. If the trick had been made for the first 100 integers we would thus have needed 10 cards (one more starting with 89, the Fibonacci number following 55 on card 9).

The theorem tells us how to build the cards. For example $40 = 34 + 5 + 1$ and $77 = 55 + 21 + 1$. Always start the decomposition of the given number with the largest included Fibonacci number, then the next largest and so on. It follows that 40 must be on cards 1, 4, and 8 (given that 1, 5, and 34 are the first numbers on these cards), while 77 must appear on cards 1, 7, and 9 (assuming in that case that the trick has been extended over 75).

CREATING MORE MYSTERIES

Now that you know how the cards are made you can make the trick seemingly more difficult by extending it to the numbers between 1 and 100. For example 99 would be on the cards starting with the Fibonacci numbers 89 and 8 and 2. With your desktop computer why not make a computer program to produce cards for numbers between 1 and 1000? You would then need cards starting with 89, 144, 233, 377, and 610. This extended trick may require however a stronger ability to add numbers mentally but you can also use a calculator to perform what would look as learned operations to discover a number randomly generated in ones mind.

This trick is not a peculiarity of the Fibonacci sequence. It is a peculiarity of every complete sequence. The sequence 1, 2, 4, 8, 16, 32, 64, ... is another such sequence [6]. Cards can thus be made with powers of two as first elements on the cards. Now it is time to use your imagination to make the guys beside you have a good time with mathematics.

REFERENCES

[1]   Bicknell-Johnson, Marjorie, Letter to the Editor, *The Fibonacci Quarterly, 24 (4)*, November 1986, page 309.

[2]   Jean, Roger V., Mathematical Approach to Pattern and Form in Plant Growth, Wiley-Interscience, New York, 1984.

[3]   Jean, Roger V., "The Fibonacci Sequence", *The UMAP Journal, 5, (1)*, pp. 23-47.

[4]   Jean, Roger V. and Johnson, Marjorie, "An Adventure into Applied Mathematics with Fibonacci Numbers", *School Science and Mathematics, 89 (6)*, October 1989, pp. 487-498.

[5]   Jean, Roger V. and Schwabe, Walter W., "Shoot Development in Plants", in UMAP Modules 1989 Tools for Teaching, pp. 169-208, 1990, UMAP Unit 702.

[6]   Kraitchik, Maurice. Mathematical Recreations, 2nd Ed. (1953): pp. 56-58, Dover Publications, N.Y.

	1
Application of Zeckendorf's Theorem	
1, 4, 6, 9, 12, 14, 17, 19, 22, 25, 27,	
30, 33, 35, 38, 40, 43, 46, 48, 51, 53,	
56, 59, 61, 64, 67, 69, 72, 74	
by Roger V. Jean	
University of Quebec	

	2
Application of Zeckendorf's Theorem	
2, 7, 10, 15, 20, 23, 28, 31, 36, 41,	
44, 49, 54, 57, 62, 65, 70, 75	
by Roger V. Jean	
University of Quebec	

	3
Application of Zeckendorf's Theorem
3, 4, 11, 12, 16, 17, 24, 25, 32, 33,
37, 38, 45, 46, 50, 51, 58, 59, 66,
67, 71, 72
by Roger V. Jean
University of Quebec

	4
Application of Zeckendorf's Theorem
5, 6, 7, 18, 19, 20, 26, 27, 28, 39,
40, 41, 52, 53, 54, 60, 61, 62, 73,
74, 75
by Roger V. Jean
University of Quebec

	5
Application of Zeckendorf's Theorem
8, 9, 10, 11, 12, 29, 30, 31, 32, 33,
42, 43, 44, 45, 46, 63, 64, 65, 66
67
by Roger V. Jean
University of Quebec

	6
Application of Zeckendorf's Theorem
13, 14, 15, 16, 17, 18, 19, 20, 47, 48,
49, 50, 51, 52, 53, 54, 68, 69, 70, 71
72, 73, 74, 75
by Roger V. Jean
University of Quebec

	7
Application of Zeckendorf's Theorem
21, 22, 23, 24, 25, 26, 27, 28, 29,
30, 31, 32, 33
by Roger V. Jean
University of Quebec

	8
Application of Zeckendorf's Theorem
34, 35, 36, 37, 38, 39, 40, 41, 42,
43, 44, 45, 46, 47, 48, 49, 50, 51,
52, 53, 54
by Roger V. Jean
University of Quebec

	9
Application of Zeckendorf's Theorem
55, 56, 57, 58, 59, 60, 61, 62, 63,
64, 65, 66, 67, 68, 69, 70, 71, 72,
73, 74, 75
by Roger V. Jean
University of Quebec

A NEW KIND OF GOLDEN TRIANGLE

Clark Kimberling

A *golden triangle* has been defined in Schoen [4] as "a triangle with two of its *sides* in the ratio ϕ:1, where ϕ is the Fibonacci Ratio, i.e., $\phi = (1+\sqrt{5})/2$." We shall exhibit another kind of triangle that deserves to be called golden, namely a triangle with two of its *angles* in the ratio ϕ:1.

The traditional golden triangle is discussed in Hoggatt [2], where, on page 23, three earlier writings on golden triangles are mentioned. Schoen's summary in [4] of two salient features of the traditional kind of golden triangle is as follows: "Let $\triangle ABC$ be a triangle whose sides are a, b, and c and let $a/b = k > 1$. Bicknell and Hoggatt [1] have shown that (1) a triangle with a side equal to b can be removed from $\triangle ABC$ to leave a triangle similar to $\triangle ABC$ if and only if $k = \phi$, and (2) a triangle similar to $\triangle ABC$ can be removed from $\triangle ABC$ to leave a triangle such that the ratios of the areas of $\triangle ABC$ and the triangle remaining is k if and only if $k = \phi$."

For purposes of motivation, consider the partitioning of a golden rectangle into an infinite set of squares: A rectangle with length L and width W satisfying $L/W = \phi$ can be partitioned into a square of sidelength W and rectangle having length W and width L-W. Since $(L-W)/W = W/L$, this smaller rectangle is also a golden rectangle. Accordingly, the partitioning process, continued inductively on the rectangle that remains after each step, partitions the original rectangle into an infinite set of squares, no two of which have the same size. An analogous process, we shall show, partitions a golden triangle of the new kind into an infinite set of golden triangles, no two of which have the same size.

G. E. Bergum et al. (eds.), Applications of Fibonacci Numbers Volume 4, 171–176.

Before turning to the process, we note that the partitioning of the golden rectangle corresponds to the continued fraction $[1,1,1,...]$ of the number ϕ. Less well known is the fact that this partitioning generalizes to arbitrary rectangles, if at each step of the process we cut away as many maximal squares as possible from the remaining rectangle.

For a golden rectangle, there is only one maximal square available at each step, but for an arbitrary rectangle, there may be more than one. Sure enough, if $[a_1,a_2,a_3,...]$ is the continued fraction for the ratio L/W of the length of the rectangle to its width, then the number of same-sized maximal squares to be removed at the nth step of the process is a_n. For an exposition about this interpretation of continued fractions, see Kimberling [3].

It is at this more general level of one-to-one correspondence between continued fractions and geometric shapes that we shall introduce a partitioning process for arbitrary triangles. It will help to state first an algorithm for the continued fraction of a number B/A greater than 1.

Theorem: Suppose $B > A > 0$. Let $P \geq 2B + A$, $A_1 = A$, $B_1 = B$, $C_1 = P - A - B$, and for $i = 2, 3, ...,$

$$A_i' = A_{i-1}$$

$$B_i' = P - B_{i-1}$$

$$C_i' = B_{i-1} - A_{i-1}$$

$$A_i = \min \{A_i', B_i', C_i'\}$$

$$C_i = \max \{A_i', B_i', C_i'\}$$

$$B_i = P - A_i - C_i$$

Let a_j be the length of the jth run of identical terms of $\{A_i\}$, excluding zero terms. Then the continued fraction $[a_1, a_2, a_3, ...]$ equals B/A.

Proof: First, assume that A and B are positive integers. Write $B_i = Q_i A_i + R_i$, $0 \leq R_i \leq A_i - 1$. Then $A_{i+1}' = A_i$, $B_{i+1}' = P - Q_i A_i - R_i$, and $C_{i+1}' = (Q_i - 1)A_i + R_i$, so that $A_{i+1} - A_i$ if and only if $Q_i \geq 2$. Accordingly, for each positive integer j, we have $A_{j+i} = A_j$ for $i = 0, 1, ..., Q_j - 1$.

The first run in the sequence $\{A_i\}$ therefore has length Q_1. The second run starts with A_{Q_1+1} and ends with $A_{Q_1+Q_2}$ and so has length Q_2. Inductively, the jth run begins with

$A_{Q_1+Q_2+\ldots Q_{j-1}+1}$ and ends with $A_{Q_1+Q_2+\ldots Q_{j-1}+Q_j}$ and so has length Q_j. Thus, the sequence $a_1, a_2, a_3, \ldots$ equals the sequence $Q_1, Q_2, Q_3, \ldots$, which, being the integers $[B_i/A_i]$ in the Euclidean algorithm for B/A, comprises the continued fraction for B/A. For irrational B/A, the proof follows straightforwardly from applying the rational case to any sequence of rationals having limit B/A.

THE PARTITIONING OF TRIANGLES

Suppose that XYZ is a triangle whose *angles* X, Y, Z satisfy $X \leq Y \leq Z$:

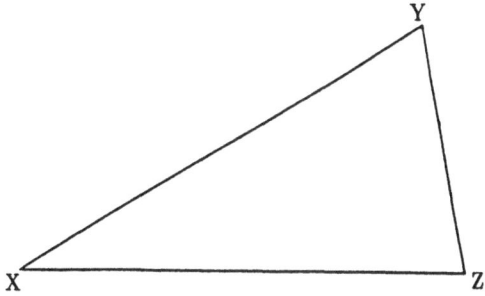

Figure 1

Inductive Step. Locate W on XY satisfying $|WZ| = |YZ|$. Triangle XYZ is thereby partitioned into two triangles: WYZ, which is isosceles, and WXZ.

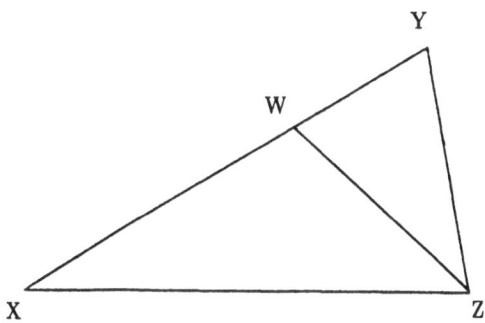

Figure 2

Triangle WYZ is isosceles, so that the inductive step, repeated (with angles always labeled so that $X \leq Y \leq Z$), partitions any given triangle ABC into a set of isosceles triangles. Following, for example, is a completed partition of a 30°-70°-80° triangle into a set of five isosceles triangles:

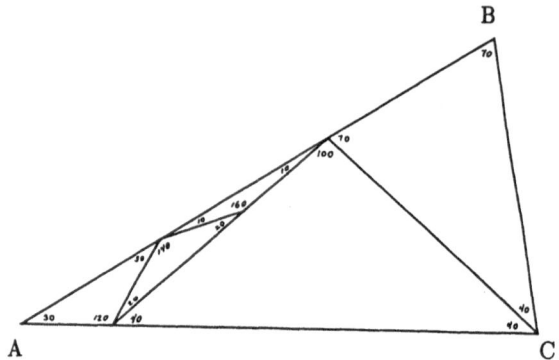

Figure 3

triangle	angles
T_1	30, 70, 80
T_2	30, 40, 110
T_3	10, 30, 140
T_4	10, 20, 150
T_5	10, 10, 160

Let T_1 be any triangle ABC whose angles satisfy $A \leq B \leq C$. For $i = 2, 3, \ldots$, let T_i be the triangle remaining after the $(i-1)^{st}$ application of the inductive step. We ask:

1. For what triangles does this partition process terminate? That is, for what given angles is T_i isosceles for some $i \geq 2$?

2. Is there a triangle for which the least angle of T_{i+1} differs from that of T_i for every i? (In Figure 3, A is the least angle of T_1 and T_2 but not of T_3.)

To answer these questions, we apply the theorem. Let $P = \pi$, so that $A_1 = A$, $B_1 = B$, and $C_1 = C$ satisfy the hypothesis. The inductive step, when applied to any triangle $T_{i-1} = A_{i-1}B_{i-1}C_{i-1}$ having $A_{i-1} \leq B_{i-1} \leq C_{i-1}$, yields a triangle T_i whose angles are $A_i' = A_{i-1}$, $B_i' = \pi - B_{i-1}$, $C_i' = B_{i-1} - A_{i-1}$. The relabeling of angles of T_i in nondecreasing order is equivalent to putting $A_i = \min\{A_i', B_i', C_i'\}$, $C_i = \max\{A_i', B_i', C_i'\}$, $B_i = \pi - A_i - C_i$. Therefore, the runs sequence of the sequence $A_1, A_2, \ldots$ is the continued fraction of the ratio B/A of angles.

The two questions are now easily answered:

1. The partition process terminates if and only if B/A is a rational number.

2. There are many such triangles. All it takes is $B = \phi A$, where $\phi = [1, 1, 1, \ldots]$ is the golden mean, and $A < \dfrac{\pi}{2 + \sqrt{5}}$.

The golden mean is the only such ratio, for any other has a continued fraction other than $[1, 1, 1, \ldots]$ and therefore, during the partitioning, at some step, has a subtriangle T_i whose least angle is also the least angle of T_{i+1}. For example, for Figure 3, $B/A = 7/3 = [2,3]$. Corresponding to the first term, 2, of the continued fraction, are the triangles T_1 and T_2 that have the same least angle, A. The least angle of T_3 is not A. Instead, and in keeping with the second term, 3, of the continued fraction, there are three T_i having the same least angle as T_3.

It will be helpful to describe the partitioning process in terms of stages. *Stage 1* consists of the inductive steps $1, 2, \ldots, a_1$; i.e., the removal of a_1 isosceles triangles from the initial triangle ABC. The least angle of each of these triangles is angle A, and the triangle remaining at the conclusion of Stage 1 is T_{a_1+1}. *Stage 2* consists of steps $a_1+1, \ldots, a_1+a_2$; i.e., the removal of a_2 isosceles triangles from triangle T_{a_1+1}, all having the same least angle. Generally, *Stage k*, for $k \geq 2$, consists of the a_k inductive steps numbered $a_{k-1}+1, \ldots, a_{k-1} + a_k$; i.e., the removal of a_k isosceles triangles from $T_{a_{k-1}+1}$.

Referring to these stages in the partitioning process and analogous stages when partitioning a rectangle into squares, we conclude with the following summary as corollaries to well-known theorems about continued fractions:

Golden Rectangle	Golden Triangle
1.	ratio of length to width: $L/W = (1 + \sqrt{5})/2$	ratio of angles: $A/B = (1 + \sqrt{5})/2$
2.	no. of congruent squares removed at each stage of partitioning: one	no. of isosceles triangles removed at each stage of partitioning: one
3.	generalization: ratio $L/W = [a_1, a_2, ...]$ corresponds to removing a_j squares at j^{th} stage of partitioning	generalization: ratio $A/B = [a_1, a_2, ...]$ corresponds to removing a_j isosceles triangles at j^{th} stage of partitioning

REFERENCES

[1] Bicknell, Marjorie and Hoggatt, Verner E. Jr., "Golden Triangles, Rectangles, and Cuboids." *The Fibonacci Quarterly, 7, No. 1* (1969) pp. 73-91.

[2] Hoggatt, Verner E., Jr. Fibonacci and Lucas Numbers. Boston: Houghton Mifflin, 1969.

[3] Kimberling, C. H., "A Visual Euclidean Algorithm," *Mathematics Teacher 72* (1983) pp. 108-109.

[4] Schoen, R., "The Fibonacci Sequence in Successive Partitions of a Golden Triangle," *The Fibonacci Quarterly 20* (1982) pp. 159-163.

TERMS COMMON TO TWO SEQUENCES
SATISFYING THE SAME LINEAR RECURRENCE

Clark Kimberling

For any linear recurrence relation over the field of real numbers, there are infinitely many recurrence sequences that obey the recurrence relation. Each such sequence is determined by a number k of initial terms, where k is the order of the recurrence relation. In this paper, we discuss the number and placement of terms that two such sequences, under a variety of additional hypotheses, can possible have in common.

As a simple example, consider the Fibonacci recurrence relation

$$u_n = u_{n-1} + u_{n-2}.$$

The sequence determined by initial terms $u_0 = 1$ and $u_1 = 2$ has zero terms in common with the sequence determined by $u_0 = 6$ and $u_1 = 42$; one term in common with the sequence determined by $u_0 = 6$ and $u_1 = 43$; and infinitely many terms in common with the sequence determined by $u_0 = 8$ and $u_1 = 13$.

There are three different possible meanings of the phrase "term common to two sequences $\{u_j\}$ and $\{v_j\}$":

(i)   a number r for which there exist indices m and n satisfying $r = u_m = v_n$;

(ii) an ordered pair (m, n) of indices such that $u_m = v_n$;

(iii) an ordered triple (r, m, n) such that $r = u_m = v_n$.

For example, meaning (i) would count the conditions "$u_2 = v_3 = 5$ and $u_4 = v_7 = 5$" as a single common term, $r = 5$, whereas (ii) and (iii) would each count two different common terms. To distinguish between (ii) and (iii) note that the conditions "$u_2 = v_3 = 5$ and $u_4 = v_7 = 6$" and "$u_2 = v_3 = 6$ and $u_4 = v_7 = 5$" are distinguishable under meaning (iii) but not meaning (ii). *It is meaning (ii) that is intended throughout this paper.*

177

G. E. Bergum et al. (eds.), Applications of Fibonacci Numbers Volume 4, 177–188.
© 1991 Kluwer Academic Publishers.

1. PRELIMINARIES

We begin with some standard definitions. A *linear homogeneous recurrence relation (with constant coefficients)* is an equation of the form

$$u_{n+k} = a_1 u_{n+k-1} + \cdots + a_k u_n, \tag{1}$$

where, for our purposes, $a_1, \cdots, a_k$ are real numbers, and $a_k \neq 0$, where $k \geq 2$.

A *linear homogeneous recurrence sequence* $u_0, u_1, u_2, \cdots$ is a sequence whose terms satisfy (1) for $n = 0, 1, 2, \ldots$. The number k is the *order* of the recurrence and of the sequence. The polynomial

$$f(x) = x^k - a_1 x^{k-1} - \cdots - a_k \tag{2}$$

is the *characteristic polynomial* of the recurrence and of the sequence.

Note that $u_{-1}, u_{-2}, \ldots$ are always definable from (1) inductively:

$$u_n = (u_{n+k} - a_1 u_{n+k-1} - \cdots - a_{k-1} u_{n+1})/a_k. \tag{3}$$

A general representation theorem. Suppose a sequence $u_0, u_1, u_2, \ldots$ satisfied the recurrence with real coefficients

$$u_{n+k} = a_1 u_{n+k-1} + \cdots + a_k u_n, \quad n \geq 0. \tag{1}$$

Let $f(x) = x^k - a_1 x^{k-1} - \cdots - a_k$ be the characteristic polynomial of the recurrence (1) and let

$$f(x) = (x - h_1)^{e_1} \cdots (x - h_c)^{e_c} (x^2 + s_1 x + t_1)^{f_1} \cdots (x^2 + s_d x + t_d)^{f_d},$$

$$\text{where } e_1 + e_2 + \cdots + e_c + 2f_1 + 2f_2 + \cdots + 2f_d = k,$$

be the factorization of $f(x)$ as a product of linear and quadratic factors irreducible over the real number field, where $x^2 + s_j x + t_j = [x - r_j(\cos\theta_j + i\sin\theta_j)][x - r_j(\cos\theta_j - i\sin\theta_j)]$ for $j = 1, 2, \ldots, d$. Then

$$u_n = \sum_{j=1}^{c} p_j(n) h_j^n + 2 \sum_{j=1}^{d} r_j^n [P_j(n) \cos n\theta_j + Q_j(n)\sin n\theta_j] \tag{4}$$

for all integers n, where the polynomials $p_j(n)$, $P_j(n)$, $Q_j(n)$ have all coefficients real. The degree of $p_j(n)$ in n is at most $e_j - 1$ for $j = 1, 2, \ldots, c$, and the degrees of $P_j(n)$ and $Q_j(n)$ in n are each at most $f_j - 1$ for $j = 1, 2, \ldots, d$. The coefficients within the polynomials $p_j(n)$, $P_j(n)$, and $Q_j(n)$ are uniquely determined by the initial values $u_0, u_1, \ldots, u_{k-1}$ of the sequence $\{u_n\}$. For a proof of this theorem, see [1]. In case the roots of $f(x)$ are distinct, the polynomials appearing in (4) are all mere constants. It is easy to prove in this case that any sequence defined by the sum (4) indeed satisfies the recurrence relation (1). Thus, (4) can be a convenient way to generate higher order recurrence sequences having various properties of interest.

The assumption $u_0 = v_0$. In order to facilitate notation, we shall, after this section, always assume, for any two sequences $U = \{u_j\}$ and $V = \{v_j\}$ that have at least one common term, that $u_0 = v_0$. In this section we shall show that under three different situations of interest, no loss in generality will result from this assumption. The three situations are (1) that the two sequences are strictly increasing, (2) that they have infinitely many terms in common, and (3) that neither (1) nor (2) holds.

First, suppose $\hat{U} = \{\hat{u}_j\}$ and $\hat{V} = \{\hat{v}_j\}$ are strictly increasing sequences. Let m be the least index i such that $\hat{u}_i = \hat{v}_j$ for some j, and let n be the least such j. Then $\hat{u}_m = \hat{v}_n$. Define new sequences $U = \{u_i\}$ and $V = \{v_i\}$ by shifting:

$u_i = \hat{u}_{i+m}$ and $v_i = \hat{v}_{i+n}$. Then $u_0 = v_0$. Suppose now for some $j \geq 1$ and $k \geq 1$ that $\hat{u}_{m+j} = \hat{v}_{n-k}$; this would force $u_j = v_{-k}$. However, since $\hat{U}$ and $\hat{V}$ are strictly increasing sequences, we would have

$$\hat{u}_{m+j} > \hat{u}_m = \hat{v}_n > \hat{v}_{n-k},$$

which shows that the undesirable condition $\hat{u}_{m+j} = \hat{v}_{n-k}$ cannot occur.

The second condition of interest is that infinitely many terms of $\hat{U}$ are also terms of $\hat{V}$. In this case, let $\hat{u}_m = \hat{v}_n$ be any common term, and put $u_i = \hat{u}_{i+m}$ and $v_i = \hat{v}_{i+n}$, so that $u_0 = v_0$. There are still clearly infinitely many pairs (i, j) of *positive* integers for which $u_i = v_j$, and any statement about the placements of these terms within the original sequences $\hat{U}$ and $\hat{V}$ easily yields a statement about the placements of corresponding terms within the original sequences $\hat{U}$ and $\hat{V}$.

Finally, if neither of the above two conditions holds, we can nevertheless assume that $u_0 = v_0$ but must then allow negative indices, as provided by the recurrence (3), so as not to lose any common terms when shifting indices. For example, if $\hat{u}_3 = \hat{v}_6$ and $\hat{u}_4 = \hat{v}_2$, we can put $u_i = \hat{u}_{i+3}$ and $v_i = \hat{v}_{i+6}$ to have $u_0 = v_0$ and $u_1 = v_{-4}$.

Common terms of the form $u_m = v_m$.   Our final preliminary consideration is with sequences $U = \{u_i\}$ and $V = \{v_i\}$ for which $u_0 = v_0$ and $u_m = v_m$ for one or several positive values of m. The question of common terms of this special type can be treated as a question about a single sequence rather than two sequences. Explicitly, the question is this: *under what conditions do* *specified values of a set of terms (such as u_0 and u_5 for a second order sequence) determine all the other terms?* This question is addressed elsewhere [2].

2. STRICTLY INCREASING SEQUENCES

Theorem 1: Suppose $U = \{u_i\}$ and $V = \{v_i\}$ are strictly increasing sequences such that (i) $u_0 = v_0$; (ii) $u_1 > 0$ and $v_1 > 0$; (iii) $u_i = au_{i-1} + bu_{i-2}$ and $v_i = av_{i-1} + bv_{i-2}$ for $i = 2, 3, \ldots$, where $a > 0$ and $b > 0$. Then, unless the sequences U and V are identical, there is at most one term of U, other than u_0, which is a term of V.

Proof: Suppose there is at least one term u_i, other than u_0, which is a term of the sequence V. Let m be the least positive index i such that $u_i = v_j$ for some j, and let n be the least such j. Then $u_m = v_n$.

Case 1: If $u_{m+1} = v_{n+1}$, then the recurrences $u_{m+1} = au_m + bu_{m-1}$ and $v_{n+1} = av_n + bv_{n-1}$ yield $u_{m-1} = v_{n-1}$, so that m = 1. Thus, $u_0 = v_{n-1}$, so that $v_{n-1} = v_0$. Since v is a strictly increasing sequence, n must equal 1. The initial conditions $u_0 = v_0$ and $u_1 = v_1$ force the sequences U and V to be identical.

Case 2: $u_{m+1} > v_{n+1}$.

First $u_m = v_n < v_{n+1}$. Next since $u_m < v_{n+1}$ and $u_{m-1} < u_m = v_n$,

$$u_{m+1} = au_m + bu_{m-1} < av_{n+1} + bv_n = v_{n+2}.$$

Finally, since $v_{n+1} < u_{m+1}$,

$$v_{n+2} = av_{n+1} + bv_n < au_{m+1} + bu_m = u_{n+2}.$$

The now established chain

$$u_m < v_{n+1} < u_{m+1} < v_{n+2} < u_{m+2}$$

and induction yield the inequalities

$$v_{n+1+i} < u_{m+1+i} < v_{n+2+i}$$

for $i = 0, 1, \ldots$, from which we conclude that every u_i for $i \geq m+1$ lies strictly between consecutive terms of the sequence V.

Case 3: $u_{m+1} < v_{n+1}.$

First, since $u_m = v_n$, it follows that $u_{m+2} < v_{n+2}$. Also, $v_n = u_m < u_{m+1}$. Next, since $v_{n-1} < v_n = u_m$, it follows that

$$v_{n+1} = av_n + bv_{n-1} < au_{m+1} + bu_m = u_{m+2}.$$

In a similar way,

$$u_{m+2} = au_{m+1} \, bu_m < av_{n+1} + bv_n = v_{n+2},$$

The now established chain

$$v_n < u_{m+1} < v_{n+1} < u_{m+2} < v_{n+2}$$

and induction yield the inequalities

$$v_{n+i} < u_{m+i} < v_{n+1+i}$$

for $i = 0, 1, \ldots,$ from which we conclude as in Case 2 that no u_i is a term of the sequence V, for any $i \geq m+1$.

The method of proof of Theorem 1 does not seem to extend directly to sequences of order 3 or higher. Yet there is ample numerical evidence to suggest the following conjecture.

Conjecture 1: Suppose k is an integer greater than 2, and $U = \{u_i\}$ and $V = \{v_i\}$ are strictly increasing sequences such that (i) $u_0 = v_0$; (ii) $u_1 > 0$ and $v_1 > 0$; (iii) U and V both satisfy the same recurrence relation (1), where $a_i > 0$ for $i = 1, ..., k$. Then there exists a positive integer B_k such that, unless $U = V$, the number of terms of U which are terms of V is $\leq B_k$.

3. DISTINCT REAL ROOTS

Theorem 2: Suppose

$$u_{n+k} = a_1 u_{n+k-1} + \cdots + a_k u_n \tag{1}$$

is a recurrence relation whose characteristic polynomial (4) has real roots with distinct absolute values $r_1, r_2, ..., r_k$. Suppose $U = \{u_i\}$ and $V = \{v_i\}$ are recurrence sequences satisfying (1) and that U satisfies no linear recurrence of order less than k. Unless $u_m = v_{m+p}$ for some integer p and all m, the sequences U and V have at most a finite number of terms in common.

Proof: Suppose, to the contrary, that there are infinitely many pairs (m_i, n_i) for which $u_{m_i} = v_{n_i}$. For each such pair of indices we have, from (4),

$$c_1 r_1^{m_i} - c_2 r_2^{m_i} + \cdots + c_k r_k^{m_i} = d_1 r_1^{n_i} + d_2 r_2^{n_i} + \cdots + d_k r_k^{n_i} \tag{5}$$

where for convenience of notation the roots r_j are indexed so that

$$|r_1| > |r_2| > \cdots > |r_k|.$$

Also, the coefficient c_i is a nonzero real number for $i = 1, 2, \ldots, k$.

Now, either $n_i \geq m_i$ for infinitely many i, or else $n_i \leq m_i$ for infinitely many i. Assume the former, noting that if only the latter holds, then the proof can be easily changed to cover that case. Dividing both sides of (5) by $r_1^{m_i}$ gives

$$c_1 + c_2 (r_2/r_1)^{m_i} + \cdots + c_k (r_k/r_1)^{m_i}$$
$$= d_1 r_1^{n_i - m_i} + d_2 r_2^{n_i - m_i} (r_2/r_1)^{m_i} + \cdots + d_k r_k^{n_i - m_i} (r_k/r_1)^{m_i}. \tag{6}$$

If $r_1 = 1$, then taking the limit of both sides of (6) as $i \to \infty$ yields $c_1 = d_1$. If $r_1 \neq 1$, then the set of numbers $n_i - m_i$ must be bounded; for if not then as $i \to \infty$, the left-hand side of (6) approaches c_1 while the right-hand side approaches 0 or else diverges, depending respectively on whether $|r_1| < 1$ or $|r_1| > 1$. Neither is possible, since c_1 is a nonzero real number. Therefore, there must be a nonnegative integer p which equals $n_i - m_i$ for infinitely many i, and the limit of (6) yields $c_1 = d_1 r_1^p$. Thus, no matter what nonzero value r_1 may have, we can cancel equal leftmost terms from the two sides of (6) to obtain

$$c_2 (r_2/r_1)^{m_i} + \cdots + c_k (r_k/r_1)^{m_i}$$
$$= d_2 r_2^{n_i - m_i} (r_2/r_1)^{m_i} + \cdots + d_k r_k^{n_i - m_i} (r_k r_1)^{m_i} \tag{7}$$

for infinitely many i. We multiply both sides of (7) by $r_1^{m_i}$ and repeat the limiting considerations that were applied earlier to (6), finding as a result that $c_2 = d_2 r_2^p$. Continuing in this manner, we conclude that $c_j = d_j r_j^p$ for $j = 1, 2, \ldots, k$, so that $u_m = v_{m+p}$ for all m.

Conjecture 2: Suppose

$$u_{n+k} = a_1 u_{n+k-1} + \cdots + a_k u_n \tag{1}$$

is a recurrence relations whose characteristic polynomial (4) has distinct real roots $r_1, r_2, \ldots, r_k$. Then there exists a positive integer B_k such that if $U = \{u_i\}$ and $V = \{v_i\}$ are sequences satisfying (1), then either $u_m = v_{m+p}$ for some integer p and all integers m, or else the number of terms of U which are terms of V is $\leq B_k$.

4. SECOND ORDER WITH A REPEATED ROOT

In this section, we assume that $k = 2$ in equation (1) and that the characteristic polynomial (2) has a repeated real root α; that is,

$$f(x) = (x - \alpha)^2 = x^2 - ax - b,$$

where $a = 2\alpha$ and $b = -\alpha^2$. We ask not only what the possibilities are for terms common to a pair of recurrence sequences, but also how to construct such pairs for each such possibility.

Sequences U and V of the form to be studied here, then, can be written out as

$$u_0 = w, \ u_1 = x, \ u_2 = 2x\alpha - w\alpha^2, \ u_3 = 3x\alpha^2 - 2w\alpha^3, \ \ldots, \ u_m = mx\alpha^{m-1} - (m-1)w\alpha^m, \ \ldots$$

$$v_0 = w, \ v_1 = y, \ v_2 = 2y\alpha - w\alpha^2, \ v_3 = 3y\alpha^2 - 2w\alpha^3, \ \ldots, \ v_n = ny\alpha^{n-1} - (n-1)w\alpha^n, \ \ldots.$$

If (m_i, n_i) are pairs of integers (not necessarily all positive) for which $u_{m_i} = v_{n_i}$ for $i = 1, 2$, then for any given α, the system of two equations

$$m_i x \alpha^{m_i - 1} - (m_i - 1)w\alpha^{m_i} = n_i y \alpha^{n_i - 1} - (n_i - 1)w\alpha^{n_i}$$

has a unique solution (x, y) given by

$$x = w[(n_2 - n_1)\alpha^{n_1 + n_2 - 1} + n_2(m_1 - 1)\alpha^{m_1 + n_2 - 1} - n_1(m_2 - 1)\alpha^{m_2 + n_1 - 1}]/D \tag{8x}$$

$$y = -w[(m_2 - m_1)\alpha^{m_1 + m_2 - 1} + m_2(n_1 - 1)\alpha^{m_2 + n_1 - 1} - m_1(n_2 - 1)\alpha^{m_1 + n_2 - 1}]/D, \tag{8y}$$

where $D = m_1 n_2 \alpha^{m_1 + n_2 - 2} - m_2 n_1 \alpha^{m_2 + n_1 - 2}$, if and only if the denominator D is not zero.

If, in addition to $u_{m_i} = v_{n_i}$ for $i = 1, 2$, there is to be a third common term $u_{m_3} = v_{n_3}$, then the corresponding system of three equations has a solution if an only if

$$\begin{vmatrix} m_1\alpha^{m_1-1} & n_1\alpha^{n_1-1} & (m_1-1)\alpha^{m_1} - (n_1-1)\alpha^{n_1} \\ m_2\alpha^{m_2-1} & n_2\alpha^{n_2-1} & (m_2-1)\alpha^{m_2} - (n_2-1)\alpha^{n_2} \\ m_3\alpha^{m_3-1} & n_3\alpha^{n_3-1} & (m_3-1)\alpha^{m_3} - (n_3-1)\alpha^{n_3} \end{vmatrix} = 0.$$

Rules of elementary linear algebra continue to apply, showing that the above determinant equals zero if and only if

$$\begin{vmatrix} m_1\alpha^{m_1-1} & n_1\alpha^{n_1-1} & \alpha^{m_1} - \alpha^{n_1} \\ m_2\alpha^{m_2-1} & n_2\alpha^{n_2-1} & \alpha^{m_2} - \alpha^{n_2} \\ m_3\alpha^{m_3-1} & n_3\alpha^{n_3-1} & \alpha^{m_3} - \alpha^{n_3} \end{vmatrix} = 0. \qquad (9)$$

The result of expanding this determinant and simplifying is the equation

$$\begin{aligned} \widehat{P}(\alpha) \equiv\ & m_1(n_2-n_3)\alpha^{m_1+n_2+n_3} + n_1(m_2-m_3)\alpha^{n_1+m_2+m_3} \qquad (10) \\ & + m_2(n_3-n_1)\alpha^{m_2+n_3+n_1} + n_2(m_3-m_1)\alpha^{n_2+m_3+m_1} \\ & + m_3(n_1-n_2)\alpha^{m_3+n_1+n_2} + n_3(m_1-m_2)\alpha^{n_3+m_1+m_2} = 0. \end{aligned}$$

Note that $\widehat{P}(\alpha)$ is a sum of six terms, each of the form $p(q-r)\alpha^{p+q+r}$. It is obvious from (9) that 1 is a root of $\widehat{P}(\alpha)$. More surprising is the fact that 1 is a *repeated* root of $\widehat{P}(\alpha)$, according to the following theorem.

Theorem 3: $\widehat{P}(\alpha)$ is divisible by $(\alpha-1)^2$. Explicitly, let J be the integer for which $\alpha^J\widehat{P}(\alpha)$ is the polynomial in α in which the least exponent of α is zero. Define $P(\alpha) = \alpha^J\widehat{P}(\alpha)$. Then there exists a polynomial $Q(\alpha)$, all of whose coefficients are integers, such that $P(\alpha) = (\alpha-1)^2 Q(\alpha)$.

(There exists a constant c such that $cQ(\alpha)$ has relatively prime coefficients and positive leading coefficient. In the sequel, we shall always take $Q(\alpha)$ to be this uniquely determined polynomial.)

Proof: A standard criterion for a polynomial $p(x)$ to have a repeated root r is that the derivative $p'(r)$ vanish. It is easy to check that $\widehat{P}'(1) = 0$, so that $P'(1) = 0$ also.

The ubiquitous repeated root $\alpha = 1$ yields pairs U and V of little interest, for if $D \neq 0$, equations (8) give $x = y$, so that U and V are identical. Other values of α lead to a variety of possibilities, as indicated by the following examples:

Example 1: What pairs of non-identical sequences $U = \{u_i\}$ and $V = \{v_i\}$ satisfying a recurrence relation $u_n = 2\alpha u_{n-1} - \alpha^2 u_{n-2}$ have $u_0 = v_0$, $u_2 = v_3$, $u_3 = v_5$, and $u_5 = v_7$? To answer this question, put $(m_1, n_1) = (2, 3)$, $(m_2, n_2) = (3, 5)$, and $(m_3, n_3) = (5, 7)$ into equation (9) to get

$$\widehat{P}(\alpha) = -10\alpha^{13} - 7\alpha^{12} - 4\alpha^{14} - 6\alpha^{11} + 12\alpha^{13} + 15\alpha^{12},$$

so that $J = -11$, and $P(\alpha) = \alpha^J \widehat{P}(\alpha) = -2(\alpha - 1)^2 (2\alpha + 3)$, so that $Q(\alpha) = 2\alpha + 3$. The root $\alpha = -3/2$ with $w = 1$ in equations (8) yields

$$u_0 = 1, u_1 = -9/16, u_2 = -9/16, u_3 = 189/64, u_4 = -243/32, \ldots$$

and

$$v_0 = 1, v_1 = -13/12, v_2 = 1, v_3 = -9/16, v_4 = -9/16, \ldots.$$

In conclusion, $v_{m+2} = u_m$ for all m, and $u_2 = u_1$.

Example 2: This time, we seek sequences U and V for which $u_0 = v_0$, $u_2 = v_3$, $u_3 = v_5$, and $u_5 = v_8$. The method of Example 1 yields $Q(\alpha) = 2\alpha^2 - \alpha + 2$, which has no real roots. Accordingly, the only solution to the problem is trivial: $U = V$.

Example 3: What sequences U and V satisfy $u_0 = v_0$, $u_2 = v_3$, $u_3 = v_5$, and $u_5 = v_9$? It can be shown that $Q(\alpha) = 8\alpha^3 - 2\alpha^2 - 3\alpha + 6$, which has only one real root α, approximately -0.95820682322962. This value of α, and only this value, provides a nontrivial pair U and V having the four prescribed common terms.

Example 4: Here, we prescribe $u_0 = v_0$, $u_2 = v_5$, $u_7 = v_7$, and $u_8 = v_4$, and find that $Q(\alpha)$ has only one real root, approximately -0.697465042951045, which leads via (8) to the trivial case $U = V$, as accounted for in [2]. (See the final paragraph of Section 1.)

Example 5: What sequences U and V satisfy $u_0 = v_0$, $u_2 = v_4$, $u_3 = v_6$, and $u_4 = v_8$? This time, $Q(\alpha)$ has only one root, namely $\alpha = -1$, which leads to sequences U and V satisfying $u_n = (-1)^n v_{2n}$, with $u_3 = 0$.

Example 6: What sequences U and V have in common the terms $u_0 = v_0$, $u_{-5} = v_4$, $u_6 = v_4$, and $u_7 = v_{-4}$? We find $Q(-1) = 0$, and $u_{6i} = v_{4i}$ and $u_{6i+1} = v_{-4i}$ for all integers i.

Example 7: As a final example, let us require that $u_0 = v_0$, $u_1 = v_2$, $u_3 = v_5$, and $u_7 = v_{10}$. Here, $Q(\alpha)$ has two roots, namely 2 and 4/5. With $\alpha = 2$, we find that

$$u_{2^n - 1} = v_{n+2^n - 1}$$

for all integers n, so that not only the required common terms occur, but also the common term $u_{15} = v_{19}$ and infinitely many others. The root $\alpha = 4/5$ yields the four required common terms. Conjecture 3 alleges that there are no others.

Conjecture 3: Suppose α is a nonzero real number, and U and V are sequences that both satisfy the recurrence relation $u_n = 2\alpha u_{n-1} - \alpha^2 u_{n-2}$. Then there exists a positive integer B such that unless U and V have infinitely many common terms, the number of common terms is $\leq$ B. Assuming this is true, we further conjecture that the least such B for all choices of real numbers α, u_0, u_1, v_0, v_1 exceeds the least such B for all choices of integers α, u_0, u_1, v_0, v_1.

Example 1 illustrates the following more general proposition:

Theorem 4: Suppose m, n, and k are integers such that $u_0 = v_0$, $u_m = v_n$, and $u_{m+k} = v_{n+k}$, where $m \neq 0$ or $n \neq 0$, and $k \neq 0$. Then $u_{m+j} = v_{n+j}$ for every integer j.

Proof: Substituting into (9) and simplifying, we find that the determinant in (9) equals

$$\alpha^{m+n+k+j} \begin{vmatrix} m & n & 1 \\ m+k & n+k & 1 \\ m+j & n+j & 1 \end{vmatrix},$$

which is easily seen to equal zero for every integer j.

We have already seen that when α is the integer 1, the sequences U and V are identical. Other integer values of α lead to more interesting results. Before turning to these, we shall prove that if all the terms of a sequence U are integers, then α itself must be an integer.

Theorem 5: If u_m is an integer for every nonnegative integer m, where $u_0 \neq 0$ and $u_m = 2\alpha u_{m-1} - \alpha^2 u_{m-2}$ for all $m \geq 2$, then α is an integer.

Proof: The equation $u_2 = 2\alpha u_1 - \alpha^2 u_0$ implies $\alpha = (u_1 \pm \sqrt{u_1^2 - u_0 u_2})/u_0$, so that α has the form $r + s\sqrt{Q}$, where r and s are rational numbers. If Q is not the square of an integer, then $\sqrt{Q}$ is irrational. The equation $u_2 = (2u_1 - u_0\alpha)\alpha$ implies, after substitution of $r + s\sqrt{Q}$ for α, that $(2u_1 s - 2u_0 rs)\sqrt{Q}$ is rational, so that either $s = 0$ or $r = u_1/u_0$. The equation $u_3 = (3u_1 - 2u_0\alpha)\alpha^2$ similarly implies that $(6rsu_1 - 6r^2su_0 - 2s^3u_0)\sqrt{Q}$ is rational, so that $s = 0$. (For if $s \neq 0$, then on substituting u_1/u_0 for r, we get $s^3 u_0 = 0$, so that $s = 0$, a contradiction.)

Thus, either α is the rational number u_1/u_0 or else $\alpha = (u_1 \pm \sqrt{u_1^2 - u_0 u_2})/u_0$, where $u_1^2 - u_0 u_2$ is the square of an integer. In case $\alpha = u_1/u_0$, it is easily found inductively that the sequence U is a geometric progression: $u_m = u_0\alpha^m$. This representation shows that α must be an integer.

For the remaining possibility, suppose $u_1^2 - u_0 u_2 = W^2$ for some nonzero integer W. Using $\alpha = (u_1 + W)/u_0$ in the recurrence relation $u_m = 2\alpha u_{m-1} - \alpha^2 u_{m-2}$, we find inductively that $u_{m+1} = \alpha^m(u_1 - mW)$, so that

$$u_{m+1} = (u_1 + W)^m(u_1 - mW)/u_0^m \qquad \text{for } m = 2, 3, \dots . \tag{11}$$

If α is not an integer, then we write $p/q = \alpha$, where p and q are relatively prime integers and $q > 1$. Then (11) implies $u_{m+1} = p^m(u_1 - mW)/q^m$, so that there exist integers k_m satisfying $k_m q^m = u_1 - mW$ for $m = 2, 3, \dots$. Thus $k_m q^m = u_1 - (m-1)W - W = k_{m-1}q^{m-1} - W$ and

$$W = q^{m-1}(k_{m-1} - qk_m) \qquad \text{for } m = 3, 4, \dots . \tag{12}$$

Since W in (12) is a fixed nonzero integer and $k_{m-1} - qk_m$ is always a nonzero integer, (12) shows that W is a nonzero multiple of q^{m-1} for all $m \geq 3$. This impossibility implies that α must be an integer.

If $\alpha = -1$, then the recurrence relation $u_m = 2\alpha u_{m-1} - \alpha^2 u_{m-2}$ can be written $u_m = -2u_{m-1} - u_{m-2}$. Induction on m yields

$$u_m = (-1)^{m-1}[m(u_0 + u_1) - u_0] \tag{13}$$

This identity will be used in the proof of the following theorem.

Theorem 6: If U and V are sequences satisfying the recurrence $u_i = -2u_{i-1} - u_{i-2}$, where $u_0 = v_0$, then a necessary and sufficient condition that $u_m = v_n$ for some nonzero index m is that one of the following two conditions holds:

  (i) $m(u_1 - v_1)/(2v_0 + 2v_1)$ is an integer, in which case
    $n = m(u_1 + v_1)/(v_0 + v_1)$;

  (ii) $[m(u_1 + v_1 + 2u_0) + v_1 - v_0]/(2v_0 + 2v_1)$ and $[2v_0 - m(u_0 + u_1)]/(v_0 + v_1)$
    are integers, in which case $n = [2v_0 - m(u_0 + u_1)]/(v_0 + v_1)$.

Proof: Using (13), we have $u_m = v_n$ if and only if

$$(-1)^{m-1}[m(u_0 + u_1) - u_0] = (-1)^{n-1}[n(v_0 + v_1) - v_0] \tag{14}$$

Case 1: Suppose $n - m + 2j$ for some integer j. Then (14) gives $n - m(u_0 + u_1)/(v_0 + v_1)$, so that

$$m(u_0 + u_1)/(v_0 + v_1) \tag{15}$$

must be an integer. Then $2j = n - m$, which equals

$$m(u_1 - v_1)/(v_0 + v_1) , \qquad (16)$$

so that $m(u_1 - v_1)/(2v_0 + 2v_1)$ must also be an integer. Conversely, if m is an integer for which $m(u_1 - v_1)/2(v_0 + v_1)$ is an integer, then the identity

$$m(u_0 + u_1)/(v_0 + v_1) = m + m(u_1 - v_1)/(v_0 + v_1)$$

shows that $m(u_0 + u_1)/(v_0 + v_1)$ is an integer, and equation (14) is satisfied using $n = m(u_0 + u_1)/(v_0 + v_1)$.

Case 2: Suppose $n = m+2j+1$ for some integer j. Then (14) gives $n = [2w - m(x+w)]/(w + y)$, so that this fraction must be an integer. Then

$$2j = n - m - 1 = [mx + (m+1)y + (2m - 1)w]/(w + y),$$

so that $[mx + (m+1)y + (2m - 1)w]/(2w + 2y)$ must be an integer. Conversely, if both fractions are integers, then clearly $u_m = v_n$.

5. OTHER CASES

Suppose U and V are second-order sequences both having characteristic polynomial $(x - \alpha)(x - \beta)$, where α and β are distinct real numbers or else complex conjugates. The possibilities for common terms in these two cases appear to be included among those in Section 4, loosely speaking.

For recurrences of higher order, in which a fuller use of equation (4) applies, the numbers of common terms, we conjecture, are still associated with specific upper bounds. More specifically, if U and V are of a specific type T (as given, for example, by the nature of the roots in (4), such as two distinct complex conjugate pairs, or a single root of multiplicity 3), and U and V do not have infinitely many common terms, then there is a maximum number B_T of terms that U and V can have in common. It would be of interest to determine B_T for various types T.

REFERENCES

[1]   Kimberling, C. "The nth Term of a General Linear Recurrence Sequence of Real Numbers." Submitted.

[2]   Kimberling, C. "Sets of Terms that Determine all the Terms of a Linear Recurrence Sequence." To appear in *The Fibonacci Quarterly*.

RECURRENCE RELATIONS IN EXPONENTIAL FUNCTIONS AND IN DAMPED SINUSOIDS AND THEIR APPLICATIONS IN ELECTRONICS

Joseph Lahr

1. INTRODUCTION

The recurrence relations in sinusoids have been exposed on the third International Conference on Fibonacci Numbers [1]. The fundamental difference equation of this paper was the relation

$$S_n - PS_{n-1} + S_{n-2} = 0, \qquad (1)$$

where the S_n are sampled values of a sinusoid, equally spaced along the time axis at a distance of a radians, and where P is the parameter of this discrete sinusoid, defined by:

$$P = 2\cos(a). \qquad (2)$$

Equation (1) was demonstrated by the following expressions:

$$\sin(an + c) = 2\cos(a)\sin[a(n-1) + c] - \sin[a(n-2) + c], \qquad (3)$$
$$\cos(an + c) = 2\cos(a)\cos[a(n-1) + c] - \cos[a(n-2) + c].$$

Similar identities may be established using the hyperbolic trigonometry:

$$\sinh(an + c) = 2\cosh(a)\sinh[a(n-1) + c] - \sinh[a(n-2) + c], \qquad (4)$$
$$\cosh(an + c) = 2\cosh(a)\cosh[a(n-1) + c] - \cosh[a(n-2) + c].$$

The recurrence relation of the hyperbolic sine and cosine is now immediately given by:

$$S_n - PS_{n-1} + S_{n-2} = 0; \qquad (5)$$
$$\text{with } P = 2\cosh(a). \qquad (6)$$

189

So we see that the recurrence relation of a sinusoid and the recurrence relation of the hyperbolic sine and cosine are identical. In the case of the hyperbolic functions P is greater than or equal 2. This fact permits to extend the method for the resolution of algebraic equations, which was presented at the just mentioned third Conference on Fibonacci Numbers [1], to the case where the roots of the algebraic equation must not be in the interval $(-2,2)$.

In electronics sinusoids are generally represented by complex exponential functions and not by trigonometric expressions. The signals defined by $\sin(an+c)$ and by $\exp[i(an+c)]$ contain the same information. So it is evident that from the point of view of electrical engineers the recurrence relations in exponential functions have the same degree of consequence than those existing in sinusoids.

2. RECURRENCE RELATIONS IN EXPONENTIAL FUNCTIONS

Suppose that the following general discrete exponential function is given:

$$E_n = K_1 \exp(an), \tag{7}$$

where a and K_1 are constants. In this paper we call K_1 the coefficient of the exponential function.

A first recurrence relation may immediately be found:

$$\boxed{E_n - UE_{n-1} = 0, \text{ with } U = \exp(a).} \tag{8}$$

This first-order linear recursion relation is named the fundamental or the first recurrence relation of a sampled exponential function, and the parameter U is called the first parameter or the ratio. From (8) and (9) we obtain:

$$U = E_n/E_{n-1}, \tag{9}$$

$$a = \ln(U), \tag{10}$$

$$K_1 = E_n/\exp(an) = E_n/U^n. \tag{11}$$

The starting point of the recursion may be chosen arbitrarily. If the determination of the coefficient K_1 is done by

$$K_1 = E_i/\exp(a) = E_i/U, \tag{12}$$

then E_i must be considered to be E_1.

The just established relations show that

$$\boxed{\text{A discrete exponential function is completely determined by two samples.}} \tag{13}$$

Using the definition of the hyperbolic sine and relation (4) we find the expression

$$\exp(an) = 2\cosh(a)\exp[a(n-1)] - \exp[a(n-2)], \tag{14}$$

which yields the following second-order linear recursion relation:

$$\boxed{E_n - PE_{n-1} + E_{n-2} = 0, \text{ with } P = 2\cosh(a),} \tag{15}$$

named the second recurrence relation of a sampled exponential function. In this paper the parameter P is called the second parameter.

The relations (1), (5) and (15) show that the sinusoids, the hyperbolic sines and cosines and the exponential functions have at least one common recurrence relation.

Let us mention some other useful and easy-to-prove relationships:

$$E_n E_{n-2} - E_{n-1}^2 = 0. \tag{16}$$

$$E_n E_{n-3} - E_{n-1} E_{n-2} = 0. \tag{17}$$

$$(E_n + E_{n-2})E_{n-2} - (E_{n-1} + E_{n-3})E_{n-1} = 0. \tag{18}$$

The repeated substitution of each sample E_i by the difference $E_i - E_{i-1}$ in the above established relations produces sets of new difference equations. If we apply this transformation to the relations (8) and (15), then we obtain the following later on useful equations.

$$E_n - (U+1)E_{n-1} + UE_{n-2} = 0. \tag{19}$$

$$E_n - (U+2)E_{n-1} + (2U+1)E_{n-2} - UE_{n-3} = 0. \tag{20}$$

$$E_n - (U+3)E_{n-1} + (3U+3)E_{n-2} - (3U+1)E_{n-3} + UE_{n-4} = 0. \tag{21}$$

$$E_n - (U+4)E_{n-1} + (4U+6)E_{n-2} - (6U+4)E_{n-3} + (4U+1)E_{n-4} - UE_{n-5} = 0. \tag{22}$$

$$E_n - (P+1)E_{n-1} + (P+1)E_{n-2} - E_{n-3} = 0. \tag{23}$$

$$E_n - (P+2)E_{n-1} + (2P+2)E_{n-2} - (P+2)E_{n-3} + E_{n-4} = 0. \tag{24}$$

$$E_n - (P+3)E_{n-1} + (3P+4)E_{n-2} - (3P+4)E_{n-3} + (P+3)E_{n-4} - E_{n-5} = 0. \tag{25}$$

$$E_n - (P+4)E_{n-1} + (4P+7)E_{n-2} - (6P+8)E_{n-3} + (4P+7)E_{n-4} - (P+4)E_{n-5}$$
$$+ E_{n-6} = 0. \tag{26}$$

3. THE SUPERPOSITION OF N EXPONENTIAL FUNCTIONS

Let us denote a first and a second exponential function by

$$A_n = K_1 \exp(an), \tag{27}$$

and by

$$B_n = K_2 \exp(bn). \tag{28}$$

The corresponding recurrence relations may be expressed by

$$A_n - UA_{n-1} = 0, \text{ and} \tag{29}$$

$$B_n - VB_{n-1} = 0, \text{ with } V = \exp(b). \tag{30}$$

The superposition of these two functions may be denoted again by E_n. Hence we can write:

$$E_n = A_n + B_n. \tag{31}$$

The determination of the corresponding recurrence relation is very simple if we use a theorem established by Maurice d'Ocagne [1], [4]. We find:

$$E_n - (U+V)E_{n-1} + UVE_{n-2} = 0. \tag{32}$$

In the same way the recurrence relation of more than two superposed exponential functions may be established. Let us define further exponential functions for instance:

$$C_n = K_3\exp(cn), \text{ and } W = \exp(c); \tag{33}$$

$$D_n = K_4\exp(dn), \text{ and } X = \exp(d). \tag{34}$$

Then in the case of the superposition of A_n, B_n and C_n the corresponding recurrence relation is given by:

$$E_n - (U+V+W)E_{n-1} + (UV+UW+VW)E_{n-2} - UVWE_{n-3} = 0. \tag{35}$$

The addition of a fourth exponential function D_n yields the following expression:

$$E_n - (U+V+W+X)E_{n-1} + (UV+UW+UX+VW+VX+WX)E_{n-2}$$
$$- (UVW+UVX+UWX+VWX)E_{n-3} + UVWXE_{n-4} = 0. \tag{36}$$

So we could continue and determine the recurrence relation of n superposed exponential functions.

The recurrence relations of superposed exponential functions permit two main statements:

| The fundamental recurrence relation of n superposed exponential functions is of the order n. | (37) |

| The coefficients of the recurrence relation of n superposed exponential functions are given by all possible combinations (the order being disregarded) of the involved parameters taken in blocks of 1, 2, 3, ... and n. The sign of the coefficients is alternating. Further it may be observed that the coefficients are identical with those of an algebraic equation having the parameters as roots. | (38) |

4. THE CASE OF AN EXPONENTIAL FUNCTION AND A SUPERPOSED
CONSTANT TERM

Suppose that a constant term k_0 is added to all samples of an exponential function.

$$E_n = K_1 \exp(an) + k_0. \tag{39}$$

The parameter U must be obtained if we subtract in relation (9) the constant k_0 from all samples.

$$U = \frac{E_n - k_0}{E_{n-1} - k_0} = \frac{E_{n-1} - k_0}{E_{n-2} - k_0}. \tag{40}$$

This identity permits the extraction of the constant term.

$$\boxed{k_0 = \frac{E_n E_{n-2} - E_{n-1}^2}{E_n - 2E_{n-1} + E_{n-2}}.} \tag{41}$$

The discussion of this relation yields the conclusion:

$$\boxed{\begin{array}{l}\text{A discrete exponential function and a superposed constant}\\ \text{term are completely determined by three samples.}\end{array}} \tag{42}$$

After the determination of k_0 it is possible to subtract k_0 from all the samples and then use the relations (9) and (11) for the calculation of U and K_1.

But the parameter U may also be determined directly. Using the equations (40) and (41) we find:

$$\boxed{U = \frac{E_n - E_{n-1}}{E_{n-1} - E_{n-2}}.} \tag{43}$$

This determination of the parameter U is independent of any constant term added to the samples of an exponential function. Comparing the relations (9) and (43) we are conducted to the following theorem:

$$\boxed{\begin{array}{l}\text{To give the formula for the determination of the parameter U of}\\ \text{a discrete exponential function the property of independence of a}\\ \text{constant term, each sample must be replaced by the difference of}\\ \text{two consecutive samples.}\end{array}} \tag{44}$$

Later on we shall see that this theorem may be generalized for any linear recurrence relation.

A transformation of equation (43) yields immediately:

$$E_n - (U+1)E_{n-1} + UE_{n-2} = 0. \tag{45}$$

This expression is the recurrence relation of the superposition of an exponential function and a constant term. Furthermore it must be observed that this equation is identical with relation (19).

The determination of the coefficient K_1 of the exponential function may be effectuated in several ways: First we have:

$$K_1 = \frac{E_1 - k_0}{U}. \tag{46}$$

After some transformations using the relations (41) and (43) we obtain a second formula:

$$K_1 = \frac{(E_2 - E_1)^3}{(E_3 - 2E_2 + E_1)(E_3 - E_2)}. \tag{47}$$

This expression has the property of independence of a constant term.

In electronics the substitution of a sample by the difference of two consecutive samples is considered as the application of a digital filter. The theory of digital filters shows that the frequency is not modified if each sample is replaced by a linear combination of samples. As the argument of an exponential function may be considered as a complex frequency, we have an other demonstration of theorem (44).

In contrast to the frequency the amplitude of a signal is not independent of the effect of a digital filter and its new value can be determined by the transfer function which may be established using the z transformation. The z transformation of a digital filter where each sample is replaced by two consecutive samples is given by

$$H(z) = 1 - z^{-1}. \tag{48}$$

The transfer function $H(a)$ is obtained if z is replaced by $\exp(i2\pi f)$ and $-$ in the case of samples coming out of exponential functions $-$ if f is substituted by $a/(i2\pi)$.

$$H(a) = 1 - \exp(-a) = \frac{U-1}{U}. \tag{49}$$

If we replace in (11) the sample by the difference of two consecutive samples and if we divide by the transfer function we obtain a third expression for K_1:

$$K_1 = \frac{E_1 - E_0}{U-1}. \tag{50}$$

If the parameter U is known then the constant term may be computed in the following way:

$$k_0 = \frac{E_n - UE_{n-1}}{1 - U}. \tag{51}$$

It may be observed that the numerator is constituted by the fundamental recurrence relation of an exponential function.

Similar relationships may be established by the reader using the second recurrence relation of a sampled exponential function (relation (15)).

Let us now consider the case of two superposed exponential functions and a constant term. On the basis of relation (32) we find, after some computations consisting in the elimination of k_0, the following equation:

$$E_n - E_{n-1} - (U+V)(E_{n-1} - E_{n-2}) + UV(E_{n-2} - E_{n-3}) = 0. \tag{52}$$

This is the recurrence relation of two superposed exponential functions and a constant term. The relation is independent of the presence of a constant term and may also be established by the substitution in equation (32) of each sample by the difference of two consecutive samples.

If the parameters U and V are known, the constant term is given by:

$$k_0 = \frac{E_n - (U+V)E_{n-1} + UVE_{n-2}}{(1-U)(1-V)}. \tag{53}$$

In reality it is sufficient to know the sum and the product of U and V.

The numerator in (53) is again constituted by the recurrence relation of two superposed exponential functions.

In the case of three superposed exponential functions and a constant term we have:

$$E_n - E_{n-1} - (U+V+W)(E_{n-1} - E_{n-2}) + (UV+UW+VW)(E_{n-2} - E_{n-3})$$
$$- UVW(E_{n-3} - E_{n-4}) = 0. \tag{54}$$

$$k_0 = \frac{E_n - (U+V+W)E_{n-1} + (UV+UW+VW)E_{n-2} - UVWE_{n-3}}{(1-U)(1-V)(1-W)}. \tag{55}$$

The conclusions of this development are:

$$\boxed{\begin{array}{l} \text{The fundamental recurrence relation of n superposed exponential} \\ \text{functions and a constant term is of the order } n+1. \end{array}} \tag{56}$$

The constant term k_0 and the fundamental recurrence relation of n superposed exponential functions are related by the following formula: (57)

$$k_0 = \frac{\text{fund. recursion of n superposed exponential functions}}{\prod_{i=1}^{n} (1 - U_i)}.$$

with $U_1 = U$; $U_2 = V$; $U_3 = W$; ...

To avoid the effect of a constant term upon the recurrence relation of superposed exponential functions, it is sufficient to replace each sample by the difference of two consecutive samples. So it is possible to determine the parameters without knowing the constant term. As soon as the parameters are known, the constant term may be computed using relation (57).   (58)

5. THE EFFECT OF THE SUBSTITUTION OF EACH SAMPLE OF A LINEAR RECURRENCE RELATION BY THE DIFFERENCE OF TWO CONSECUTIVE SAMPLES

Let us consider a general linear recurrence relation of order $n - 1$ and let us subtract from this recurrence relation the same relation but with subscripts shifted by one in the negative direction.

$$a_n H_n + a_{n-1} H_{n-1} + \cdots + a_2 H_2 + a_1 H_1 = 0. \tag{59}$$

$$- a_n H_{n-1} - a_{n-1} H_{n-2} - \cdots - a_2 H_1 - a_1 H_0 = 0. \tag{60}$$

This yields a new recursion of order n:

$$a_n H_n + (a_{n-1} - a_n) H_{n-1} + \cdots + (a_1 - a_2) H_1 - a_1 H_0 = 0. \tag{61}$$

The same equation may be obtained if we substitute in relation (59) each sample H_i by the expression $H_i - H_{i-1}$.

This substitution has the special effect that the new recurrence relation is independent of a constant term added to each sample. The relations (19) and (23) belong to this category. The demonstration is very easy. In fact we change nothing in the substitution-expression if we add a constant term k_0 to each sample since:

$$(H_i + k_0) - (H_{i-1} + k_0) = H_i - H_{i-1}. \tag{62}$$

We may continue with this transformation and we may apply the substitution a second time. That signifies that we replace in (59) each sample

$$H_i \text{ by } (H_i - H_{i-1}) - (H_{i-1} - H_{i-2}), \text{ or by } (H_i - 2H_{i-1} + H_{i-2}). \tag{63}$$

If we add to each sample H_i in the substitution-formula (63) the expression

$$k_0 + k_1 i, \tag{64}$$

then we observe that this operation does not change the substitution. The recurrence relation established by this substitution is independent of the addition of the expression $k_0 + k_1 i$. The relations (20) and (24) belong to this category. They are recurrence relations of the general term:

$$E_n = K \exp(an) + k_0 + k_1 n. \tag{65}$$

It is obvious that k_0 and k_1 may have any value.

We can continue and apply a third substitution. The corresponding substitution-formula is then given by:

$$(H_i - 3H_{i-1} + 3H_{i-2} - H_{i-3}), \tag{66}$$

and the new recurrence relation is independent of the expression

$$k_0 + k_1 i + k_2 i^2. \tag{67}$$

The relations (21) and (25) possess the same property, and the corresponding general term is represented by:

$$E_n = K \exp(an) + k_0 + k_1 n + k_2 n^2. \tag{68}$$

Also k_2 may admit any value.

Two theorems can be announced after these considerations.

If in a linear recurrence relation of order n each sample H_i is m-times replaced by $H_i - H_{i-1}$ or by $H_i - H_{i+1}$, then the new recurrence relation is of the order $n + m$ and is independent of the addition of

$$\sum_{j=0}^{m-1} k_j i^j$$

to each sample. k_0, k_1, $k_3 \ldots k_{m-1}$ having arbitrary values.

(69)

The coefficients of all the substitution-formulae correspond to the coefficients of Pascal's triangle.

(70)

The determination of the parameter U is not affected by m substitutions of each sample by the difference of two consecutive samples. The computation of the coefficient K_1 however must be modified using the z transformation. For m substitutions we have:

$$H(z) = (1-z^{-1})^m.$$
(71)

The corresponding transfer function is given by:

$$H(a) = \frac{(U-1)^m}{U^m}.$$
(72)

And for K_1 we obtain using (11):

$$K_1 = \frac{\sum_{k=0}^{m} (-1)^k \binom{m}{k} E_{m-k}}{(U-1)^m}.$$
(73)

6. A FIRST APPLICATION IN ELECTRONICS CONSISTING IN THE DETERMINATION OF THE PARAMETERS AND COEFFICIENTS OF SUPERPOSED EXPONENTIAL FUNCTIONS

Complex exponential functions are generally used for the description of signals appearing in telecommunication systems and data transmission equipments. Since the relations established in this paper are independent of the discrimination whether the parameters and the coefficients of exponential functions are real or complex, they may be very useful in the analysis of electronic signals.

In order to illustrate this statement let us consider the case of three superposed exponential functions and let us try to determine all the characteristics of these functions.

In a first step the parameters are determined. With relation (35) we may write:

$$E_n - (U+V+W)E_{n-1} + (UV+UW+VW)E_{n-2} - UVWE_{n-3} = 0,$$
$$E_{n-1} - (U+V+W)E_{n-2} + (UV+UW+VW)E_{n-3} - UVWE_{n-4} = 0,$$
$$E_{n-2} - (U+V+W)E_{n-3} + (UV+UW+VW)E_{n-4} - UVWE_{n-5} = 0.$$
(74)

The samples being known, this system of linear equations yields the different combinations of the parameters U, V and W. Using the well-known theorem of Vieta we may establish an algebraic equation of degree 3, the roots of which are the desired parameters.

$$x^3 - (U+V+W)x^2 + (UV + UW + VW)x - UVW = 0.$$
(75)

For the determination of the coefficients of the involved exponential functions we must know a sample of each of these functions. In this case the corresponding samples are A_1, B_1 and C_1; they are given by the solutions of the following linear system.

$$A_1 + B_1 + C_1 = E_1,$$
$$UA_1 + VB_1 + WC_1 = E_2,$$
$$U^2A_1 + V^2B_1 + W^2C_1 = E_3.$$
(76)

The principal determinant of this system is the well-known Vandermonde determinant, and the solutions are given by

$$A_1 = \frac{E_3 - E_2(V + W) + E_1(VW)}{(U - V)(U - W)}, \tag{77}$$

$$B_1 = \frac{E_3 - E_2(U + W) + E_1(UW)}{(V - U)(V - W)}, \tag{78}$$

$$C_1 = \frac{E_3 - E_2(U + V) + E_1(UV)}{(W - U)(W - V)}. \tag{79}$$

We observe that the numerator of these expressions corresponds to the recurrence relation of two superposed exponential functions. This fact permits a general formulation for the samples number 1.

> If n exponential functions are superposed, the sample number 1
> of the exponential function i is given by a fraction where the
> numerator corresponds to the recurrence relation of $n - 1$
> superposed exponential functions affected with all parameters
> different from U_i and where the denominator is given by
>
> $$\prod (U_i - U_k),$$
>
> with $k = 1, 2, 3, \ldots i - 1, i + 1, \ldots n - 1, n.$ (80)

After these calculations the coefficients can immediately be determined:

$$K_1 = \frac{A_1}{U}, \; K_2 = \frac{B_1}{V}, \; K_3 = \frac{C_1}{W}, \; \cdots \tag{81}$$

In addition to this generalization let us indicate two results of the case of two superposed exponential functions. The sum and the product of the parameters are then given by:

$$U + V = \frac{E_n E_{n-3} - E_{n-1} E_{n-2}}{E_{n-1} E_{n-3} - E_{n-2}^2} \quad \text{and} \tag{82}$$

$$U V = \frac{E_n E_{n-2} - E_{n-1}^2}{E_{n-1} E_{n-3} - E_{n-2}^2}. \tag{83}$$

Conclusions of this chapter:

> n superposed discrete exponential functions
> are completely determined by 2n samples.   (84)

> The main mathematical operations consist in the resolution of a system of n linear equations and of the determination of the n roots of an algebraic equation.   (85)

If we are in the presence of a constant term the following procedure must be applied:

1)   build up a new sequence of samples N_i by the substitution $N_i = E_i - E_{i-1}$,

2)   then use the same algorithm as in the case without a constant term (only for the determination of the parameters U_i).

Finally we may conclude:

> n superposed discrete exponential functions and a constant term are completely determined by $2n + 1$ samples.   (86)

7. A SECOND APPLICATION IN ELECTRONICS CONSISTING IN THE DETERMINATION OF THE PARAMETERS OF A DAMPED SINUSOID

Let S_n and E_n be the samples of a discrete sinusoid respectively of a discrete exponential function. Let D_n be the samples of a damped sinusoid defined by:

$$D_n = S_n E_n. \tag{87}$$

Since $S_n = PS_{n-1} - S_{n-2}$ and $E_n = UE_{n-1}$ we find immediately:

$$D_n = PUD_{n-1} - U^2 D_{n-2}. \tag{88}$$

This is the fundamental recurrence relation of a damped sinusoid. If the samples are known, the parameters P and U may be determined on the basis of the following linear system:

$$D_n - PUD_{n-1} + U^2 D_{n-2} = 0, \tag{89}$$
$$D_{n-1} - PUD_{n-2} + U^2 D_{n-3} = 0.$$

Hence we obtain:

$$U = \sqrt{\frac{D_{n-1}^2 - D_n D_{n-2}}{D_{n-2}^2 - D_{n-1} D_{n-3}}} \quad \text{and} \tag{90}$$

$$P = \frac{D_n + U^2 D_{n-2}}{UD_{n-1}}. \tag{91}$$

It may be noticed that the relations (83) and (90) contain the same recurrence expression.

Further we observe that relation (91) may immediately be derived from relation (1) if we substitute

$$S_{n-i} \text{ by } U^i D_{n-i}. \tag{92}$$

This relationship is really evident if we consider the following reflections:

The discrete sinusoid is defined by

$$S_n = A\sin(2\pi f T_s n + c), \tag{93}$$

where A signifies the peak amplitude, f the frequency, T_s the sampling period and c the phase angle. Since the discrete exponential function is defined by $E_n = K\exp(an)$, the product $D_n = S_n E_n$ let the amplitude A and the coefficient K merge into a unique constant A_u. For this reason it is not possible to separate A and K, if only the samples D_i are known. This explains why we could use for D_n also the following definition:

$$D_n = A_u\sin(2\pi f T_s n + c)\exp(an). \tag{94}$$

By this way we have immediately:

$$S_n = \frac{D_n}{U^n} = U^{-n}D_n. \tag{95}$$

The demonstration of (91) and (92) is now perfect.

The frequency, the amplitude and the phase angle may then be determined with relations published in [1].

$$f = \frac{\cos^{-1}(P/2)}{2\pi T_s}. \tag{96}$$

$$A_u = 2U^{-n}\sqrt{\frac{D_n^2 - D_{n-1}D_{n+1}}{4 - P^2}}. \tag{97}$$

$$c = \tan^{-1}\left[\frac{UD_0\sqrt{4 - P^2}}{2D_1 - UPD_0}\right]. \tag{98}$$

This development allows the conclusion:

$$\boxed{\text{A discrete damped sinusoid is completely determined by four samples.}} \tag{99}$$

If we have a damped sinusoid and a constant term, then according to theorem (69) it is sufficient to replace in the relations (90) and (91) each sample by the difference of two consecutive samples.

After the determination of the parameters P and U, the constant term is given by the following relation established using the fundamental recurrence relation of a damped sinusoid.

$$k_0 = \frac{D_n - PUD_{n-1} + U^2 D_{n-2}}{1 - PU + U^2}.$$

(100)

Attention: In this equation the original samples must be used, and not those transformed by the substitution of D_i by $D_i - D_{i-1}$.

Finally we can say:

> A damped sinusoid and a constant term
> is completely determined by five samples.

(101)

REFERENCES

[1]   Lahr, J.H.G. "Recurrence Relations in Sinusoids and Their Applications to Spectral Analysis and to the Resolution of Algebraic Equations". Applications of Fibonacci Numbers, Vol. 3, (1989) edited by G. E. Bergum, A. N. Philippou and A. F. Horadam, Kluwer Academic Press, ISBN 0-7923-0523-X.

[2]   Lahr, J.H.G. "High Resolution Spectral Analysis Using Recurrence Relations". Recent Advances in Fourier Analysis and Its Applications, NATO ASI Series, Series C, Vol. 315, (1990) edited by J. S. Byrnes and Jennifer L. Byrnes, Kluwer Academic Publishers, ISBN 0-7923-0875-1.

[3]   Lahr, J. H. G. "Theorie Elektrischer Leitungen unter Anwendung und Erweiterung der Fibonacci-Funktion". Dissertation ETH Nr. 6958, (1981) Zürich.

[4]   d'Ocagne, M. "Mémoire sur les Suites Récurrentes". Journal de l'Ecole Polytechnique 64 (1894), Paris, pp. 151-224.

SOME BASIC PROPERTIES OF THE FIBONACCI LINE-SEQUENCE

Jack Y. Lee

1. INTRODUCTION

We define Fibonacci ray-sequences for the non-negative suffix and Fibonacci line-sequences for the unrestricted suffix. The properties of the latter were developed by Horadam [4]. We further define line-sequences as congruent if they are composed of the same numbers.

For example, the neighboring pairs [1, 0] and [0, 1] will generate, separately, the congruent line-sequences

$$F_{1,0} : \ldots, \; -3, \, 2, \; -1, \, [1, \, 0], \, 1, \, 1, \, 2, \ldots \tag{1.1}$$

and

$$F_{0,1} : \ldots, \, 2, \; -1, \, 1, \, [0, \, 1], \, 1, \, 2, \, 3, \ldots \tag{1.2}$$

according to the recurrence relation

$$u_n = u_{n-2} + u_{n-1}. \tag{1.3}$$

It is the purpose of this paper to examine some relations among these congruent line-sequences.

2. THE NATURE OF THE LINE-SEQUENCE

Let us consider a set $\{F\}$ of such line-sequences. An element in this set can be expressed in general by

203

G. E. Bergum et al. (eds.), Applications of Fibonacci Numbers Volume 4, 203–214.
© 1991 *Kluwer Academic Publishers.*

$$F_{u_0, u_1}: \ldots, u_{-2}, u_{-1}, [u_0, u_1], u_2, u_3, \ldots \tag{2.1}$$

which may or may not be the Fibonacci line-sequence. The u's are, for our present work, restricted to integers.

We shall say that two line-sequences are equal if their corresponding numbers are equal. By "corresponding", we mean matching of the generating numbers. Thus, according to the convention given by (2.1),

$$F_{u_0, u_1} = F_{u'_0, u'_1} \tag{2.2}$$

iff

$$u_n = u'_n, \quad n = \text{integer}. \tag{2.3}$$

We now state two theorems, the full ramifications of which have been explored by Stein [7].

Theorem 1. A pair of numbers generates one and only one line-sequence.

Theorem 2. Two line-sequences are equal iff their generators are equal.

In addition to the preceding Theorems, we need the following definitions.

1.   Addition of two line-sequences in {F} is found by adding their corresponding numbers. The result is of course another line-sequence with the same recurrence relation. Hence it is an element in {F}. That is,

$$F_{u_0, u_1} = F_{u'_0, u'_1} + F_{u''_0, u''_1} \tag{2.4}$$

where

$$u_n = u'_n + u''_n, \quad n = 0, 1. \tag{2.5}$$

2.   The multiplication of a line-sequence in {F} by a scalar is obtained by multiplying every number in the line-sequence with the scalar. The result is of course another line-sequence with the same recurrence relation and hence is in {F}. That is,

$$F_{u_0, u_1} = k \, F_{u'_0, u'_1} \tag{2.6}$$

where

$$u_n = k \, u'_n, \quad k = \text{scalar}, \quad n = 0, 1. \tag{2.7}$$

We shall call this operation a dilation and say that the resulting line-sequence is "similar" to the original line-sequence.

3. The inner product of two line-sequences in $\{F\}$, denoted by $(F_{u_0, u_1}, F_{u_0', u_1'})$, is found by multiplying the corresponding generating numbers and then adding the products together. The result is a scalar. Namely,

$$(F_{u_0, u_1}, F_{u_0', u_1'}) = u_0 u_0' + u_1 u_1' . \tag{2.8}$$

See also equation (5) of Catlin [3].

Two line-sequences are said to be orthogonal if their inner product is zero. The "length" of a line-sequence is the square root of its inner product with itself.

4. Obviously the line-sequence generated by $u_0 = u_1 = 0$ is a sequence of zeros. This is the identity element with respect to addition.

We are now ready to declare the obvious. Since all the rules for a vector space are obviously satisfied [2], the set $\{F\}$ of line-sequences (2.1) is a vector space and the line-sequences $F_{1,0}$ and $F_{0,1}$ in (1.1) and (1.2) form a set of basis vectors which spans the two-dimensional vector space of the "line-sequential vectors". Furthermore, any arbitrary line-sequential vector in $\{F\}$ can then be expressed as a certain linear superposition of $F_{1,0}$ and $F_{0,1}$ so that

$$F_{u_0, u_1} = u_0 F_{1,0} + u_1 F_{0,1}. \tag{2.9}$$

In particular, the Lucas line-sequence can be simply expressed as

$$F_{2,1} = 2 F_{1,0} + F_{0,1}. \tag{2.10}$$

Obviously the following symmetry properties hold:

$$F_{-i, j} = -F_{i, -j}, \tag{2.11}$$

and

$$F_{-i, -j} = -F_{i, j}. \tag{2.12}$$

We therefore say that the set $\{F\}$ is of odd parity. This is displayed in Figure 1 for the Fibonacci line-sequences and their dilations. Notice that the addition of any two neighboring horizontal line-sequences produces the third, which is a dilation of the Fibonacci line-sequence, according to (2.4) and (2.6). Similarly for the vertical line-sequences. Note also that the symmetry properties (2.11) and (2.12) are displayed.

.	.	.	.	.	.	.	.	.	.	.	.	.	.	.
.	.	.	.	.	.	.	.	.	.	.	.	.	.	.
.	.	25	-15	10	-5	5	0	5	5	10	15	25	.	.
.	.	-15	9	-6	3	-3	0	-3	-3	-6	-9	-15	.	.
.	.	10	-6	4	-2	2	0	2	2	4	6	10	.	.
.	.	-5	3	-2	1	-1	0	-1	-1	-2	-3	-5	.	.
.	.	5	-3	2	-1	1	0	1	1	2	3	5	.	.
.	.	0	0	0	0	0	0	0	0	0	0	0	.	.
.	.	5	-3	2	-1	1	0	1	1	2	3	5	.	.
.	.	5	-3	2	-1	1	0	1	1	2	3	5	.	.
.	.	10	-6	4	-2	2	0	2	2	4	6	10	.	.
.	.	15	-9	6	-3	3	0	3	3	6	9	15	.	.
.	.	25	-15	10	-5	5	0	5	5	10	15	25	.	.
.	.	.	.	.	.	.	.	.	.	.	.	.	.	.
.	.	.	.	.	.	.	.	.	.	.	.	.	.	.

Figure 1: The Parity Chart

We shall say that the parity chart is "flat" if the numbers in the four quadrants are the same barring the signs. For the Fibonacci line-sequence, this is characterized by the four ones at the corners of the innermost square. As we shall see later, the Fibonacci line-sequence is not the only one with a flat parity chart.

Combining (1.3) and (2.4), we have

$$F_{u_0, u_1} + F_{u_1, u_2} = F_{u_2, u_3}.$$ (2.13)

This is the recurrence relation of the line-sequences. It also shows the translation property of the generators, and consequently the numbers in the sequences, which keep shifting under the addition of any two neighboring line-sequences. This is displayed in Figure 2 for the Fibonacci line sequences.

.	.	.	.	.	.	.	.	.	.	.	.
.	.	.	.	.	.	.	.	.	.	.	.
.	.	34	- 21	13	- 8	5	- 3	2	- 1	1	0	.	.
.	.	- 21	13	- 8	5	- 3	2	- 1	1	0	1	.	.
.	.	13	- 8	5	- 3	2	- 1	1	0	1	1	.	.
.	.	- 8	5	- 3	2 .	- 1	1	0	1	1	2	.	.
.	.	5	- 3	2	- 1	1	0	1	1	2	3	.	.
.	.	- 3	2	- 1	1	0	1	1	2	3	5	.	.
.	.	2	- 1	1	0	1	1	2	3	5	8	.	.
.	.	- 1	1	0	1	1	2	3	5	8	13	.	.
.	.	1	0	1	1	2	3	5	8	13	21	.	.
.	.	0	1	1	2	3	5	8	13	21	34	.	.
.	.	.	.	.	.	.	.	.	.	.	.
.	.	.	.	.	.	.	.	.	.	.	.

Figure 2: The Translation Chart

Notice that the addition of any two neighboring line-sequences, horizontal or vertical, produces the third, with successive shifting of the corresponding generators, and hence the numbers, in accordance with (2.13).

3. THE TRANSLATION PROPERTIES

We shall now investigate the translation properties of the line-sequence. Let us define an operation T_i on an arbitrary number u_n in a line-sequence by

$$T_i u_n = u_{n+i},$$ (3.1)

where i = integer. Since the effect of T_i is only to shift the number along the line-sequence by shifting the index n of the number, and the sequence of the index n is none other than the sequence of integers, it follows that the set $\{T_i\}$ is a cyclic group of infinite order [6].

Using (2.1) and (3.1), we must have

$$T_i F_{u_0, u_1} = F_{u_0+i, u_1+i}.$$ (3.2)

Since the new pair of generating numbers is a pair of neighboring numbers in the same line-sequence, the new line-sequence must be congruent to the original one. Thus we have

Theorem 3. Two line-sequences F_{u_0, u_1} and $F_{u'_0, u'_1}$ are congruent iff there exists a translation T_i such that

$$T_i F_{u_0, u_1} = F_{u'_0, u'_1} . \tag{3.3}$$

Note that the translation relation (3.2) is displayed in Figure 2. Also note the presence of the tree structures in the Translation Chart but observe that the nodes are joined into a network rather than spreading out into a tree form.

Observe that addition of the numbers below the diagonal of zeros in the Translation Chart, Figure 2, results in the sequences discussed in [1]. That is, addition of numbers below the diagonal zeros in this chart results in the sequence (1, 2, 4, 7, 12, 20, . . .). If the row of numbers second to the last row is left out in the addition, the result will be (1, 1, 3, 5, 9, 15, 25, . . .).

4. GENERAL SECOND ORDER LINE-SEQUENCES

Now, we extend our investigation to line-sequences $\{G\}$ with the following general form of a second order recurrence relation. That is,

$$u_n = cu_{n-2} + bu_{n-1} \tag{4.1}$$

where b, c are non-zero integers. Again, the set $\{G\}$ of line-sequences forms a vector space with the following set of basis vectors:

$$G_{1,0}: \ldots, (c+b^2)/c^2, -b/c, [1, 0], c, cb, c(c+b^2), \ldots \tag{4.2}$$

and

$$G_{0,1}: \ldots, (c+b^2)/c^3, -b/c^2, 1/c, [0, 1], b, c+b^2, \ldots . \tag{4.3}$$

Note that, unlike the pair $F_{1,0}$ and $F_{0,1}$, which are congruent, the pair $G_{1,0}$ and $G_{0,1}$ are not. In fact $G_{1,0}$ contains the neighboring pair $[0, c]$ so it is congruent to $G_{0,c}$ which in turn is a dilation of $G_{0,1}$ by a factor c. When $c = 1$, we do have congruence.

We now seek an expression for a general term in (4.3), which reduces to Binet's formula when (4.3) reduces to (1.2). This expression is composed of the roots of the equation, [5],

$$q^2 - bq - c = 0. \tag{4.4}$$

Let A and B represent these two roots where

$$A = [b + (b^2 + 4c)^{\frac{1}{2}}] / 2 \tag{4.5}$$

and

$$B = [b - (b^2 + 4c)^{\frac{1}{2}}] / 2 \tag{4.6}$$

then

$$u_n = (A^n - B^n) / (A - B). \tag{4.7}$$

We assume that $A - B \neq 0$. Otherwise, u_n has to be expressed in the expanded form. The corresponding conjugate numbers are given by

$$v_n = A^n + B^n. \tag{4.8}$$

The following parity relation among the numbers in (4.3) can be easily shown to hold. That is,

$$u_{-n} = (-1)^{n+1} u_n / c^n. \tag{4.9}$$

Also, it is easy to show that the parity relation

$$v_{-n} = (-1)^n v_n / c^n \tag{4.10}$$

holds for the corresponding conjugate numbers.

In (4.5) to (4.8), we have $u_0 = 0$, $v_0 = 2$, $u_1 = 1$ and $v_1 = b$. Thus the sequence conjugate to $G_{0,1}$ is none other than $G_{2,b}$, where

$$G_{2,b} = 2 G_{1,0} + b G_{0,1}. \tag{4.11}$$

Substituting (4.2) and (4.3) into (4.11), we obtain the explicit expression

$$G_{2,b} : \ldots, (2c+b^2)/c^2, \ -b/c, \ [2, \ b], \ 2c+b^2, \ b(3c+b^2), \ldots \tag{4.12}$$

for the numbers in the conjugate line-sequence.

By (4.2), all line sequences with c=1 possess a flat parity chart.

5. SPECIAL CASES, c = ±1

Now, by discussing some special cases of the recurrence relation (4.1), we shall discover some surprisingly simple yet interesting facts. We first discuss the case of c = 1 in (4.1). Here, (4.1) reduces to

$$u_n = u_{n-2} + bu_{n-1}. \tag{5.1}$$

The basis vectors (4.2) and (4.3) are then

$$G_{1,0} : \ldots, 1+b^2, \ -b, \ [1, \ 0], \ 1, \ b, \ 1+b^2, \ldots \tag{5.2}$$

and

$$G_{0,1} : \ldots, 1+b^2, \ -b, \ 1, \ [0, \ 1], \ b, \ 1+b^2, \ldots . \tag{5.3}$$

Hence, by putting c = 1, we not only obtain an integer number sequence as compared to a fraction number sequence but also find that the two basis vectors become a pair of congruent line-sequences. If we now put b = 1, we obtain the Fibonacci line-sequence. Thus, we see that the Fibonacci sequence is not the only one having a pair of congruent basis vectors. From (4.12), the corresponding conjugate sequence is the Lucas line-sequence. For b = 2, we obtain the Pell line-sequence. The corresponding conjugate sequence is the conjugate Pell line-sequence (with $u_0 = u_1 = 2$).

Letting c = −1 in (4.1), the recurrence relation becomes

$$u_n = -u_{n-2} + bu_{n-1}. \tag{5.4}$$

The basis vectors are

$$G_{1,0} : \ldots, (b^2-1), \ b, \ [1, \ 0], \ -1, \ -b, \ -(b^2-1), \ldots \tag{5.5}$$

and

$$G_{0,1} : \ldots, -(b^2-1), \ -b, \ -1, \ [0, \ 1], \ b, \ (b^2-1), \ldots . \tag{5.6}$$

These two line-sequences are not congruent because of the presence of the negative sign in the dilation factor. Letting $b = 1$ in (5.6), we obtain

$$G_{0,1}: \ldots, 0, -1, -1, 0, 1, 1, 0, -1, -1, 0, 1, 1, \ldots, \qquad (5.7)$$

a repeating sequence with a period of 6 digits.

Putting $b = -1$ in (5.6), we obtain

$$G_{0,1}: \ldots, 0, 1, -1, 0, 1, -1, 0, 1, -1, 0, 1, -1, \ldots, \qquad (5.8)$$

a repeating sequence with a period of 3 digits.

The corresponding conjugate sequences, according to (4.12), are

$$G_{2,1}: \ldots, -2, -1, 1, 2, 1, -1, -2, -1, 1, \ldots \qquad (5.9)$$

and

$$G_{2,-1}: \ldots, 2, -1, -1, 2, -1, -1, 2, -1, -1, \ldots, \qquad (5.10)$$

which are repeating sequences with periods of 6 and 3 digits respectively.

6. SPECIAL CASES, $c = -b^2/4$

With $c = -b^2/4$, the recurrence relation (4.1) becomes

$$u_n = -(b^2/4)\, u_{n-2} + b\, u_{n-1}. \qquad (6.1)$$

The basis vectors (4.2) and (4.3) are

$$G_{1,0}: \ldots 12/b^2, \ 4/b, \ [1,\ 0], \ -b^2/4, \ -b^3/4, \ -3b^4/16, \ldots \qquad (6.2)$$

and

$$G_{0,1}: \ldots -48/b^4, \ -16/b^3, \ -4/b^2, \ [0,\ 1], \ b, \ 3b^2/4, \ldots . \qquad (6.3)$$

The corresponding conjugate line-sequence, according to (4.12) is

$$G_{2,b}: \ldots 8/b^2, \ 4/b, \ [2,\ b], \ b^2/2, \ b^3/4, \ldots . \qquad (6.4)$$

Letting $b = 2$ and $c = -1$ in (6.3) we have

$$G_{0,1}: \ \ldots -3, \ -2, \ -1, \ [0, \ 1], \ 2, \ 3, \ldots \tag{6.5}$$

which is none other than the sequence of integers. The corresponding conjugate line-sequence, according to (4.12) becomes

$$G_{2,2}: \ \ldots, \ 2, \ 2, \ 2, \ 2, \ 2, \ 2, \ldots, \tag{6.6}$$

an infinite sequence of twos!

7. SPECIAL CASES, $c = b^2/4$

With $c = b^2/4$, the recurrence relation (4.1) becomes

$$u_n = (b^2/4) \ u_{n-2} + b \ u_{n-1}. \tag{7.1}$$

The basis vectors (4.2) and (4.3) are

$$G_{1,0}: \ \ldots, \ 20/b^2, \ -4/b, \ [1, \ 0], \ b^2/4, \ b^3/4, \ 5b^4/16, \ldots \tag{7.2}$$

and

$$G_{0,1}: \ \ldots, \ 80/b^4, \ -16/b^3, \ 4/b^2, \ [0, \ 1], \ b, \ 5b^2/4, \ldots . \tag{7.3}$$

The corresponding conjugate line-sequence, according to (4.12) is

$$G_{2,b}: \ \ldots, \ 24/b^2, \ -4/b, \ [2, \ b], \ 3b^2/2, \ 7b^2/4, \ldots . \tag{7.4}$$

Letting $b = 2$, and $c = 1$ in (7.2) and (7.3) we obtain the congruent Pell line-sequences

$$G_{1,0}: \ \ldots, \ 5, \ -2, \ [1, \ 0], \ 1, \ 2, \ 5, \ldots \tag{7.5}$$

and

$$G_{0,1}: \ \ldots, \ 5, \ -2, \ 1, \ [0, \ 1], \ 2, \ 5, \ldots . \tag{7.6}$$

Observe that (7.4) reduces to the corresponding conjugate Pell line-sequence

$$G_{2,2}: \ \ldots, \ 6, \ -2, \ [2, \ 2], \ 6, \ 14, \ldots . \tag{7.7}$$

We summarize the preceding findings in the following table

c	b	$b^2/4$	basis 2	conjugate
			Fibonacci	Lucas
1	1	1/4	2, -1, 1, 0, 1, 1, 2,	3, -1, 2, 1, 3, 4,
			Pell	conj. Pell
1	2	1	5, -2, 1, 0, 1, 2, 5,	-14, 6, -2, 2, 2, 6, 14,
			period 6	period 6
-1	1	1/4	0, -1, -1, 0, 1, 1,	-2, -1, 1, 2, 1, -1,
			tertiary	period 3
-1	-1	1/4	0, 1, -1, 0, 1, -1,	2, -1, -1, 2, -1, -1,
			integer	constant
-1	2	1	-3, -2, -1, 0, 1, 2, 3,	2, 2, 2, 2, 2, 2,
			jigsaw	binary
-1	-2	1	-3, 2, -1, 0, 1, -2, 3	2, -2, 2, -2, 2, -2,

Table 1: Special Integer Number Line-Sequences

ACKNOWLEDGEMENT

The author wishes to express his gratitude to Prof. R. Stowasser for his inspiring seminars during an NSF Summer Program held in the City College of the City University of New York, 1989. The author also wishes to express his appreciation for an NSF budget sponsored by the Project Leadership under the directorship of Profs. Frank R. Pomilla and T. Y. Dai of the York College of the City University of New York, 1989.

REFERENCES

[1]   Bicknell-Johnson, Marjorie and Bergum, Gerald E., "The Generalized Fibonacci Numbers $\{C_n\}$, $C_n = C_{n-1} + C_{n-2} + k$." A. N. Philippou et al. (eds.), Applications of Fibonacci Numbers, Kluwer Academic Publishers (1988): p. 193.

[2]   Bowen, Ray M. and Wang, C. C., "Introduction to Vectors and Tensors." Plenum Press, NY (1976): pp. 40-59.

[3]   Catlin, P. A., "On the Multiplication of Recurrences," *The Fibonacci Quarterly*, Vol. 12,
      No. 4 (1974): pp. 365-368.

[4]   Horadam, A. F., "Basic Properties of a Certain Generalized Sequence of Numbers." *The
      Fibonacci Quarterly, 3* (1965): pp. 161-176.

[5]   Horadam, A. F. and Shannon, A. G., "Asveld's Polynomials $p_j(n)$." A. N. Philippou, et.
      al. (eds.), <u>Applications of Fibonacci Numbers</u>, Kluwer Academic Publishers (1988): p. 163.

[6]   Magnus, W., Karrass, A., and Solitar, D., "Combinatorial Group Theory." Interscience
      Publishers (1966): p. 2.

[7]   Stein, S. K., "The Intersection of Fibonacci Sequences." *Michigan Math., J., 9* (1962): pp.
      399-402.

DE MOIVRE-TYPE IDENTITIES FOR THE TETRABONACCI NUMBERS

Pin-Yen Lin

1. INTRODUCTION

It is well known that, for $x^2 - x - 1 = 0$, the two roots are $(1 \pm \sqrt{5})/2$, and that

$$\left(\frac{1 \pm \sqrt{5}}{2}\right)^n = \frac{L_n \pm \sqrt{5}F_n}{2} \tag{1}$$

where L_n are the Lucas numbers and F_n the Fibonacci numbers. Identities (1) are called "de Moivre-type identities" [1].

Lin [4] showed that for $x^3 - x^2 - x - 1 = 0$ there are similar results. Here, we will solve the quartic equation $x^4 - x^3 - x^2 - x - 1 = 0$ and expand the nth power of each root to get five kinds of Tetrabonacci numbers.

2. DE MOIVRE-TYPE IDENTITIES FOR THE TETRABONACCI NUMBERS

Using Ferrari's method [3], we can get four roots of the quartic equation $x^4 - x^3 - x^2 - x - 1 = 0$:

$$r_1 = \tfrac{1}{4}\left\{1 + A + \sqrt{11 - A^2 + \tfrac{26}{A}}\right\} \tag{2}$$

$$r_2 = \tfrac{1}{4}\left\{1 + A - \sqrt{11 - A^2 + \tfrac{26}{A}}\right\} \tag{3}$$

$$r_3 = \tfrac{1}{4}\left\{1 - A + \sqrt{11 - A^2 - \tfrac{26}{A}}\right\} \tag{4}$$

$$r_4 = \tfrac{1}{4}\left\{1 - A - \sqrt{11 - A^2 - \tfrac{26}{A}}\right\} \tag{5}$$

where $A = \sqrt{\dfrac{11 + 2\sqrt[3]{-260 + 12\sqrt{1689}} - 2\sqrt[3]{260 + 12\sqrt{1689}}}{3}}$

215

G. E. Bergum et al. (eds.), Applications of Fibonacci Numbers Volume 4, 215–218.
© 1991 Kluwer Academic Publishers.

Let $B = \sqrt{11 - A^2 + \frac{26}{A}}$ for equation (2) and (3) or $B = \sqrt{11 - A^2 - \frac{26}{A}}$ for equation (4) and (5); using the identity $44A^2 - 4A^4 + \frac{676}{A^2} = 460$, we have

$$r_1 = \tfrac{1}{4}\{1 + \tfrac{4}{4}A + \tfrac{8}{8}B\},$$

$$r_1^2 = \tfrac{1}{4}\{3 + \tfrac{2}{4}A + \tfrac{4}{8}B + \tfrac{52}{8A} + \tfrac{4}{8}AB\},$$

$$r_1^3 = \tfrac{1}{4}\{7 + \tfrac{9}{4}A + \tfrac{7}{8}B + \tfrac{39}{8A} + \tfrac{3}{8}AB + \tfrac{13B}{8A} + \tfrac{A^2B}{8} - \tfrac{A^3}{8}\},$$

$$r_1^4 = \tfrac{1}{4}\{15 + \tfrac{15}{4}A + \tfrac{19}{8}B + \tfrac{91}{8A} + \tfrac{7}{8}AB + \tfrac{13B}{8A} + \tfrac{A^2B}{8} - \tfrac{A^3}{8}\},$$

$$r_1^5 = \tfrac{1}{4}\{26 + \tfrac{30}{4}A + \tfrac{38}{8}B + \tfrac{182}{8A} + \tfrac{14}{8}AB + \tfrac{26B}{8A} + \tfrac{2A^2B}{8} - \tfrac{2A^3}{8}\},$$

$$r_1^6 = \tfrac{1}{4}\{51 + \tfrac{56}{4}A + \tfrac{68}{8}B + \tfrac{364}{8A} + \tfrac{28}{8}AB + \tfrac{52B}{8A} + \tfrac{4A^2B}{8} - \tfrac{4A^3}{8}\},$$

$$r_1^7 = \tfrac{1}{4}\{99 + \tfrac{110}{4}A + \tfrac{132}{8}B + \tfrac{676}{8A} + \tfrac{52}{8}AB + \tfrac{104B}{8A} + \tfrac{8A^2B}{8} - \tfrac{8A^3}{8}\},$$

$$r_1^8 = \tfrac{1}{4}\{191 + \tfrac{211}{4}A + \tfrac{257}{8}B + \tfrac{1313}{8A} + \tfrac{101}{8}AB + \tfrac{195B}{8A} + \tfrac{15A^2B}{8} - \tfrac{15A^3}{8}\}.$$

The coefficients in the above equations yield five kinds of Tetrabonacci sequences which we denote by R_n, S_n, T_n, U_n, and V_n. The first ten numbers of these sequences are shown in the following Table (as well as Tetrabonacci W_n and P_n).

n	0	1	2	3	4	5	6	7	8	9	10
R_n	4	1	3	7	15	26	51	99	191	367	708
S_n	4	3	7	14	28	52	101	195	376	724	1396
T_n	0	1	1	2	4	8	15	29	56	108	208
U_n	4	2	9	15	30	56	110	211	407	784	1512
V_n	8	4	7	19	38	68	132	257	495	952	1836
W_n	0	1	2	4	7	14	27	52	100	193	372
P_n	0	1	2	3	6	12	23	44	85	164	316

By induction, we establish that

$$r_1^n = \tfrac{1}{4}\{R_n + \tfrac{U_n}{4}A + \tfrac{V_n}{8}B + \tfrac{13S_{n-1}}{8}\tfrac{1}{A} + \tfrac{S_{n-1}}{8}AB + \tfrac{13T_{n-2}}{8}\tfrac{B}{A} + \tfrac{T_{n-2}}{8}A^2B - \tfrac{T_{n-2}}{8}A^3\}. \quad (6)$$

Using the same method, we obtain

$$r_2^n = \tfrac{1}{4}\{R_n + \tfrac{U_n}{4}A - \tfrac{V_n}{8}B + \tfrac{13S_{n-1}}{8}\tfrac{1}{A} - \tfrac{S_{n-1}}{8}AB - \tfrac{13T_{n-2}}{8}\tfrac{B}{A} - \tfrac{T_{n-2}}{8}A^2B - \tfrac{T_{n-2}}{8}A^3\}, \quad (7)$$

$$r_3^n = \tfrac{1}{4}\{R_n - \tfrac{U_n}{4}A + \tfrac{V_n}{8}B - \tfrac{13S_{n-1}}{8}\tfrac{1}{A} - \tfrac{S_{n-1}}{8}AB - \tfrac{13T_{n-2}}{8}\tfrac{B}{A} - \tfrac{T_{n-2}}{8}A^2B + \tfrac{T_{n-2}}{8}A^3\}, \quad (8)$$

$$r_4^n = \tfrac{1}{4}\{R_n - \tfrac{U_n}{4}A - \tfrac{V_n}{8}B - \tfrac{13S_{n-1}}{8}\tfrac{1}{A} + \tfrac{S_{n-1}}{8}AB + \tfrac{13T_{n-2}}{8}\tfrac{B}{A} + \tfrac{T_{n-2}}{8}A^2B + \tfrac{T_{n-2}}{8}A^3\}. \quad (9)$$

Hence we find that r_1^n, r_2^n, r_3^n, and r_4^n can be expressed in terms of R_n, U_n, V_n, S_{n-1}, and T_{n-2}, so we have formulas equivalent to equation (1) for the Tetrabonacci numbers.

3. BINET'S FORMULA FOR R_n, S_n, T_n, U_n, AND V_n

From Spickerman [5], we can obtain Binet's formula for R_n, S_n, T_n, U_n, and V_n. That is,

$$R_n = r_1^n + r_2^n + r_3^n + r_4^n, \tag{10}$$

$$S_n = \frac{r_1^{n+2}(4r_1 - 1)}{(r_1 - r_2)(r_1 - r_3)(r_1 - r_4)} + \frac{r_2^{n+2}(4r_2 - 1)}{(r_2 - r_1)(r_2 - r_3)(r_2 - r_4)}$$

$$+ \frac{r_3^{n+2}(4r_3 - 1)}{(r_3 - r_1)(r_3 - r_2)(r_3 - r_4)} + \frac{r_4^{n+2}(4r_4 - 1)}{(r_4 - r_1)(r_4 - r_2)(r_4 - r_3)}, \tag{11}$$

$$T_n = \frac{r_1^{n+3}}{(r_1 - r_2)(r_1 - r_3)(r_1 - r_4)} + \frac{r_2^{n+3}}{(r_2 - r_1)(r_2 - r_3)(r_2 - r_4)}$$

$$+ \frac{r_3^{n+3}}{(r_3 - r_1)(r_3 - r_2)(r_3 - r_4)} + \frac{r_4^{n+3}}{(r_4 - r_1)(r_4 - r_2)(r_4 - r_3)}, \tag{12}$$

$$U_n = \frac{r_1^{n+1}(4r_1^2 - 2r_1 + 3)}{(r_1 - r_2)(r_1 - r_3)(r_1 - r_4)} + \frac{r_2^{n+1}(4r_2^2 - 2r_2 + 3)}{(r_2 - r_1)(r_2 - r_3)(r_2 - r_4)}$$

$$+ \frac{r_3^{n+1}(4r_3^2 - 2r_3 + 3)}{(r_3 - r_1)(r_3 - r_2)(r_3 - r_4)} + \frac{r_4^{n+1}(4r_4^2 - 2r_4 + 3)}{(r_4 - r_1)(r_4 - r_2)(r_4 - r_3)}, \tag{13}$$

and

$$V_n = \frac{r_1^{n+1}(8r_1^2 - 4r_1 - 5)}{(r_1 - r_2)(r_1 - r_3)(r_1 - r_4)} + \frac{r_2^{n+1}(8r_2^2 - 4r_2 - 5)}{(r_2 - r_1)(r_2 - r_3)(r_2 - r_4)}$$

$$+ \frac{r_3^{n+1}(8r_3^2 - 4r_3 - 5)}{(r_3 - r_1)(r_3 - r_2)(r_3 - r_4)} + \frac{r_4^{n+1}(8r_4^2 - 4r_4 - 5)}{(r_4 - r_1)(r_4 - r_2)(r_4 - r_3)}. \tag{14}$$

Equation (10) is the Waring formula [3].

4. SOME PROPERTIES OF R_n, S_n, T_n, U_n, AND V_n

As Ian Bruce shows in [2], using the Tetrabonacci sequence definition as equation (15), some interesting results can be derived. We have found the following:

$$R_n = R_{n-1} + R_{n-2} + R_{n-3} + R_{n-4} \tag{15}$$

(S_n, T_n, U_n, V_n, P_n, and W_n have the same definition.)

$$P_n = T_n + T_{n-1} \tag{16}$$

$$W_n = T_n + T_{n-1} + T_{n-2} \tag{17}$$

$$R_n = T_n + 2T_{n-1} + 3T_{n-2} + 4T_{n-3} \tag{18}$$

$$S_n = 4T_{n+1} - T_n \tag{19}$$

$$U_n = 4T_n - 2T_{n-1} + 3T_{n-2} \tag{20}$$

$$V_n = 8T_{n+1} - 4T_n - 5T_{n-1} \tag{21}$$

$$T_n^2 - T_{n-1}^2 = P_n W_{n-2} \tag{22}$$

$$2(T_{n-1}^2 + T_n^2) = P_n^2 + W_{n-2}^2 \tag{23}$$

$$W_{n-2}P_n + W_{n-3}P_{n-1} = T_n^2 - T_{n-2}^2 \tag{24}$$

$$\sum_{i=1}^{n} W_i = T_{n+2} - 1 \tag{25}$$

$$\sum_{i=1}^{n} T_i = \frac{1}{3}(T_{n+2} + 2T_n + T_{n-1} - 1) \tag{26}$$

$$\sum_{i=1}^{n} P_i = \frac{1}{3}(2T_{n+2} + T_n + 2T_{n-1} - 2) \tag{27}$$

$$\sum_{i=0}^{n} S_i = \frac{1}{3}(4T_{n+3} - T_{n+2} + 8T_{n+1} + 2T_n - T_{n-1} - 3) \tag{28}$$

REFERENCES

[1]   Bicknell, M. and Hoggatt, V. E. Jr., eds., A Primer for the Fibonacci Numbers, Santa Clara, CA, The Fibonacci Association, 1972, p. 45, B-10.

[2]   Bruce, Ian, "A Modified Tribonacci Sequence," The Fibonacci Quarterly 22, No. 3 (1984): pp 244-246.

[3]   Dickson, L.E., First Course in the Theory of Equations, Chicago. 1921. (Chinese Translation)

[4]   Lin, Pin-Yen, "De Moivre-Type Identities for the Tribonacci Numbers," The Fibonacci Quarterly 26, No. 2 (1988): pp. 131-134.

[5]   Spickerman, W. R., "Binet's Formula for the Tribonacci Sequence," The Fibonacci Quarterly 20, No. 2 (1982): pp. 118-120.

[6]   Spickerman, W. R. and Joyner, R. N., "Binet's Formula for the Recursive Sequence of Order K," The Fibonacci Quarterly 21, No. 4 (1984): pp. 327-331.

TWO GENERALIZATIONS OF GOULD'S STAR OF DAVID THEOREM

Calvin Long and Shiro Ando

1. INTRODUCTION

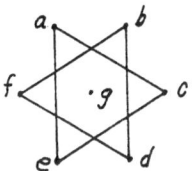

Let $a, b, c, d, e,$ and f denote the hexagon of elements immediately surrounding any given element g in Pascal's triangle. Since the first paper by Hoggatt and Hansel [8] showing that $ace = bdf$ and hence that $abcdef = k^2$ for some integer k, a spate of papers examining the fascinating properties of these arrays and their generalizations has appeared. Among the more surprising of these is the GCD Star of David theorem that $(a, c, e) = (b, d, f)$, conjectured by H. W. Gould [3] and proved and/or generalized by Hillman and Hoggatt [4] and [5], Strauss [13], Singmaster [12], Hitotumatu and Sato [6] and [7], Sato [10] and [11], and Ando and Sato [1] and [2]. In this paper we further generalize the result along lines suggested by Long and Hoggatt [9, p 78] by showing that

$$(a_1,\ a_3,\ \cdots,\ a_{17}) = (a_2,\ a_4, \cdots,\ a_{18})$$

where $a_1,\ a_2,\ \cdots,\ a_{18}$ are the eighteen adjacent binomial coefficients in the regular hexagon of coefficients centered on any particular coefficient $\binom{n}{r}$ and that

$$(b_1,\ b_3,\ \cdots,\ b_{11}) = t \cdot (b_2,\ b_4,\ \cdots,\ b_{12})$$

G. E. Bergum et al. (eds.), Applications of Fibonacci Numbers Volume 4, 219–230.

where b_1, b_2, $\cdots$, b_{12} are the twelve adjacent binomial coefficients in the regular hexagon centered at $\binom{n}{r}$ with $t = 1$ if r or $n - r = s$ is even, $t = 2$ if r and s are odd and $r \equiv 3(\bmod\ 4)$ or $s \equiv 3(\bmod\ 4)$, and $t = 4$ if $r \equiv s \equiv 1(\bmod\ 4)$.

In fact, these theorems are special cases of more general conjectures which we are, as yet, unable to prove. In particular, if a_1, a_2, $\cdots$, a_{2m} are the coefficients in a regular hexagon of binomial coefficients with edges along the rows and main diagonals of Pascal's triangle and with an even number of coefficients per edge, we conjecture that

$$(a_1, a_3, \cdots, a_{2m-1}) = (a_2, a_4, \cdots, a_{2m}).$$

For regular hexagons with an odd number of coefficients per side, we conjecture that

$$(a_1, a_3, \cdots, a_{2m-1}) = t \cdot (a_2, a_4, \cdots, a_{2m}).$$

where t depends only on m and is independent of the location of the hexagon in Pascal's triangle.

2. NOTATION AND PRELIMINARY RESULTS

Throughout the paper small Latin letters will always denote integers. Let $r + s = n$ as above, with $r \geq 3$ and $s \geq 3$. Set $A = \binom{n}{r}$ and, for simplicity, set

$$(h, k) = \binom{n + h + k}{r + h}.$$

Define the distance, $d(A, B)$, between A and the coefficient $B = \binom{n'}{r'}$ by

$$d(A, B) = \tfrac{1}{2}(|n - n'| + |r - r'| + |s - s'|)$$

where $r' + s' = n'$.

Let p be a prime. For any rational number α there exists a unique integer $v = v(\alpha)$ such that $\alpha = p^v a/b$ where $(a, p) = (b, p) = 1$. If $v(n) = e$ then $p^e||n; i.e.,\ p^e|n$ and $p^{e+1} \nmid n$. Moreover, it is clear that

$$v(1) = 0, \tag{1}$$
$$v(\alpha\beta) = v(\alpha) + v(\beta), \tag{2}$$
$$v(\alpha/\beta) = v(\alpha) - v(\beta), \tag{3}$$
$$v(\alpha \pm \beta) \geq \min(v(\alpha), v(\beta))\ \forall \alpha,\beta, \tag{4}$$
$$v(\alpha \pm \beta) = \min(v(\alpha), v(\beta))\ \text{if}\ v(\alpha) \neq v(\beta). \tag{5}$$

Finally, if $m = m_1 m_2 \cdots m_k$, then

$$(m_1, m_2, \cdots, m_k) = \prod_{p \mid m} p^{\min(v(m_1), \cdots, v(m_k))} . \tag{6}$$

3. THE MAIN RESULTS

The 18 binomial coefficients at distance 3 from A form a regular hexagon with four entries per side and centered at A. Taking these entries alternately, we form two sets $S_1 = S_1(n, r, s)$ and $S_2 = S_2(n, r, s)$ as follows:

$S_1 = S_1(n, r, s)$	$S_2 = S_2(n, r, s)$
$(-3, 0) = \frac{r(r-1)(r-2)}{n(n-1)(n-2)} A$	$(-2, -1) = \frac{r(r-1)s}{n(n-1)(n-2)} A$
$(-1, -2) = \frac{rs(s-1)}{n(n-1)(n-2)} A$	$(0, -3) = \frac{s(s-1)(s-2)}{n(n-1)(n-2)} A$
$(1, -3) = \frac{s(s-1)(s-2)}{n(n-1)(r+1)} A$	$(2, -3) = \frac{s(s-1)(s-2)}{n(r+1)(r+2)} A$
$(3, -3) = \frac{s(s-1)(s-2)}{(r+1)(r+2)(r+3)} A$	$(3, -2) = \frac{s(s-1)(n+1)}{(r+1)(r+2)(r+3)} A$
$(3, -1) = \frac{s(n+1)(n+2)}{(r+1)(r+2)(r+3)} A$	$(3, 0) = \frac{(n+1)(n+2)(n+3)}{(r+1)(r+2)(r+3)} A$
$(2, 1) = \frac{(n+1)(n+2)(n+3)}{(r+1)(r+2)(s+1)} A$	$(1, 2) = \frac{(n+1)(n+2)(n+3)}{(r+1)(s+1)(s+2)} A$
$(0, 3) = \frac{(n+1)(n+2)(n+3)}{(s+1)(s+2)(s+3)} A$	$(-1, 3) = \frac{r(n+1)(n+2)}{(s+1)(s+2)(s+3)} A$
$(-2, 3) = \frac{r(r-1)(n+1)}{(s+1)(s+2)(s+3)} A$	$(-3, 3) = \frac{r(r-1)(r-2)}{(s+1)(s+2)(s+3)} A$
$(-3, 2) = \frac{r(r-1)(r-2)}{n(s+1)(s+2)} A$	$(-3, 1) = \frac{r(r-1)(r-2)}{n(n-1)(s+1)} A$

It is easy to see that the product of the elements in S_1 is equal to the product of the elements in S_2. Moreover, if we let $gcdS_i$ and $lcmS_i$ respectively denote the greatest common divisor and the least common multiple of the elements in S_i for each i, we have the following.

Theorem 1: Let $n, r, s, A, S_1,$ and S_2 be as above. Then

$$gcd S_1 = gcd S_2. \tag{7}$$

On the other hard

$$\ell cm S_1 = \ell cm S_2 \tag{8}$$

does not necessarily hold.

Proof: That (8) does not necessarily hold is most easily shown by a counter example. For $n = 7$, $r = 3$, and $s = 4$, we note that $\ell cm S_1 = 2^3 \cdot 3^2 \cdot 5 \cdot 7 = 2520$ and $\ell cm S_2 = 1260$.

We now turn to the proof of (7). Let p be any prime and for convenience set $v((a,b)) = v(a,b)$. Also, set

$$v_i = v_i(p) = \min_{(a,b) \in S_i} \{v(a,b)\}.$$

Clearly, we must show that $v_1 = v_2$ for all p. To do this, we show that $v_1 \geq v_2$. Then, since $S_1(n,r,s) = S_2(n,s,r)$, we have by symmetry that $v_2 \geq v_1$ and hence that $v_1 = v_2$ as required. For each element $(a,b) \in S_1$, we show that if $v_1 = v(a,b)$ then $v_1 \geq v_2$. Actually it suffices to show this only for $(a,b) = (-3, 0)$, $(-1, -2)$, and $(1, -3)$ since the cyclic permutation $(r,s,-n-1)$ carries each element in the hexagon into the element (or its negative) located 120° around the hexagon. Thus, $(r,s,-n-1)$, applied repeatedly, carries the first three elements of S_1 into the second three, the second three into the third three, and the third three back into the first three; and the same is true of S_2. In particular, the same argument that proves the result for $(a,b) = (-3, 0)$ can be repeated with r replaced by s, s by $-n-1$, and $-n-1$ by r to give the proof for $(a,b) = (3, -3)$, and so on.

For $(a,b) \in S_i$, let $u((a,b)) = u(a,b) = v(a,b) - v(A)$ and let $u_i = v_i - v(A)$ for each i. Assume that $v_1 < v_2$ or, equivalently, that $u_1 < u_2$.

Case 1: $v_1 = v(-3, 0) = v\left(\dfrac{r(r-1)(r-2)}{n(n-1)(n-2)} A\right)$

Since $(-3, 1) \in S_2$, we have by assumption that

$$u(-3, 1) = v\left(\frac{r(r-1)(r-2)}{n(n-1)(s+1)}\right) > u_1 = v\left(\frac{r(r-1)(r-2)}{n(n-1)(n-2)}\right).$$

Hence, $0 \leq v(s+1) < v(n-2)$ and $p \mid (n-2)$. If $v(s) \leq v(r-2)$, then

$$u(-2, -1) = v\left(\frac{r(r-1)s}{n(n-1)(n-2)}\right) \leq v\left(\frac{r(r-1)(r-2)}{n(n-1)(n-2)}\right) = u_1$$

contrary to assumption since $(-2, -1) \in S_2$. Hence, $v(s) > v(r-2) \geq 0$ and $v(r-2) = v(n-2)$ by (5) since $s+r-2 = n-2$. All this implies that $v(s) \geq 2$ and that $u_1 = v(r/n)$ since $p \nmid (n-1)(r-1)$. Now assume that p is odd. Then we also have that $p \nmid rn$ and hence $u_1 = 0$. Since $(-1, 3) \in S_2$,

$$v\left(\frac{r(n+1)(n+2)}{(s+1)(s+2)(s+3)}\right) \geq u_2 > u_1 = 0. \tag{9}$$

Since $v(s) > 0$ and p is odd, $p \nmid (s+1)(s+2)$. Similarly, $p \nmid r(n+2)$. Hence, (9) implies that $v(n+1) > v(s+3)$. But then, by (5), $v(s+3) = v(r-2)$ and it follows that

$$u(-3, 3) = v\left(\frac{r(r-1)(r-2)}{(s+1)(s+2)(s+3)}\right) = 0 = u_1$$

contrary to assumption since $(-3, 3) \in S_2$. For $p = 2$, it follows from above that r, s, and n must all be even. If $v(n) > v(r)$, then (5) implies that $v(r) = v(s) \geq 2$. Hence, $v(r-2) = v(s-2) = 1$ since every second even integer is divisible by at most 2^1. But then

$$u(0, -3) = v\left(\frac{s(s-1)(s-2)}{n(n-1)(n-2)}\right) = v\left(\frac{r(r-1)(r-2)}{n(n-1)(n-2)}\right) = u(-3, 0) = u_1$$

which is again a contradiction since $(0, -3) \in S_2$. Therefore, $v(n) \leq v(r)$ and

$$u_1 = v\left(\frac{r(r-1)(r-2)}{n(n-1)(n-2)}\right) \geq 0 \tag{10}$$

since $v(r-2) = v(n-2)$ from above. If $v(s) \leq v(r+2)$, then $r+1$ and $r+3$ are both odd and

$$u(3, -2) = v\left(\frac{s(s-1)(n+1)}{(r+1)(r+2)(r+3)}\right) \leq 0.$$

But, since $(3, -2) \in S_2$, this with (10) implies $u_2 \leq u_1$ which is again a contradiction. Thus, $v(s) > v(r+2)$ and so $v(r+2) = v(n+2)$ by (5). Hence,

$$u(3, 0) = v\left(\frac{(n+1)(n+2)(n+3)}{(r+1)(r+2)(r+3)}\right) = 0 \leq u_1$$

since $(n+1)$, $(n+3)$, $(r+1)$, and $(r+3)$ are all odd, and this is again a contradiction. Thus,

$v_1 \geq v_2$ for all p and the proof for Case 1 is complete.

Case 2: $v_1 = v(-1, -2) = v\left(\dfrac{rs(s-1)}{n(n-1)(n-2)} A\right)$

We proceed somewhat more briefly. If $v(s-1) \geq v(r-1)$, then

$$u(-2, -1) = v\left(\frac{r(r-1)s}{n(n-1)(n-2)}\right)$$

$$\leq v\left(\frac{rs(s-1)}{n(n-1)(n-2)}\right) = u(-1, -2) = u_1$$

which is a contradiction since $(-2, -1) \in S_2$. Therefore, $v(s-1) < v(r-1)$ and hence $p \mid (r-1)$. Similar consideration of $(0, -3)$ leads to the inequality $v(r) < v(s-2)$ so that $p \mid (s-2)$. Hence, it follows that $p \mid (n-3)$. Now suppose that p is odd. Then $p \nmid r(r+1)(r+3)(n-1)(n-2)(n+1)$ and assuming that $v(s-2) \leq v(r+2)$ leads to the contradiction $u(2, -3) \leq u_1$. Hence, $v(s-2) > v(r+2)$ and it follows from (5) that $v(r+2) = v(n)$. But this implies that $u(3, -2) = u_1$ which is again a contradiction. If $p = 2$, then r and n are odd and s is even. But then the assumption $v(r-1) \leq v(s)$ implies the contradiction $u(-3, 1) \leq u_1$. Hence, $v(r-1) > v(s)$, $v(s) = v(n-1)$ by (5), and $u_1 = 0$. Now the assumption $v(n+1) \leq v(s+2)$ implies the contradiction $u(-1,3) \leq 0 = u_1$ and so $v(n+1) > v(s+2)$ and hence $v(s+2) = v(r-1)$ by (5). But this leads to the contradiction $u(-3, 3) = 0 = u_1$. Thus, finally, $v_1 < v_2$ is false and $v_1 \geq v_2$ for all p in this case as required.

Case 3: $v_1 = v(1, -3) = v\left(\dfrac{s(s-1)(s-2)}{n(n-1)(r+1)} A\right)$

By assumption, $u(0, -3) > u(1, -3)$ and $u(2, -3) > u(1, -3)$. But these imply respectively that $v(r+1) > v(n-2)$ and $v(r+2) < v(n-1)$. In turn, these imply that $p \mid (r+1)$, $p \mid (n-1)$ and hence $p \mid (s-2)$. If $v(s-2) \geq v(r+1)$, it follows that $u_1 = u(1, -3) \geq u(-1, -2) = u_1$ since $u_1 = \min_{(a,b) \in S_1}\{u(a,b)\}$. Hence, this reduces to Case 2. If $v(s-2) < v(r+1)$, then $v(s-2) = v(n-1)$ by (5). Also, $u(3, -2) > u_1$ implies that $v(n+1) > v(r+3)$ so that $p \mid (n+1)$. But since $p \mid (n-1)$, this implies that $p = 2$. Therefore, r and n must be odd and s even. But then $v(s-2) > 0$ and hence, from above, $v(r+1) \geq 2$. Similarly $2 \mid (r+3)$ and so $v(n+1) \geq 2$. But these last two inequalities imply that

$$v(r-1) = v(r+3) = v(n-1) = v(n+3) = 1. \tag{11}$$

If $v(s) \geq v(r+1)$, then, from above, $u_1 \geq 0$ and, from (11), $u(-3, 1) = 0$. But this contradicts the initial assumption that $v_1 < v_2$ and hence $u_1 < u_2$. Therefore, $v(s) < v(r+1)$ so that $v(s) = v(n+1)$ by (5). But then, by (11),

$$u(3, 0) = v\left(\frac{n+1}{r+1}\right) = v\left(\frac{s}{r+1}\right) = u(1, -3) = u_1$$

since, from above, $v(s-2) = v(n-1)$. This is again a contradiction and so $v_1 \geq v_2$ in this case as well. This completes the proof.

Now consider the 12 binomial coefficients at distance 2 from $A = \binom{n}{r}$. These form a regular hexagon with three entries per side and centered at A. Taking these elements alternately, we form two sets $T_1 = T_1(n,r,s)$ and $T_2 = T(n,r,s)$ as follows:

$T_1 = T_1(n,r,s)$	$T_2 = T_2(n,r,s)$
$(-2,0) = \frac{r(r-1)}{n(n-1)} A$	$(-1, -1) = \frac{rs}{n(n-1)} A$
$(0, -2) = \frac{s(s-1)}{n(n-1)} A$	$(1, -2) = \frac{s(s-1)}{n(r+1)} A$
$(2, -2) = \frac{s(s-1)}{(r+1)(r+2)} A$	$(2, -1) = \frac{s(n+1)}{(r+1)(r+2)} A$
$(2, 0) = \frac{(n+1)(n+2)}{(r+1)(r+2)} A$	$(1,1) = \frac{(n+1)(n+2)}{(r+1)(s+1)} A$
$(0, 2) = \frac{(n+1)(n+2)}{(s+1)(s+2)} A$	$(-1,2) = \frac{(n+1)r}{(s+1)(s+2)} A$
$(-2, 2) = \frac{r(r-1)}{(s+1)(s+2)} A$	$(-2,1) = \frac{r(r-1)}{n(s+1)} A$

For this decomposition, it is clear that the products of the elements in the two sets are not the same. However, $gcdT_1$ is sometimes equal to $gcdT_2$ though not always. In fact, we have the following theorem.

Theorem 2: For T_1 and T_2 as just described

$$gcdT_1 = t \cdot gcdT_2 \tag{12}$$

where $t = 1$ if r or s is even, $t = 2$ if r and s are odd and $r \equiv 3(\text{mod } 4)$ or $s \equiv 3(\text{mod } 4)$, and $t = 4$ if $r \equiv s \equiv 1(\text{mod } 4)$.

Example 1: For $r = s = 2$ and $n = 4$, $T_1 = \{1, 1, 1, 15, 15, 1\}$ and $T_2 = \{2, 1, 5, 20, 5, 1\}$. Thus, $gcd T_1 = gcd T_2 = 1$ and $t = 1$.

Example 2: For $r = 3$, $s = 5$ and $n = 8$, $T_1 = \{6, 20, 56, 252, 120, 8\}$ and $T_2 = \{15, 35, 126, 210, 36, 7\}$. Thus, $gcd T_1 = 2$, $gcd T_2 = 1$, and $t = 2$.

Example 3: For $r = s = 5$ and $n = 10$, $T_1 = \{56, 56, 120, 792, 792, 120\}$ and $T_2 = \{70, 84, 330, 924, 330, 84\}$ Thus, $gcd T_1 = 8$, $gcd T_2 = 2$, and $t = 4$.

Proof of Theorem 2: We use the same notation as in Theorem 1 and we begin by showing that $u_1 \geq u_2$ and hence $v_1 \geq v_2$. This need be done only for the case $u_1 = u(-2, 0)$ since the case $u_1 = u(0, -2)$ follows by symmetry and the other cases all follow by 120° rotations.

Assume that $u_1 < u_2$. Then $u(-1, -1) > u(-2, 0)$ and $u(-2, 1) > u(-2, 0)$, and these imply that $v(s) > v(r-1)$ and $v(s+1) < v(n-1)$. Using (5), we have from the first of these inequalities that $v(n-1) = v(r-1)$ and hence that

$$v(s) > v(r-1) = v(n-1) > v(s+1) \geq 0. \tag{13}$$

Thus,

$$p \mid (n-1), p \mid (r-1), p \mid s, \text{ and } p \nmid rn(s-1)(s+1) \tag{14}$$

From (13) and (14), it follows that

$$u_1 = v\left(\frac{r(r-1)}{n(n-1)}\right) = 0. \tag{15}$$

For p odd, we further have that $p \nmid (n+1)(s+2)$ and this, with (14), implies that $u(-1, 2) = 0$ in contradiction to the assumption that $u_1 < u_2$. For $p = 2$, the inequality $v(r+1) \geq v(s)$ together with (14) implies that $u(1, -2) \leq 0$ again in contradiction to $u_1 < u_2$. Therefore, $v(r+1) < v(s)$ so $v(r+1) = v(n+1)$ by (5). But then $u(1, 1) = 0$ since $2 \nmid n$ implies that $2 \nmid (n+2)$ and $2 \mid s$ implies that $2 \nmid (s+1)$. But this again contradicts $u_1 < u_2$. Therefore, $u_1 \geq u_2$ and so $v_1 \geq v_2$ for all p.

Now assume that $v_2 < v_1$ and hence $u_2 < u_1$. This time we have to check both $u_2 = u(-1, -1)$ and $u_2 = u(1, -2)$ since we cannot use symmetry. However, the other cases will follow from these two by 120° rotations.

Case 1: $u_2 = u(-1, -1) = v\left(\frac{rs}{n(n-1)}\right)$, p odd

From $u_2 < u(-2, 0)$ and $u_2 < u(0, -2)$ respectively, we obtain $v(s) < v(r-1)$ and $v(r) < v(s-1)$. Thus $p \mid (r-1), p \mid (s-1), p \mid (n-2)$, and hence

$$p \nmid rsn(r+1)(n-1)(n+2).$$ (16)

If $v(s-1) \leq v(r+2)$, it follows from (16) that

$$u_1 \leq u(2, -2) = v\left(\frac{s(s-1)}{(r+1)(r+2)}\right) \leq 0 = u_2$$

in contradiction to the assumption that $u_2 < u_1$. Therefore, $v(s-1) > v(r+2)$ and so $v(r+2) = v(n+1)$ by (5). But again, this implies the contradiction

$$u_1 \leq u(2, 0) = v\left(\frac{(n+1)(n+2)}{(r+1)(r+2)}\right) = 0 = u_2.$$

Therefore, in this case $u_1 \leq u_2$ and $v_1 \leq v_2$ as desired.

Case 2: $u_2 = u(1, -2) = v\left(\frac{s(s-1)}{n(r+1)}\right)$, p odd

From $u_2 < u(0, -2)$ and $u_2 < u(2, -2)$ respectively, it follows that $v(n-1) < v(r+1)$ and $v(r+2) < v(n)$. Thus, $p \mid (r+1), p \mid n, p \mid (s-1)$, and hence

$$p \nmid rs(r-1)(r+2)(n-1)(n+1)(n+2).$$ (17)

If $v(r+1) \leq v(s-1)$, it follows from (17) that

$$u_1 \leq u(-2,0) = v\left(\frac{r(r-1)}{n(n-1)}\right) \leq v\left(\frac{s(s-1)}{n(r+1)}\right) = u_2$$

in contradiction to the assumption $u_2 < u_1$. Therefore, $v(r+1) > v(s-1)$ and so $v(s-1) = v(n)$ by (5). But then

$$u_1 \leq u(2, 0) = v\left(\frac{(n+1)(n+2)}{(r+1)(r+2)}\right) = v\left(\frac{s(s-1)}{n(r+1)}\right) = u_2,$$

again a contradiction. Therefore, $u_2 \geq u_1$ and $v_2 \geq v_1$ in this case as well, and hence for all odd p.

Case 3: $p = 2$

Suppose first that $u_2 = u(-1, -1)$. From $u_2 < u(0, -2)$ and $u_2 < u(-2,0)$ respectively, we obtain $v(r) < v(s-1)$ and $v(s) < v(r-1)$. Thus, $2 \mid (s-1), 2 \mid (r-1), 2 \mid (n-2)$, and hence r and s are odd and n is even. Similarly, if $u_2 = u(1, -2)$, the inequalities $u_2 < u(0, -2)$ and $u_2 < u(2, -2)$ respectively imply that $v(n-1) < v(r+1)$ and $v(r+2) < v(n)$.

Thus, $2 \mid (r+1), 2 \mid n, 2 \mid (s-1)$ and so, again, r and s are odd and n is even. Since these are the only two cases that need to be considered, it follows that if either r or s is even then $u_1 \leq u_2$ and hence $u_1 = u_2$ and $v_1 = v_2$. Thus, in this case, equality holds for all p and $gcdT_2 = t \cdot gcdT_2$ with $t = 1$ as claimed.

If r and s are both odd and $r \equiv 3 \pmod{4}$ or $s \equiv 3 \pmod{4}$, then, since $(r+1)+(s+1) = n+2$, one of the following must hold.

$$r \equiv s \equiv 3 \pmod{4}, \ n \equiv 2 \pmod{4} \tag{18}$$
$$r \equiv 3 \pmod{4}, \ s \equiv 1 \pmod{4}, \ n \equiv 0 \pmod{4} \tag{19}$$
$$r \equiv 1 \pmod{4}, \ s \equiv 3 \pmod{4}, \ n \equiv 0 \pmod{4} \tag{20}$$

Since all lead to the same conclusion, we consider only (18). The argument for the other two cases is the same and is omitted. From (18) it follows that the values of u for T_1 and T_2 are as follows.

T_1	T_2
$u(-2,0) = 1 - v(n)$	$u(-1,-1) = -v(n)$
$u(0,-2) = 1 - v(n)$	$u(1,-2) = -v(r+1)$
$u(2,-2) = 1 - v(r+1)$	$u(2,-1) = -v(r+1)$
$u(2,0) > 1 - v(r+1)$	$u(1,1) = v(n+2) - v(r+1) - v(s+1)$
$u(0,2) > 1 - v(s+1)$	$u(-1,2) = -v(s+1)$
$u(-2,-2) = 1 - v(s+1)$	$u(-2,1) = -v(s+1)$

Moreover, since $n + 2 = (r+1) + (s+1)$, it follows from (4) that $v(n+2) \geq \min\{v(r+1), v(s+1)\}$ and hence that $u(1, 1) = v(n+2) - v(r+1) - v(s+1) \geq -\max\{v(r+1), v(s+1)\}$. Therefore, setting $\bar{v} = \max\{v(r+1), v(s+1), v(n)\}$, we have that $u_1 = 1 - \bar{v}$ and $u_2 = -\bar{v}$. Thus, $u_1 = 1 + u_2$, $v_1 = 1 + v_2$, and, since $v_1(p) = v_2(p)$ for all odd p, $gcdT_1 = t \cdot gcdT_2$ with $t = 2$ as claimed.

Finally, if $r \equiv s \equiv 1(\text{mod } 4)$, then $n \equiv 2(\text{mod } 4)$ and we show, as in the preceding paragraph, that $u_1 = 1$, $u_2 = -1$ and $v_1 = v_2 + 2$. Then, since $v_1 = v_2$ for all odd p, it follows that $gcdT_1 = t \cdot gcdT_2$ with $t = 4$ and this completes the proof.

Finally, note that, for $n = 6$, $r = 2$ and $s = 4$, we find that $lcmT_1 = 2^3 \cdot 3 \cdot 5 \cdot 7$ and $lcmT_2 = 2^3 \cdot 5 \cdot 7$.   Also,   for $n = 7$, $r = 3$ and $s = 4$, $lcmT_1 = 2^6 \cdot 3^2 \cdot 5 \cdot 7$ and $lcmT_2 = 2^3 \cdot 3^2 \cdot 5 \cdot 7$. Thus, no result corresponding to Theorem 2 appears to hold for least common multiples.

REFERENCES

[1]   Ando, S. and Sato, D. "A GCD Property on Pascal's Pyramid and the corresponding LCM Property on the Modified Pascal Pyramid." Applications of Fibonacci Numbers, Edited by G.E. Bergum, A.N. Philippou and A.F. Horadam, Kluwer Academic Publishers, (1988) pp. 7-14.

[2]   Ando, S. and Sato, D. "Translatable and Rotatable Configurations Which Give Equal Product, Equal GCD and Equal LCM Properties Simultaneously." Applications of Fibonacci Numbers, Edited by G.E. Bergum, A.N. Philippou and A.F. Horadam, Kluwer Academic Publishers, (1988) pp. 15-26.

[3]   Gould, H. W. "A New Greatest Common Divisor Property of The Binomial Coefficients." The Fibonacci Quarterly 10 (1972): pp. 579-584, 628.

[4]   Hillman, A. P. and Hoggatt, V. E. Jr. "A Proof of Gould's Pascal Hexagon Conjecture." The Fibonacci Quarterly 10 (1972): pp. 565-568, 598.

[5]   Hillman, A. P. and Hoggatt, V. E. Jr. "Exponents of Primes in Generalized Binomial Coefficients." Journal fur die Reine und Angewandte Mathematik. 262-263 (1973): pp. 375-380.

[6]   Hitotumatu, S. and Sato, D. "Star of David Theorem (1)." The Fibonacci Quarterly 13 (1975): p. 70.

[7]   Hitotumatu, S. and Sato, D. "Expansion of the Star of David Theorem." Preliminary report (II). Notices of the A. M. S. 22 (1975): p. A-377.

[8]   Hoggatt, V. E. Jr. and Hansell, W. "The Hidden Hexagon Squares." The Fibonacci Quarterly 9, (1971): pp. 120, 133.

[9]   Long, C. T. and Hoggatt, V. E. Jr. "Sets of Binomial Coefficients with Equal Products." The Fibonacci Quarterly 12 (1974): pp. 71-79.

[10]   Sato, D. "An Algorithm to Expand the Star of David Theorem (GCD Properties of
        Binomial Coefficients)." Preliminary report. *Notices of the A.M.S. 22* (1975): p. A-296.

[11]   Sato, D. "Expansion of the Star of David Theorem of H. W. Gould and David
        Singmaster." Preliminary report (I), *Notices of the A.M.S. 22* (1975): p. A-377.

[12]   Singmaster, D. "Notes on Binomial Coefficients: IV-Proof of a Conjecture of Gould on
        the GCD's of Two Triples of Binomial Coefficients." *The Fibonacci Quarterly 11* (1973):
        pp. 282-284.

[13]   Straus, E. G. "On the Greatest Common Divisor of Some Binomial Coefficients." *The
        Fibonacci Quarterly 11* (1973): pp. 25-26.

ON TRIANGULAR LUCAS NUMBERS

Ming Luo

1. INTRODUCTION

In the paper [3], we have proved that the only triangular numbers (i.e., the positive integers of the form $\frac{1}{2}m(m+1)$) in the Fibonacci sequence

$$u_{n+2} = u_{n+1} + u_n, \ u_0=0, \ u_1=1$$

are $u_{\pm 1}=u_2=1$, $u_4=3$, $u_8=21$ and $u_{10}=55$. This verifies a conjecture of Vern Hoggatt [2]. In this paper we shall find all triangular numbers in the Lucas sequence

$$v_{n+2} = v_{n+1} + v_n, \ v_0=2, \ v_1=1,$$

where n ranges over all integers.

Theorem 1: The complete set of triangular Lucas numbers is $v_1=1$, $v_{\pm 2}=3$ and $v_{\pm 18}=5778$.

The method used is as in [3]. It is sufficient to find all n's such that $8v_n+1$ is a perfect square. In section 2 we shall prove that there is exactly one value in each of the five residue classes $n \equiv 1$, ± 2, ± 18 (mod 8800) such that $8v_n+1$ is a perfect square, which are $n=1$, ± 2, ± 18. To do this we use a Jacobi symbol criterion established in [3], which is as follows:

G. E. Bergum et al. (eds.), Applications of Fibonacci Numbers Volume 4, 231–240.
© 1991 *Kluwer Academic Publishers.*

<u>Criterion:</u> If a, n are positive integers such that $n \equiv \pm 2 \pmod 6$, $(a, v_n)=1$, then

$$\left(\frac{\pm 4au_{2n} + 1}{v_{2n}}\right) = -\left(\frac{8au_n \pm v_n}{64a^2 + 5}\right).$$

In section 3 we use congruence calculations to prove that for all other residue classes of n modulo 8800 $8v_n+1$ cannot be a perfect square. Thus we conclude that $v_1=1$, $v_{\pm 2}=3$ and $v_{\pm 18}=5778$ are the only triangular Lucas numbers.

Moreover, since $x=\pm u_n$, $y=\pm v_n$ are the complete set of solutions of the Diophantine equations $5x^2 - y^2 = \pm 4$, the condition $v_n=\frac{1}{2}z(z+1)$ is equivalent to finding all integer solutions of the two Diophantine equations $20x^2 - z^2(z+1)^2 = \pm 16$, i.e., finding all integer points on these two elliptic curves. Hence we obtain:

<u>Theorem 2:</u> (a) The Diophantine equation $20x^2 - y^2(y+1)^2=16$ has only the integer solutions $(\pm x, y)=(1, -2)$, $(1, 1)$.

(b) The Diophantine equation $20x^2 - y^2(y+1)^2= -16$ has only the integer solutions $(\pm x, y)=(1, -3)$, $(1, 2)$, $(2584, -108)$, $(2584, 107)$.

2. SOME LEMMAS OBTAINED BY THE CRITERION

We use the following well known formulae about Fibonacci and Lucas numbers:

$$v_{-n} = (-1)^n v_n; \tag{1}$$
$$2v_{m+n} = 5u_m u_n + v_m v_n; \tag{2}$$
$$v_n^2 - 5u_n^2 = 4(-1)^n; \tag{3}$$
$$v_{2kt+n} \equiv (-1)^t v_n \pmod{v_k}; \tag{4}$$

(see [1]), where n, m, t denote integers and $k \equiv \pm 2 \pmod 6$.

<u>Lemma 1:</u> If $n \equiv 1 \pmod{2^4 \cdot 5 \cdot 11}$, then $8v_n+1$ is a perfect square only for $n=1$.

<u>Proof:</u> If $n \neq 1$, put $n=1+2 \cdot 3^r \cdot 5 \cdot 11m$, where $r \geq 0$, $3 \nmid m$ and $m \equiv \pm 8 \pmod{24}$ which are equivalent to $m \equiv 8, 16, 32, 40, 56, 64 \pmod{72}$.

If $2|r$, let

$$\ell = 3^r \cdot 5 \cdot 11, \qquad k = m, \qquad \text{if } m \equiv 32, 56 \pmod{72};$$
$$\ell = 3^r \cdot 11, \qquad k = 5m, \qquad \text{if } m \equiv 40, 64 \pmod{72};$$
$$\ell = 3^r \cdot 5, \qquad k = 11m, \qquad \text{if } m \equiv 8 \pmod{72};$$
$$\ell = 3^r, \qquad k = 5 \cdot 11m, \qquad \text{if } m \equiv 16 \pmod{72};$$

if $2{\nmid}r$, let

$$\ell = 3^r \cdot 5 \cdot 11, \qquad k = m, \qquad \text{if } m \equiv 16, 40 \pmod{72};$$
$$\ell = 3^r \cdot 11, \qquad k = 5m, \qquad \text{if } m \equiv 8, 32 \pmod{72};$$
$$\ell = 3^r \cdot 5, \qquad k = 11m, \qquad \text{if } m \equiv 64 \pmod{72};$$
$$\ell = 3^r, \qquad k = 5 \cdot 11m, \qquad \text{if } m \equiv 56 \pmod{72}.$$

These permit us to write $n = 1 + 2\ell k$ such that one of the following cases holds:

Case 1: $\ell \equiv 1 \pmod{4}$ and $k \equiv 16$ or $40 \pmod{72}$;

Case 2: $\ell \equiv -1 \pmod{4}$ and $k \equiv 32$ or $56 \pmod{72}$.

In the first case, using (4) and (2), we obtain

$$8v_n + 1 \equiv 8v_{2k+1} + 1 \equiv 4(5u_{2k}u_1 + v_{2k}v_1) + 1 \equiv 20u_{2k} + 1 \pmod{v_{2k}}.$$

Using the Criterion, we get

$$\left(\frac{8v_n+1}{v_{2k}}\right) = \left(\frac{20u_{2k}+1}{v_{2k}}\right) = -\left(\frac{40u_k+v_k}{5 \cdot 321}\right).$$

Since $2|k$ implies $v_k \equiv \pm 2 \pmod{5}$, so we further get

$$\left(\frac{8v_n+1}{v_{2k}}\right) = \left(\frac{40u_k+v_k}{321}\right).$$

Take modulo 321 to the sequence $\{40u_k + v_k\}$, we obtain its residue sequence with period 72, and $k \equiv 16, 40 \pmod{72}$ imply $40u_k + v_k \equiv 278, 194 \pmod{321}$ respectively. Since

$$\left(\frac{278}{321}\right) = \left(\frac{194}{321}\right) = -1,$$

so that $8v_n + 1$ is not a perfect square in this case.

For the second case, similarly we have

$$8v_n + 1 = -8v_{2k+1} + 1 \equiv -4(5u_{2k}u_1 + v_{2k}v_1) + 1 \equiv -20u_{2k} + 1 \pmod{v_{2k}}$$

and, by the Criterion,

$$\left(\frac{8v_n+1}{v_{2k}}\right) = \left(\frac{-20u_{2k}+1}{v_{2k}}\right) = -\left(\frac{40u_k - v_k}{5 \cdot 321}\right) = \left(\frac{40u_k - v_k}{321}\right).$$

The residue sequence of $\{40u_k - v_k\}$ modulo 321 has period 72 and $k \equiv 32, 56 \pmod{72}$ imply $40u_k - v_k \equiv 127, 43 \pmod{321}$ respectively. Since

$$\left(\frac{127}{321}\right) = \left(\frac{43}{321}\right) = -1,$$

so that $8v_n + 1$ is also not a perfect square in this case.

Finally, when $n=1$ we have $8v_n + 1 = 3^2$, which completes the proof. $\square$

Lemma 2: If $n \equiv \pm 2 \pmod{2^5}$, then $8v_n + 1$ is a perfect square only for $n = \pm 2$.

Proof: If $n \equiv 2 \pmod{2^5}$ and $n \neq 2$, then put $n = 2 + 2 \cdot 3^r \cdot k$, where $3 \nmid k$ and $k \equiv \pm 16 \pmod{48}$. Using (4) we obtain

$$8v_n + 1 \equiv -8v_2 + 1 \equiv -23 \pmod{v_k}.$$

Take modulo 23 to $\{v_k\}$ we get its residue sequence with period 48 and that both $k \equiv \pm 16 \pmod{48}$ imply $v_k \equiv -1 \pmod{23}$. Hence,

$$\left(\frac{8v_n+1}{v_k}\right) = \left(\frac{-23}{v_k}\right) = \left(\frac{v_k}{23}\right) = \left(\frac{-1}{23}\right) = -1,$$

$8v_n + 1$ is not a perfect square.

If $n \equiv -2 \pmod{2^5}$ and $n \neq -2$, then by (1), $8v_n + 1 = 8v_{-n} + 1$. Since $-n \equiv 2 \pmod{2^5}$ and $-n \neq 2$, we see, according to the argument above, $8v_n + 1$ is not a perfect square.

Finally, when $n = \pm 2$, we have $8v_n + 1 = 5^2$, which completes the proof. $\square$

Lemma 3: If $n \equiv \pm 18 \pmod{2^3 \cdot 5^2 \cdot 11}$, then $8v_n + 1$ is a perfect square only for $n = \pm 18$.

Proof: If $n \equiv 18 \pmod{2^3 \cdot 5^2 \cdot 11}$ and $n \neq 18$, put $n = 18 + 2 \cdot 3^r \cdot \ell \cdot 5^2 \cdot 11 \cdot 2^t$, where $2 \nmid \ell$, $t \geq 2$. Let $k = 2^t$, $5 \cdot 2^t$, $11 \cdot 2^t$, $5^2 \cdot 2^t$ or $5 \cdot 11 \cdot 2^t$, which will be determined later. For each case, we have, by (4),

$$8v_n + 1 \equiv -8v_{18} + 1 \equiv -46223 \pmod{v_k}.$$

Take modulo 46223 to $\{v_n\}$, our calculations show that the sequence of residues has period 5436 and when n is in one of the following thirty residue classes modulo 5436, whose set is denoted by A,

$$
\begin{aligned}
A = \quad &\{4, 32, 64, 80, 128, 304, 340, 380, 440, 512, 644, 680, \\
&760, 1024, 1280, 1364, 2276, 2432, 2576, 2740, 2888, \\
&2908, 3040, 3164, 3316, 3440, 4024, 4096, 4292, 4868\},
\end{aligned}
$$

then $\left(\dfrac{v_n}{46223}\right) = -1$, so that $8v_n+1$ is not a perfect square.

Thus it suffices to choose k such that k is congruent to one of the integers in A. Note that the residue sequence of $\{2^t\}$ modulo 5436 has period 30. If we let

$$
k = \begin{cases}
2^t & \text{if } t \equiv 2, 5, 6, 7, 9, 10, 12, 16, 19, 21, 25, 27, 28, 29 \ (\text{mod } 30) \\
5 \cdot 2^t & \text{if } t \equiv 0, 4, 8, 13, 14, 15, 17, 18, 20, 22, 24 \ (\text{mod } 30) \\
11 \cdot 2^t & \text{if } t \equiv 1, 23 \ (\text{mod } 30) \\
5^2 \cdot 2^t & \text{if } t \equiv 11 \ (\text{mod } 30) \\
5 \cdot 11 \cdot 2^t & \text{if } t \equiv 3, 26 \ (\text{mod } 30) \quad ,
\end{cases}
$$

then it is easy to check that k satisfies our requirement. So it is verified that $8v_n+1$ is not a perfect square.

If $n \equiv -18 \ (\text{mod } 2^3 \cdot 5^2 \cdot 11)$ and $n \neq -18$, then, similarly to the proof of the case $n \equiv -2 \ (\text{mod } 2^5)$, we can show that $8v_n+1$ is not a perfect square.

Finally, the cases $n = \pm 18$ give $8v_n+1 = 215^2$, which completes our proof. $\square$

The three lemmas immediately imply the following result:

Corollary 1: Assume that $n \equiv 1, \pm 2, \pm 18 \ (\text{mod } 2^5 \cdot 5^2 \cdot 11)$, then $8v_n+1$ is a perfect square only for $n=1, \pm 2, \pm 18$ respectively. $\square$

3. SOME LEMMAS OBTAINED BY CONGRUENCE CALCULATIONS

The lemmas in this section provide a system of necessary conditions for $8v_n+1$ to be a perfect square. We prove them mainly by the following process of calculation: First we study $\{8v_n+1\}$ modulo a_1. We get a sequence with period k_1 (with respect to n), in which we eliminate every residue class modulo k_1 of n, for which $8v_n+1$ is a quadratic nonresidue modulo a_1. Next we take modulo a_2 to $\{8v_n+1\}$ and get a sequence with period k_2. For our purpose,

a_2 will be chosen in such a way so that $k_1|k_2$. Then we eliminate every residue class modulo k_2 of n from those left in the preceding step for which $8v_n+1$ is a quadratic nonresidue modulo a_2. We repeat this procedure until we reach the desired results.

Remark: To obtain the period 4ℓ in the residue sequence of $\{8v_n+1\}$ modulo a, we generally take a to be a prime factor of v_ℓ according to (4), and the calculation may be carried out directly from the recurrence relation

$$8v_{n+2}+1=(8v_{n+1}+1)+(8v_n+1)-1 \ .$$

Lemma 4: If $8v_n+1$ is a perfect square, then $n \equiv 1, \ \pm 2, \ \pm 18 \pmod{2^5 \cdot 5}$.

Proof: (i) Modulo 11. The sequence of residues of $\{8v_n+1\}$ has period 10. We can eliminate $n \equiv 0, 4, 6, 7 \pmod{10}$ since they imply, respectively, $8v_n+1 \equiv 6, 2, 2, 2, \pmod{11}$, all of which are quadratic nonresidues modulo 11. So there remain $n \equiv \pm 1, \ \pm 2, 3, 5, \pmod{10}$

For brevity, in what follows we shall omit the sentences about periods, which can be inferred from the moduli whose residue classes are eliminated, e.g., mod 10 in the above step. Since the reasons for eliminations are always that $8v_n+1$ is a quadratic nonresidue modulo a prime, these sentences are also omitted.

(ii) Modulo 5. Eliminate $n \equiv 0, 3 \pmod 4$, which imply $8v_n+1 \equiv \pm 2 \pmod 5$, so there remain $n \equiv 1, \ \pm 2, 5, 9, 13, \pmod{20}$.

(iii) Modulo 41. Eliminate $n \equiv 5, 9, 13, 21, 33 \pmod{40}$, which imply, respectively, $8v_n+1 \equiv 7, 35, 28, 34, 15 \pmod{41}$, so there remain $n \equiv 1, \ \pm 2, \ \pm 18, 25, 29 \pmod{40}$.

(iv) Modulo 2161. Eliminate $n \equiv 22, 29, 38, 41, 42, 58, 65, 69 \pmod{80}$, which imply, respectively, $8v_n+1 \equiv 1319, 1593, 2138, 2154, 2138, 1319, 2055, 570 \pmod{2161}$, so there remain $n \equiv 1, \ \pm 2, \ \pm 18, 25 \pmod{80}$.

(v) Modulo 47. Eliminate $n \equiv 9, 17 \pmod{32}$, which imply, respectively, $8v_n+1 \equiv 45, 40 \pmod{47}$. This leads to eliminating $n \equiv 81, 105 \pmod{160}$, so there remain $n \equiv 1, \ \pm 2, \pm 18, 25, \pm 62, \pm 78 \pmod{160}$.

(vi) Modulo 1601. Eliminate $n \equiv 25, \ \pm 78 \pmod{160}$, which imply, respectively, $8v_n+ 1 \equiv 451, 1578 \pmod{1601}$, so there remain $n \equiv 1, \ \pm 2, \ \pm 18, \ \pm 62 \pmod{160}$.

(vii)   Now we eliminate $n \equiv \pm 62$ (mod 160) by the following calculation. Note that the cases $n \equiv \pm 62$ (mod 160) are equivalent to $n \equiv \pm 62$, ± 98 (mod 320). If $n \equiv \pm 98$ (mod 320), then let $n = \pm 98 + 320k = \pm 2 + 2(10k \pm 3) \cdot 16$, by (4), we have

$$8v_n + 1 \equiv -8v_{\pm 2} + 1 \equiv -23 \ (\text{mod } v_{16}).$$

Thus,

$$\left(\frac{8v_n + 1}{v_{16}}\right) = \left(\frac{-23}{2207}\right) = -1,$$

$8v_n + 1$ is not a perfect square in these cases.

If $n \equiv \pm 62$ (mod 320), then let $n = \pm 62 + 320k$, we obtain, by (4),

$$8v_n + 1 \equiv 8v_{\pm 62} + 1 \equiv 72497608814425 \ (\text{mod } v_{80}).$$

Note that $v_{80} = 523613963978820127 = 2207 \cdot 23725145626561$.   Our calculations show that the Jacobi symbol

$$\left(\frac{72497608814425}{23725145626561}\right) = -1,$$

so that $8v_n + 1$ is not a perfect square in these cases.   Hence we can eliminate $n \equiv \pm 62$ (mod 160) and finally there remain $n \equiv 1$, ± 2, ± 18 (mod 160), which completes our proof. $\square$

Lemma 5: If $8v_n + 1$ is a perfect square, then $n \equiv 1$, ± 2, ± 18 (mod $2^2 \cdot 5^2$).

Proof: (i) If $8v_n + 1$ is a perfect square then Lemma 4 follows that $n \equiv 1$, ± 2, ± 18 (mod 40), which are equivalent to $n \equiv 1$, 41, 81, 121, 161, ± 2, ± 18, ± 22, ± 38, ± 42, ± 58, ± 62, ± 78, ± 82, ± 98 (mod 200).

Modulo 151.   Eliminate $n \equiv 11$, 21, ± 8, ± 22 (mod 50), which imply, respectively, $8v_n + 1 \equiv 83$, 113, 75, 27 (mod 151). Thus we can eliminate $n \equiv 121$, 161, ± 22, ± 42, ± 58, ± 78 (mod 200), so there remain $n \equiv 1$, 41, 81, ± 2, ± 18, ± 38, ± 62, ± 82, ± 98 (mod 200).

(ii)   Modulo 401.   Eliminate $n \equiv 81$, ± 38, ± 82, ± 98 (mod 200), which imply, respectively, $8v_n + 1 \equiv 207$, 309, 293, 378 (mod 401).

Modulo 570601.   Eliminate $n \equiv 41$ (mod 200), which implies $8v_n + 1 \equiv -2182$ (mod 570601).

Hence there remain $n \equiv 1$, ± 2, ± 18, ± 62 (mod 200).

(iii) The cases $n \equiv \pm 62 \pmod{200}$ are equivalent to $n \equiv \pm 62, \pm 138 \pmod{400}$. By modulo 7 we eliminate $n \equiv \pm 6 \pmod{16}$, which imply $8v_n + 1 \equiv 5 \pmod 7$. So we can eliminate $n \equiv \pm 138 \pmod{400}$. Thus it suffices to eliminate $n \equiv \pm 62 \pmod{400}$.

Let $n = \pm 62 + 400k$, then, by (4), we obtain

$$8v_n + 1 \equiv 8v_{\pm 62} + 1 \equiv 72497608814425 \pmod{v_{100}}.$$

Since $v_{100} = 792070839848372253127$, our calculations show that

$$\left(\frac{72497608814425}{792070839848372253127} \right) = -1,$$

so that $8v_n + 1$ is not a perfect square in these cases.

Hence, finally there remain $n \equiv 1, \pm 2, \pm 18 \pmod{200}$, which completes our proof. □

Lemma 6: If $8v_n + 1$ is a perfect square, then $n \equiv 1, \pm 2, \pm 18 \pmod{2^3 \cdot 11}$.

Proof: From the second step of the proof of Lemma 4 it is readily seen that if $8v_n + 1$ is a perfect square then $n \equiv 1, 2 \pmod 4$, which are equivalent to $n \equiv 1, 5, 9, 13, 17, 21, 25, 29, 33, 37, 41, \pm 2, \pm 6, \pm 10, \pm 14, \pm 18, 22 \pmod{44}$.

(i) Modulo 89. Eliminate $n \equiv 9, 13, 25, 41, \pm 6, \pm 10, \pm 14, 22 \pmod{44}$, which imply, respectively, $8v_n + 1 \equiv 75, 75, 58, 58, 56, 6, 70, 74 \pmod{89}$

Modulo 199. Eliminate $n \equiv 7, 15, 21 \pmod{22}$, which imply, respectively, $8v_n + 1 \equiv 34, 167, 192 \pmod{199}$, so that we can eliminate $n \equiv 21, 29, 37 \pmod{44}$.

Hence there remain $n \equiv 1, 5, 17, 33, \pm 2, \pm 18 \pmod{44}$. These are equivalent to $n \equiv 1, 5, 17, 33, 45, 49, 61, 77, \pm 2, \pm 18, \pm 26, \pm 42 \pmod{88}$.

(ii) Modulo 3. Eliminate $n \equiv 5 \pmod 8$, which implies $8v_n + 1 \equiv 2 \pmod 3$, so we can eliminate $n \equiv 5, 45, 61, 77 \pmod{88}$.

Modulo 43. Eliminate $n \equiv 33, 49, \pm 26, \pm 42 \pmod{88}$, which imply, respectively, $8v_n + 1 \equiv 2, 42, 2, 20 \pmod{43}$.

Modulo 307. Eliminate $n \equiv 17 \pmod{88}$, which implies $8v_n + 1 \equiv 18 \pmod{307}$.

Finally, there remain $n \equiv 1, \pm 2, \pm 18 \pmod{88}$, which completes the proof. □

Lemma 7: If $8v_n+1$ is a perfect square, then $n \equiv 1$, ± 2, $\pm 18 \pmod{2^2 \cdot 5 \cdot 11}$.

Proof: It is readily seen that $n \equiv 1$, $\pm 2 \pmod{20}$ from Lemma 4 and $n \equiv 1$, ± 2, $\pm 18 \pmod{44}$ from Lemma 6, these imply simultaneously $n \equiv 1$, ± 2, ± 18, ± 42, $\pm 62 \pmod{220}$.

Modulo 331. Eliminate $n \equiv \pm 42$, $\pm 62 \pmod{110}$ which imply, respectively, $8v_n+1 \equiv 292$, $263 \pmod{331}$, so that there remain $n \equiv 1$, ± 2, $\pm 18 \pmod{220}$, which completes the proof. $\square$

From Lemmas 4 to 7 we can derive the following result:

Corollary 2: If $8v_n+1$ is a perfect square, then $n \equiv 1$, ± 2, $\pm 18 \pmod{2^5 \cdot 5^2 \cdot 11}$.

Proof: Suppose that $8v_n+1$ is a perfect square, it follows from Lemmas 4 to 7 that n must satisfy the following congruences simultaneously:

$$\begin{cases} n \equiv c_1 \pmod{2^5 \cdot 5} \\ n \equiv c_2 \pmod{2^3 \cdot 5^2} \\ n \equiv c_3 \pmod{2^3 \cdot 11} \\ n \equiv c_4 \pmod{2^2 \cdot 5 \cdot 11} \end{cases} \qquad c_1, c_2, c_3, c_4 \in \{1, \pm 2, \pm 18\} .$$

Since the greatest common divisor of the last two moduli is 44 and $|c_3 - c_4| \leq 36$, we get $c_3 = c_4$. Note that the least common multiple of the last two moduli is $2^3 \cdot 5 \cdot 11$, the congruences are reduced to

$$\begin{cases} n \equiv c_1 \pmod{2^5 \cdot 5} \\ n \equiv c_2 \pmod{2^3 \cdot 5^2} \\ n \equiv c_3 \pmod{2^3 \cdot 5 \cdot 11} \end{cases} \qquad c_1, c_2, c_3, \in \{1, \pm 2, \pm 18\} .$$

Now, the three moduli have the greatest common divisor 40 and the least common multiple $2^5 \cdot 5^2 \cdot 11$. With the reason as above, we have $c_1 = c_2 = c_3$, and therefore $n \equiv 1$, ± 2, $\pm 18 \pmod{2^5 \cdot 5^2 \cdot 11}$, which completes our proof. $\square$

4. MAIN RESULTS

Proof of Theorem 1: It is easy to see that v_n is triangular if and only if $8v_n+1$ is square. By Corollary 2 this implies $n \equiv 1$, ± 2, $\pm 18 \pmod{2^5 \cdot 5^2 \cdot 11}$, moreover, by Corollary 1, we get $n=1$, ± 2, ± 18. In fact, $v_1 = 1 \cdot 2/2$, $v_{\pm 2} = 2 \cdot 3/2$, $v_{\pm 18} = 107 \cdot 108/2$. The proof is complete. $\square$

Finally, we prove the equivalent Diophantine equation results of Theorem 1.

Proof of Theorem 2: (a) According to (3) and the explanation at the end of section 1, the equation

$$20x^2 - y^2(y+1)^2 = 16 \tag{5}$$

implies that $\frac{1}{2}y(y+1)=v_n$ and n is odd. Therefore it follows from Theorem 1 that $\frac{1}{2}y(y+1)=1$, so that $y=-2$ or 1, which give all the integer solutions of equation (5).

(b) Similarly, the equation

$$20x^2 - y^2(y+1)^2 = -16 \tag{6}$$

implies that $\frac{1}{2}y(y+1)=v_n$ and n is even, so that $\frac{1}{2}y(y+1)=3$ or 5778 by Theorem 1. Hence we get $y=-3, 2, -108$ or 107, which give all the integer solutions of equation (6). $\square$

ACKNOWLEDGMENT

The author would like to thank Professor Sun Qi for his warm encouragement and the referee for the helpful comments.

REFERENCES

[1] Cohn, J. H. E. "On Square Fibonacci Numbers." *J. London Math. Soc. 39* (1964): pp. 537-541.

[2] Guy, R. K. Unsolved Problems in Number Theory. New York: Springer-Verlag, 1981, p. 106.

[3] Luo, Ming. "On Triangular Fibonacci Numbers." *The Fibonacci Quarterly, 27.2* (1989): pp. 98-108.

A FAST ALGORITHM OF THE CHINESE REMAINDER THEOREM AND ITS APPLICATION TO FIBONACCI NUMBERS

Kenji Nagasaka, Jau-Shyong Shiue and Chung-Wu Ho

The Chinese Remainder Theorem is an old but important theorem in the elementary number theory which can be stated as follows:

The Chinese Remainder Theorem: Let m_1, m_2, ..., m_k be positive integers which are relatively prime in pairs. Then there exists a unique integer x (mod M) satisfying the system of congruences:

$$
\begin{cases}
x \equiv a_1 \pmod{m_1}, \\
x \equiv a_2 \pmod{m_2}, \\
\quad --- \qquad --- \\
x \equiv a_k \pmod{m_k},
\end{cases}
\tag{1.1}
$$

where $M = m_1 m_2 \cdots m_k$.

The number of arithmetic operations in order to get a unique integer x (mod M) satisfying (1.1) can be estimated from the Lame's theorem of the Euclidian algorithm by at most $5 \cdot \sum_{i=1}^{k} \log_{10}\{\min(M_i, m_i)\}$ divisions and the same number of multiplications and additions, where $M_i = M/m_i$ for i = 1, 2, ..., k.

(∗) The research of the second author was partially supported by the University Research Council of the University of Nevada, Las Vegas.

G. E. Bergum et al. (eds.), Applications of Fibonacci Numbers Volume 4, 241–246.
© 1991 *Kluwer Academic Publishers.*

Now we state our fast algorithm of the Chinese Remainder Theorem as Theorem 1 and shall show the efficiency of it in terms of the number of arithmetic operations required.

Theorem 1: Under the same assumptions as in the Chinese Remainder Theorem, the system of congruences (1.1) is equivalent to the following single linear congruence

$$(\sum_{i=1}^{k} b_i M_i)x \equiv \sum_{i=1}^{k} a_i b_i M_i \pmod{M}, \tag{1.2}$$

where b_i's are arbitrary integers coprime to m_i's, respectively, and $M_i = M/m_i$ for $i = 1, 2, \cdots$, k with $M = \prod_{i=1}^{k} m_i$.

Lemma 1: The system of congruences (1.1) is equivalent to the following system of congruences:

$$\begin{cases} M_1 x \equiv a_1 M_1 \pmod{M} \\ M_2 x \equiv a_2 M_2 \pmod{M} \\ \quad --- \\ M_{k-1} x \equiv a_{k-1} M_{k-1} \pmod{M} \\ (\sum_{i=1}^{k} b_i M_i) x \equiv \sum_{i=1}^{k} a_i b_i M_i \pmod{M}, \end{cases} \tag{1.3}$$

where b_i is relatively prime to m_i for $i = 1, 2, \cdots, k$.

Proof of Lemma 1: We show the necessity. Suppose x is a solution of (1.1). Then, for $i = 1, 2, \ldots, k$,

$$x - a_i = c_i m_i,$$

with an integer c_i. Multiplying the above identity with M_i for each i, we get for $i = 1, 2, \ldots, k$,

$$M_i x - M_i a_i = c_i m_i M_i = c_i M,$$

which is rewritten as the congruence

$$M_i x \equiv M_i a_i \pmod{M} \tag{1.4}$$

for $i = 1, 2, \cdots, k$. The first $k-1$ congruences in (1.3) are identical with those of (1.4) for $i = 1, 2, \cdots, k-1$. The last congruence of (1.3) can be obtained as a linear combination of (1.4).

Conversely, we assume that x is a solution of the system of congruences. Since M_i and m_i are coprime, (1.4) can be reduced to

$$x \equiv a_i \pmod{m_i},$$

for $i = 1, 2, \cdots, k-1$.

Subtracting the linear combination of the first $k-1$ congruences in (1.3), we have

$$b_k M_k x \equiv a_k b_k M_k \pmod{M}.$$

Since M_k and b_k are relatively prime to m_k, we conclude that

$$x \equiv a_k \pmod{m_k}. \qquad \text{<Q.E.D. of Lemma 1>}$$

Lemma 2: Under the same assumptions as in Theorem 1 and in Lemma 1, M is relatively prime to $\sum_{i=1}^{k} b_i M_i$.

Proof of Lemma 2: Let g be the greatest common divisor of $\sum_{i=1}^{k} b_i M_i$ and M and let p be a prime factor of g. Then p divides $\sum_{i=1}^{k} b_i M_i$ and $M = \sum_{i=1}^{k} m_i$. The divisibility of M by p implies that p divides only one of m_i, say m_j, since $(m_i, m_j) = 1$ if $i \neq j$. Then p divides all M_i but M_j, since $M_j = M/m_j$. Together with the divisibility of $\sum_{i=1}^{k} b_i M_i$ by p, we derive that p divides $b_j M_j$ and consequently divides b_j. Hence p divides the greatest common divisor of m_j and b_j which is equal to one from the assumption.

$$\text{<Q.E.D. of Lemma 2>}$$

Proof of Lemma 1: By Lemma 1, we know that the system of congruences (1.1) is equivalent to the system (1.3). From Lemma 2, the last congruence has a unique solution, say $x_0 \mod M$. Thus the system (1.3) has a unique solution x_0, which implies that x_0 is the unique solution of the system of congruences (1.1), since the Chinese remainder theorem assures the unique existence of the solution of the system (1.1) mod M.

$$\text{<Q.E.D. of Theorem 1>}$$

Applying Theorem 1 to the Fibonacci numbers, we have:

Corollary: Let u_{n-1}, u_n and u_{n+1} be three successive terms of the Fibonacci numbers. Then the system of congruences

$$\begin{cases} x \equiv a \pmod{u_{n-1}} \\ x \equiv b \pmod{u_n} \\ x \equiv c \pmod{u_{n+1}} \end{cases}$$

has a unique solution

$$x \equiv (-1)^{n+1} \left[a\, u_n u_{n+1} - (b u_{n-1} u_{n+1} + c u_{n-1} u_n) \right] \pmod{u_{n-1} u_n u_{n+1}}.$$

Proof: It is known that $(u_m, u_n) = u_{(m,n)}$. Thus we have $(u_{n-1}, u_n) = (u_n, u_{n+1}) = u_1 = 1$ and $(u_{n-1}, u_{n+1}) = u_1$ or $u_2 = 1$. Now

$$u_n u_{n+1} x \equiv a\, u_n u_{n+1} \pmod{u_{n-1} u_n u_{n+1}}$$
$$u_{n-1} u_{n+1} x \equiv b\, u_{n-1} u_{n+1} \pmod{u_{n-1} u_n u_{n+1}}$$
$$u_{n-1} u_n x \equiv c\, u_{n-1} u_n \pmod{u_{n-1} u_n u_{n+1}},$$

then, by Theorem 1,

$$\left[u_n u_{n+1} - u_{n-1}(u_{n+1} + u_n) \right] x \equiv a\, u_n u_{n+1} - (b\, u_{n-1} u_{n+1} + c\, u_{n-1} u_n) \pmod{u_{n-1} u_n u_{n+1}},$$

or

$$(u_n u_{n+1} - u_{n-1} u_{n+2})\, x \equiv a\, u_n u_{n+1} - (b\, u_{n-1} u_{n+1} + c\, u_{n-1} u_n).$$

Since

$$u_n u_{n+1} - u_{n-1} u_{n+2} = (-1)^{n-1},$$

we then obtain

$$(-1)^{n-1} x \equiv a\, u_n u_{n+1} - (b\, u_{n-1} u_{n+1} + c\, u_{n-1} u_n) \pmod{u_{n-1} u_n u_{n+1}}.$$

Hence

$$x \equiv (-1)^{n+1} \left[a\, u_n u_{n+1} - (b\, u_{n-1} u_{n+1} + c\, u_{n-1} u_n) \right] \pmod{u_{n-1} u_n u_{n+1}}.$$

<Q.E.D. of Corollary>

We can now compute the number of arithmetic operations required by our algorithm with that of the traditional one in the solution of the Chinese Remainder Theorem. To find the solution by the traditional method, we need, $5 \cdot \log_{10}\{\min (M_i, m_i)\}$ divisions and the same number of multiplications for each congruence:

$$M_i z_i \equiv 1 \pmod{m_i}, \; i = 1, 2, \ldots, k.$$

Assuming that $\min(M_i, m_i) = m_i$ for each i, which would be true in most cases, we find that the total number of multiplications and the total number of divisions needed to solve all k such congruences are each $5 \cdot \sum_{i=1}^{k} \log_{10} m_i$.

On the other hand, in our algorithm, we need to solve only a single congruence (1.2). Without loss of generality, assume that

$$M_1 \geq M_2 \geq M_3 \geq \ldots \geq M_k.$$

Choosing $b_i = (-1)^{i+1}$ for each i, we can be sure that

$$\sum_{i=1}^{k} b_i M_i \le M_1 = m_2 m_3 \cdots m_k.$$

Thus, the number of multiplications and divisions required in our algorithm is at most

$$5 \cdot \log_{10}\{ \min(M_1, M) \}$$
$$= 5 \cdot \log_{10} M_1 = 5 \cdot \sum_{i=2}^{k} \log_{10} m_1 .$$

Therefore, our algorithm requires at least $5 \cdot \log_{10} m_1$ fewer number of divisions and $5 \cdot \log_{10} m_1$ fewer number of multiplications than that of the traditional algorithm. In fact, for a given system of congruences, we may do better by choosing the integers $b_1, b_2, ..., b_k$ to make the sum $\sum b_i M_i$ smaller than M_1. Since M_i's are relatively prime, it may even be possible to choose the b_i's so that $\sum b_i M_i = 1$, making the Euclidian algorithm entirely unnecessary.

We recall the generalized Chinese remainder theorem [3].

The Generalized Chinese Remainder Theorem: A necessary and sufficient condition that the system of congruences (1.1) be solvable is that, for each $1 \le i < j \le k$,

$$a_i - a_j \equiv 0 \pmod{(m_i, m_j)}.$$

Any two solutions are congruent $\mathrm{mod}[m_1, m_2, \cdots, m_k]$, where $[a, b]$ denotes the least common multiple of two positive integers a and b.

Tracing the argument in the proof of Theorem 1, we get the following:

Theorem 2: Under the same assumptions as in the generalized Chinese remainder theorem, the system of congruences (1.1) is equivalent to the following single linear congruence

$$\left(\sum_{i=1}^{k} b_i M_i \right) x \equiv \sum_{i=1}^{k} a_i b_i M_i \pmod{M},$$

where b_i's are arbitrary integers coprime to μ_i's, respectively, and $M_i = M/\mu_i$, for $i = 1, 2, \cdots, k$ with $M = \prod_{i=1}^{k} \mu_i = [m_1, m_2, \cdots, m_k]$.

Here we put

$$M = [m_1, m_2, \cdots, m_k]$$

the least common multiple of m_i's. Then we can express M in the form

$$M = \prod_{i=1}^{k} \mu_i,$$

of relatively prime factors, including unity, such that μ_i divides m_i for $i = 1, 2, \cdots, k$.

REFERENCES

[1]   Dickson, E., History of the Theory of Numbers, vol. II, Diophantine Analysis. Chelsea
      (1971).

[2]   Knuth, D., The Art of Computer Programming, vol. 2, Seminumerical Algorithms.
      Addition-Wesley (1969).

[3]   Rosen, K., Elementary Number Theory and its Applications. Addison-Wesley (1984).

GENERATING THE PYTHAGOREAN TRIPLES
VIA SIMPLE CONTINUED FRACTIONS

A. G. Schaake and J. C. Turner

1. INTRODUCTION

Perhaps the most famous equation in mathematics is $A^2 + B^2 = C^2$, which is known as Pythagoras' equation. The problem of finding positive integer solutions for it is an ancient one indeed. It is known to have been tackled successfully in the Babylonian period of more than 3500 years ago, for a clay tablet bearing a list of solutions in sexagesimal notation, dated at circa 1600 B.C., is held in the G.A. Plimpton collection at Columbia University, New York. The methods used by the Babylonians for solving this equation in positive integers have not come down to us.

The first recorded formulae for families of solutions are credited to Pythagoras (B.C. 582-500) and Plato (B.C. 428-348?). Any integer solution (A, B, C) is now known as a Pythagorean triple; if g.c.d. $(A, B, C) = 1$, it is said to be a primitive triple. There is an infinity of different primitive triples; and any multiple of a primitive triple is also a Pythagorean triple. Formulae for the whole set of triples were given by Euclid, in his famous treatise *Elements* (ca B.C. 300), and by Diophantus, in his Arithmetica (ca 250 A.D.).

Since then, countless major and minor mathematicians have written on properties of the triples, usually dealing with special subfamilies of them. For example the first Proposition in Leonardo Pisano's (Fibonacci's) book *Liber Quadratorum (Book of Squares)*, which was published in 1225 A.D., shows how to compute Pythagorean triples from sums of consecutive odd integers. After Diophantus, we do not believe there was presented an essentially new method for

G. E. Bergum et al. (eds.), Applications of Fibonacci Numbers Volume 4, 247–256.
© 1991 *Kluwer Academic Publishers*.

solving $A^2 + B^2 = C^2$, giving all solutions and providing means for classifying them in useful ways, until Gauss (1777-1855 A.D.) introduced his ideas on algebraic integers in the field $Z(\sqrt{-1})$. In 1912 A.D. P. Lambert solved $A^2 + B^2 = C^2$ by use of numbers $e+fi$, in Nouv. Ann. Math. (4), 12.

The method given by Diophantus, is essentially that which is given today in most textbooks on number theory (see [6], for example). It is embodied in the following theorem:

Theorem: All the primitive-triple solutions of the equation $x^2 + y^2 = z^2$ for which y is an even integer are given by the formula

$$x = u^2 - v^2, \ y = 2uv, \ z = u^2 + v^2,$$

where u, v is any pair of coprime natural numbers with $u > v$ and such that one of the pair is odd and the other is even.

In recent years we have been developing a new theory of braiding, and one of the principal tools used for modelling regular braids [1,2] is a binary tree with rational numbers assigned to its nodes [7]. Root-to-node paths in this tree are found to correspond with simple continued fractions (SCFs) of the rational numbers at the nodes. We discovered by studying these paths that we could relate SCFs and their convergents to solutions of certain families of quadratic Diophantine equations. In particular, we found that by these methods we could solve Pythagoras' equation, and then classify the set of solutions in terms of the related SCFs and their properties.

An extensive search of the literature on the subject, and communications with number theorists, encourages us to believe that our methods are new; this is surprising, in view of the antiquity of the problem and the vast literature upon it.

The properties of rational numbers which we discovered from the tree used in the braiding models also appear to be new results. We have applied them to study other quadratic form problems, and to generalize Euclid's algorithm and the concept of prime number; this work is described in research reports [3] and [4].

In this paper we give a brief description of our method for generating all the Pythagorean triples from a class of simple continued fractions. We give an algorithm for determining what kind of a triple will arise from a given SCF in this class. Finally we show how to use parities of the terms in SCFs to provide a way to partition the whole family of Pythagorean triples.

For further details and proofs of our methods, the reader is referred to [3,4]. A discursive account of the methods only, written for mathematics students and teachers as well as for professional mathematicians, is presented in our book, *A New Chapter for Pythagorean Triples* [5].

2. THE NEW METHOD FOR PYTHAGOREAN TRIPLES

The new method for solving $A^2 + B^2 = C^2$ is as follows. First the equation is written in the form $(nx)^2 + (ny)^2 = (nz)^2$, where y is even and $n=$g.c.d.(A, B, C). Then it is transformed into the form $ps=qr-1$, with the following conditions being imposed:

$$r<s \text{ (except when } r = s = 1) \text{ and } p<q;$$
$$r<p \text{ (except when } r = p = 1) \text{ and } s<q.$$

In [3] we give full reasons and details for imposing the conditions.

Given these conditions, we can deduce that r/s and p/q must be the penultimate and final convergents respectively of a certain type of simple continued fraction. Thus we can relate Pythagorean triples to SCFs.

We shall define below the class of all those SCFs which generate Pythagorean triples. A member of this class will be called a PSCF.

First we give step-by-step details of the transformations, and then show the PSCFs determined by them.

Starting Equation

$$(nx)^2 + (ny)^2 = (nz)^2, \text{ with } y \text{ even.}$$

Transformations

The equation can be written in one of the following forms:

I. $\quad (ny + 1)(ny - 1) = (nz + nx)(nz - nx) - 1$

II. $\quad (nx + 1)(nx - 1) = (nz + ny)(nz - ny) - 1$

III. $\quad \left(n\frac{y}{2} + 1\right)\left(n\frac{y}{2} - 1\right) = \left(n\frac{z}{2} + n\frac{x}{2}\right)\left(n\frac{z}{2} - n\frac{x}{2}\right) - 1.$

Comparisons

Each of the three forms I, II and III may be compared with the equation

$$p \cdot s = q \cdot r - 1.$$

This leads to the following possible transformations and the consequent Pythagorean equation $a^2 + b^2 = c^2$.

$$
\text{I}^* \quad
\begin{cases}
\begin{cases}
r = nz - nx \; ; & p = ny + 1 \\
s = ny - 1 \; ; & q = nz + nx
\end{cases} \\[2ex]
\begin{cases}
r = nz - nx \; ; & p = ny - 1 \\
s = ny + 1 \; ; & q = nz + nx
\end{cases}
\end{cases}
\left.\begin{array}{l}
\\
a = 2nx = q - r \\[1ex]
b = 2ny = p + s \\[1ex]
c = 2nz = q + r \\
\end{array}\right\}
$$

$$
\text{II}^* \quad
\begin{cases}
\begin{cases}
r = nz - ny \; ; & p = nx + 1 \\
s = nx - 1 \; ; & q = nz + ny
\end{cases} \\[2ex]
\begin{cases}
r = nz - ny \; ; & p = nx - 1 \\
s = nx + 1 \; ; & q = nz + ny
\end{cases}
\end{cases}
\left.\begin{array}{l}
\\
a = 2nx = p + s \\[1ex]
b = 2ny = q - r \\[1ex]
c = 2nz = q + r \\
\end{array}\right\}
$$

$$
\text{III}^* \quad
\begin{cases}
\begin{cases}
r = n\left(\frac{z}{2} - \frac{x}{2}\right) \; ; & p = n\frac{y}{2} + 1 \\
s = n\frac{y}{2} - 1 \; ; & q = n\left(\frac{z}{2} + \frac{x}{2}\right)
\end{cases} \\[2ex]
\begin{cases}
r = n\left(\frac{z}{2} - \frac{x}{2}\right) \; ; & p = n\frac{y}{2} - 1 \\
s = n\frac{y}{2} + 1 \; ; & q = n\left(\frac{z}{2} + \frac{x}{2}\right)
\end{cases}
\end{cases}
\left.\begin{array}{l}
\\
a = nx = q - r \\[1ex]
b = ny = p + s \\[1ex]
c = nz = q + r \\
\end{array}\right\}
$$

Note that in each of the cases I*, II* and III* it is possible to interchange the p and s transformations, but not the q and r ones, due to the conditions on p, q, r and s. In every case, the above transformations impose the condition that $|p - s| = 2$. This leads to a simple continued fraction for p/q (see [3] for proof) of the following two forms:

(1) $\quad p \, / \, q = [0;U,V,\tilde{U}]$

(2) $\quad p \, / \, q = [0;U,\tilde{V},\tilde{U}]$

where $\quad U = u_1, u_2, \ldots, u_i$, with $u_k = 1,2,3,\ldots$

$\left.\begin{array}{l} V = v,1,j,(v+1) \\ \tilde{V} = (v+1),j,1,v \end{array}\right\}$ with $v = 0,1,2,3,\ldots$ and $j = 0,1,2,\ldots$

$\tilde{U} = u_i, u_{i-1}, \ldots, u_2, u_1$.

For brevity we write $UV\tilde{U}$ and $U\tilde{V}\tilde{U}$ for these quasi-symmetric continued fractions representing p/q, and we call them Pythagorean simple continued fractions (PSCFs).

The parameter j is related to the g.c.d. of (a, b, c) in the following ways:

I*   g.c.d.$(a, b, c) = 2n = j + 1$ hence $n = \frac{j+1}{2}$, hence j has to be odd;

II*   g.c.d.$(a, b, c) = 2n = 2(j + 1)$ hence $n = j + 1$;

III*   g.c.d.$(a, b, c) = n = j + 1$.

Furthermore, note that j may be zero. When $j = 0$ the rule, for using the PSCF (for calculating its convergents), is to contract the centre portion thus:

$$V = v, 1, 0, (v+1)$$
$$= v, [1 + (v+1)]$$
$$= v, (v+2);$$

and similarly for $\tilde{V}$.

Example:

Suppose the PSCF is the following:

$$[0; 5, 1, 3, 1, 1, 5, 2, 3, 1, 5]$$

We can show (see the decision tree later) that this PSCF belongs to Method III* with $n = j + 1$. It also belongs to Method I* with $2n = j + 1$.

Note that $U = 5, 1, 3$ and $V = 1, 1, 5, 2$, hence $v = 1$ and $j = 5$, thus $j + 1 = 6$.

Working out the convergents of this PSCF we find that:

the penultimate one is $r/s = 486/2809$

and the final one is $p/q = 2807/16224$.

From the formulae given under I* or III* we obtain the Pythagorean triple

$$\left. \begin{array}{l} a = q - r = 15738 \\ b = p + s = 5616 \\ c = q + r = 16710 \end{array} \right\} \quad \text{g.c.d.} = j + 1 = 6$$

The primitive triple is therefore (dividing by 6):

$$x = 2623, \quad y = 936, \quad z = 2785.$$

3. CLASSIFYING THE PSCFs

In this section we show how PSCFs may be classified into five types, which are distinguishable by means of conditions placed on their elements. Any PSCF may be contracted (or expanded) by certain rules (proved in [3]) *without changing its type*. We shall only use the contraction rule here, which is as follows:

Contraction rule, to be applied to U and $\tilde{U}$ of the PSCF

(i)   Write down under the elements of string U the parity of each element: e for
        even, o for odd.

(ii)   Discard any number of e's which are consecutive left-most elements.

(iii)   If the left-most element is o, contract by discarding the second element, and
        adding the first and third, using parity addition, thus: $o + o = e$ and $o + e = o$.

(iv)   Repeat from (ii) until no further action is possible.

(v)   The above contraction operations are to be applied simultaneously from the left
        on U, and from the right on $\tilde{U}$, so that the PSCF (contracted) remains quasi-
        symmetric always.

It will be evident that after contractions have ceased we shall have just one of the following
terminal strings, say U^*, for the contracted U (and $\tilde{U}^*$ for $\tilde{U}$):

(a) $U^* = \emptyset$;   (b) $U^* = o$;   (c) $U^* = o,o$;   (d) $U^* = o,e$

Example: (the rules applied in order are i, ii, iii, ii, iii, ii)

$U =$	2,	3,	7,	9,	13,	7,	39,	4,	2,	1,	2
	e,	o,	o,	o,	o,	o,	o,	e,	e,	o,	e
		o,	o,	o,	o,	o,	o,	e,	e,	o,	e
			e,	o,	o,	o,	o,	e,	e,	o,	e
				o,	o,	o,	o,	e,	e,	o,	e
					e,	e,	e,	e,	o,	e
						o,	e

Therefore $U^* = o,e$; and simultaneously $\tilde{U}^* = e,o$.

We are now in a position to divide all the PSCFs into five types in the following way:

(i)   first divide them into two classes according to their U^* forms, and the v-values
        within the V-strings;

(ii)   next divide one class into three types, and the other into two types, according to
        the values of j in their V-strings.

This procedure is clearly defined by the following decision tree.

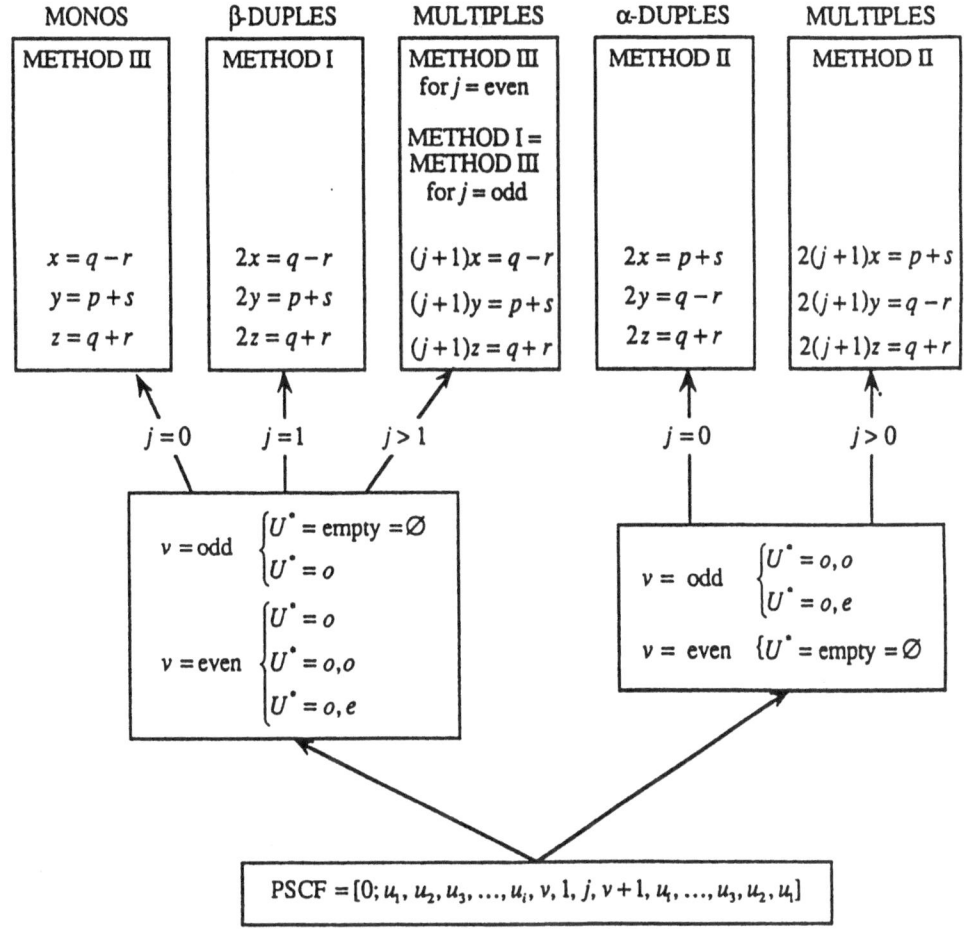

DECISION TREE

Notice from the top boxes of the tree how the PSCFs have been sorted out into five distinct types. The final characteristics of these are two-fold, namely by transformation method (I, II, III, used on a $a^2 + b^2 = c^2$) and by values of the g.c.d.(a, b, c). For the latter, we use the description **mono** when a triple with g.c.d. $= 1$ results from a PSCF (i.e. a *primitive triple*); **duple** when a triple with g.c.d. $= 2$ results; and **multiple** when a triple with g.c.d. > 2 results.

Example:

Suppose that the PSCF is:

$$[0; 4, 5, 3, 1, 2, 13, 4, 7, 8, 9, 1, 13, 10, 8, 7, 4, 13, 2, 1, 3, 5, 4]$$

Hence $U = 4, 5, 3, 1, 2, 13, 4, 7, 8$, and $V = 9, 1, 13, 10 = v,1,j,(v + 1)$. Thus $j = 13$ and $v = 9 =$ odd. Using the contraction rule gives $U^* = \emptyset = \tilde{U}^*$.

Entering the Decision Tree at the root, we use information $v =$ odd, $U^* = \emptyset$ to step left; then since $j > 1$ we step right. Thus the PSCF belongs to the type which is a multiple ($j = 13$, so the g.c.d. of the resulting Pythagorean triple is 14), and belongs to Method I $\equiv$ Method III (for j is odd).

4. A TREE CLASSIFICATION OF THE PRIMITIVE TRIPLES

In the previous section we showed that those PSCFs which have parity properties as given in the left-hand lower box of the decision tree, will, with $j = 0$ and using the Method III transformation, generate all the "monos", which are the primitive Pythagorean triples.

We shall now present a tree, whose nodes after the root bear a U-string in parity form. We call this the *Mono Primitive Pythagorean Triple Tree*. The parity U-string on each node (together with the root form 0,2 for the common V-string) determines a general family or subclass of primitive triples.

For example, the middle branch from the root goes to the node labeled o, o; this indicates the general family of the monotype $UV\tilde{U} = [0; o, o, 0, 2, o, o]$. An example member of this family is:

$$\text{PSCF} = [0;1,3,0,2,3,1]$$

which gives by Method III the primitive triple

$$x = 9, y = 40, z = 41.$$

We have proved in [3] that each node beyond the root determines a subclass of the nontrivial primitive triples, which is mutually exclusive of any other node subclass. And, moreover, that the whole class of nontrivial primitive triples is generated through the union of the node subclasses.

Thus the tree effects a classification of the nontrivial primitive Pythagorean triples. The diagram of the tree is on the following page.

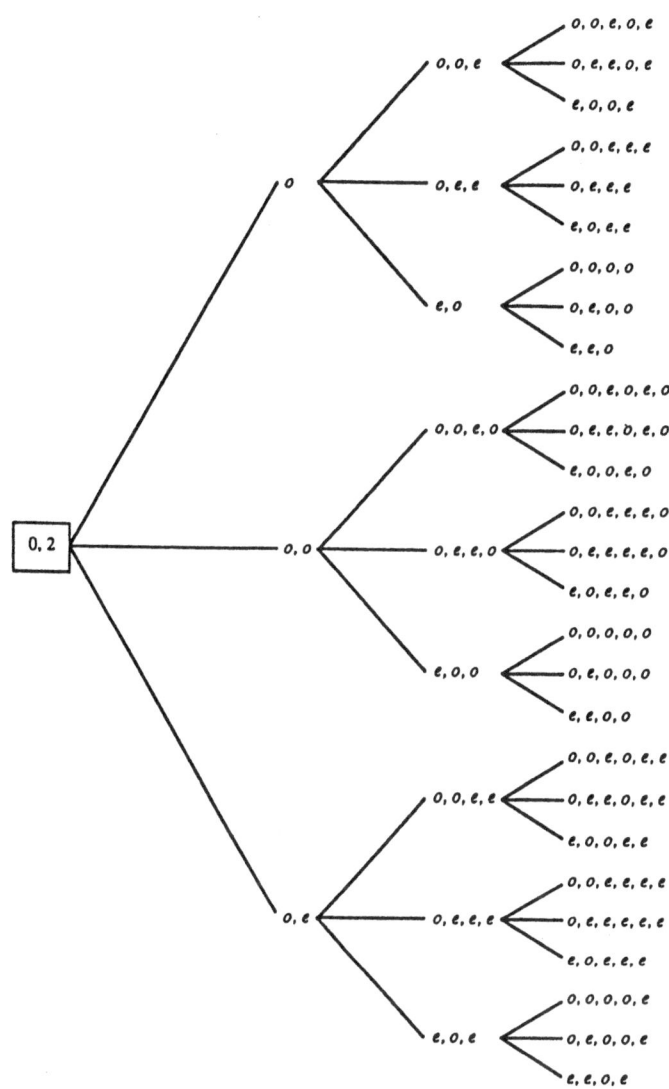

THE MONO PRIMITIVE PYTHAGOREAN TRIPLE TREE

U-string

Formulae for each subclass.

With a little patience, and a lot of algebra, general formulae for computing (x, y, z)-values in each of the node-subclasses can be computed. We give just one example, for the $UV\tilde{U} = [0; o, o, 0, 2, o, o]$ family.

Writing $2m-1$ and $2n-1$ for the *odd* terms we see that the general PSCF is of the form $[0; 2m-1, 2n-1, 0, 2, 2n-1, 2m-1]$. Deriving the final two convergents of this, and applying the formulae of Method III, gives the required formulae for the primitive triples in the subclass, viz:

$$\begin{cases} x = [2n(2m-1) + 1]^2 - (2n)^2 \\ y = 2[2n(2m-1) + 1]2n \\ z = [2n(2m-1) + 1]^2 + (2n)^2 \end{cases}$$

where m and n are positive integers.

REFERENCES

[1]   Schaake, A. G., Turner, J. C. and Sedgwick, D. A. Braiding-Regular Knots. University of Waikato, Hamilton, New Zealand, (1988): 117 pp.

[2]   Schaake, A. G. and Turner, J. C. A New Theory of Braiding (RR 1/1). Research Report No. 165, University of Waikato, Hamilton, New Zealand, (1988): 42 pp.

[3]   Schaake, A. G. and Turner, J. C. New Methods for Solving Quadratic Diophantine Equations (Parts I and II). Research Report No. 192, University of Waikato, Hamilton, New Zealand, (1989): 99 pp.

[4]   Schaake, A. G. and Turner, J. C. Generalizing Euclid's Algorithm, via the Regular and Moebius Knot Trees. Order-n Arithmetics. Research Report No. 196, University of Waikato, Hamilton, New Zealand, (1990): 61 pp.

[5]   Schaake, A. G. and Turner, J. C. A New Chapter for Pythagorean Triples. University of Waikato, Hamilton, New Zealand, (1989): 155 pp.

[6]   Sierpinski, W. Pythagorean Triangles. Publ: Graduate School of Science, Yeshiva University, New York, 1962, p. 9.

[7]   Turner, J. C. "Three Number Trees - Their Growth Rules and Related Number Properties." Applications of Fibonacci Numbers, Volume 3. Edited by G.E. Bergum, et. al., Kluwer Academic Pub. (1990): pp. 335-350.

Note: The publications [1,2,3,4,5] may be obtained by writing to the Registrar (Book Sales), University of Waikato, Hamilton, New Zealand.

ON THE MOEBIUS KNOT TREE AND EUCLID'S ALGORITHM

A. G. Schaake and J. C. Turner

1. INTRODUCTION

In Eudlid's Elements, Book VII, proposition 2, (circa 300 B.C.) there appears a method for determining the greatest common divisor of two integers, which is now known simply as Euclid's algorithm. It is a fundamental theorem of arithmetic. It supplies also the terms (partial quotients) of the simple continued fraction for the ratio of the two integers. Since then the theory of simple continued fractions has steadily grown, and the fractions have found countless applications in number theory and other branches of mathematics. There is a huge literature upon them.

In [1], [2] we give details of a new theory of braiding which we are developing. The principal mathematical tools of this theory are modular arithmetic and simple continued fractions and number trees. In order to model the process of evolution of an infinite class of single-string regular cylindrical braids, we were led to the study of a binary tree graph upon whose nodes were arranged the set of rational numbers, each number being in lowest form. This tree has two roots and a trunk-node. It was found that the terms of the simple continued fraction $[u_0; u_1, u_2, \ldots, u_k, 1]$ of any chosen rational number $\frac{p}{q}$, on some node of the tree, corresponded to the number of steps to be taken from the trunk, branching first right for u_0 steps, then left for u_1 steps, and so on alternately, until the node carrying $\frac{p}{q}$ was reached.

To each rational number there corresponds uniquely a single-string, cylindrical, regular braid. In order to construct the braid which corresponds to $\frac{p}{q}$, one must, in effect, construct in turn each of the braids corresponding to the convergents of $[u_0; u_1, u_2, \ldots, u_k, 1] = \frac{p}{q}$.

G. E. Bergum et al. (eds.), Applications of Fibonacci Numbers Volume 4, 257–270.

Thus there is a direct correspondence between braiding operations, and movements along paths in the tree. The relations between the rational numbers on the nodes of the tree constitute what we now call **order-1 arithmetic**. In [1] we called the tree a Regular Knot Tree (RKT). In this paper it becomes the first of an infinite sequence of rational number trees which we designate by n-RT, with $n = 1,2,3,\ldots$. Thus the 1-RT is the RKT.

A single-string, cylindrical, regular braid is tied around the outer surface of a cylinder. We discovered a way to tie a single-string braid in the form of a Moebius band (see [3] for details). Such braids we call **Moebius braids**. It turns out that there is again an evolutionary process for constructing an infinite class of single-string Moebius braids; and again one Moebius braid corresponds to each rational number, *but the latter is not necessarily in lowest form*. Again the process can be represented by paths in a rational number tree (which we call the regular Moebius knot tree, RMKT), and the braids are governed by an arithmetic determined by this tree. Thus we were led to an **order-2** arithmetic, and a second-order rational number tree **2-RT**. By considering paths from the trunk to the nodes of the 2-RT we obtained a generalization of the classical simple continued fraction.

It was then a natural process for us to generalize the above trees, and their associated arithmetics and continued fractions, to obtain the **order-n** arithmetics, the n-RTs, and nth order continued fractions which we designate by $[u_0; u_2, u_2, \ldots, u_k, 1]_n$.

This paper begins by showing the regular knot tree and the regular Moebius knot tree. Then the n-RT tree is shown, and used to introduce generalizations of Euclid's algorithm, and of simple continued fractions and their convergents.

The paper goes on to show how this leads to generalizations of other fundamental concepts of "ordinary" arithmetic, such as the *g.c.d.* of two numbers, coprimeness, and Euler's totient function.

Almost the whole of current number theory is founded on Euclid's algorithm, and the unique factorization property of integers. We believe that a study of the order-n arithmetics introduced in this paper, which enable an entirely new view of the rational numbers to be taken, with the generalized concept of coprimeness and its consequences playing a crucial role, will lead to much new insight into the theory of numbers. Our hope is that old, long-standing problems will thereby be solved; and a broad new theory of numbers will emerge.

2. THE RKT AND THE RMKT

The Regular Knot Tree (called the RKT) represents the evolution of the members of the class of regular knots (see [1,2,3] for definitions and details). It is a **weighted rooted tree** consisting of a set of **nodes** weighted by rational numbers p/q in lowest form, connected by **arcs**, with no circuits. This tree has two base points, weighted 0/1 and 1/0, which are called the **roots**. The node immediately above these is called the **trunk** and is weighted 1/1.

An important property [1] of the RKT is that the set of weights on its nodes is the set of rationals Q^+, in lowest form. Indeed, we can say that in terms of paths described upwards from the trunk, the RKT represents the whole set of real numbers (each uniquely) in the interval $(0,\infty)$. Finite paths correspond to ratios, and infinite paths to irrationals.

Rules and relationships concerning this tree are contained within the generalization to the n-RT, which we are to treat in this paper. Many properties of the RKT (which is the 1-RT) are given in [1,2,3].

The first five levels of the RKT are shown in Figure 1.

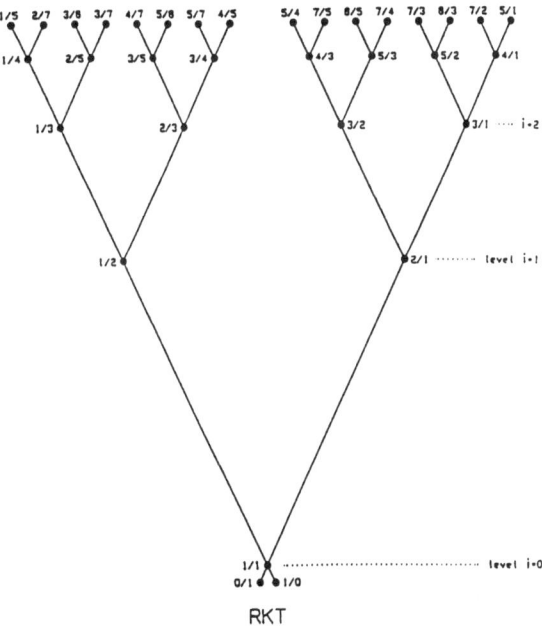

Figure 1

The Regular Moebius Knot Tree (called the RMKT) represents the evolution of regular braids which can be tied in the form of a Moebius band (see [3] for details). The roots of this tree are weighted 0/2 and 2/0, and the trunk node is weighted 1/1. It will be seen, in Figure 2, that in the RMKT not all the rational numbers representing the node weights are in the lowest form; the significance of this will become apparent later. It is still true, however, that the set of node weights is one-to-one with Q^+. A picture of a Moebius braid belonging to the tree, which corresponds to the node weighted 11/79, is shown after the tree.

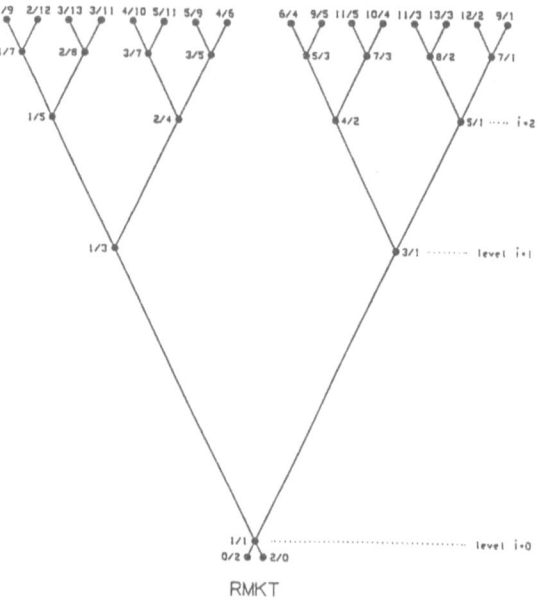

Figure 2: Regular Moebius Knot Tree

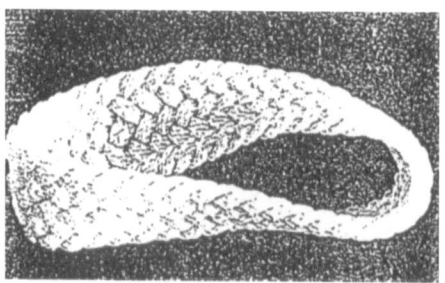

A Regular Moebius Knot

11-parts / 79-bights   Headhunter coding (pattern)

3. THE GENERAL RATIONAL NUMBER TREE (n-RT)

The general rational number tree of order-n.

As with the RKT and RMKT, we can "grow" a binary tree and assign rational numbers to its nodes according to generalized growth rules, beginning with the assignation of $0/n$ and $n/0$ to the pair of roots. We retain $1/1$ to assign to the trunk node.

The resulting tree (see Fig. 3 for a diagram of the first five levels) we call the rational number tree of order-n, and designate it by n-RT.

The assignation of rational numbers to nodes above the trunk is made by computing mediant sums (see Figure 3) in the same way as is used for the RKT (i.e. the 1-RT) and the RMKT (i.e. the 2-RT); the reader may verify this from the diagram.

It may be shown that for each $n = 1,2,3,\ldots$ the set of node weights in n-RT is one-to-one with Q^+.

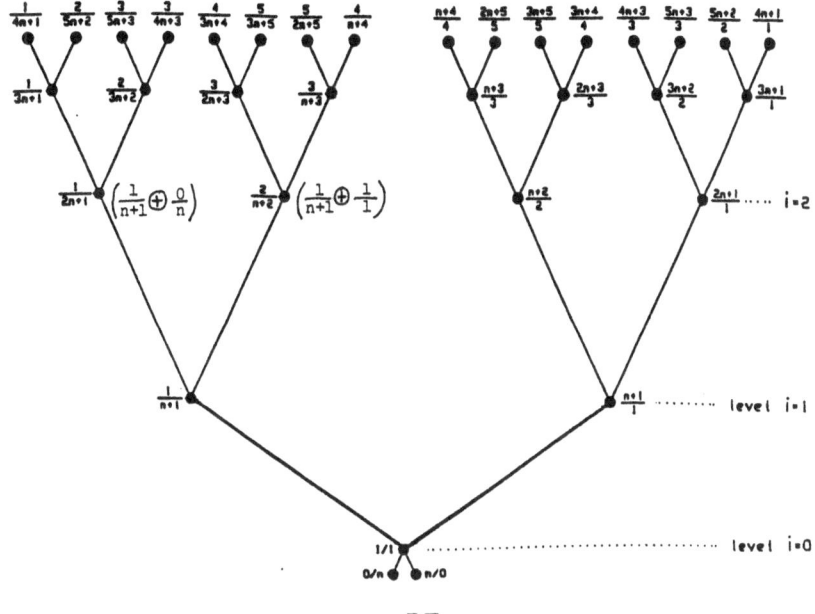

$$n\text{-RT}$$

Figure 3: General Rational Number Tree

Node-number assignation by mediant sums: $\frac{a}{b} \oplus \frac{c}{d} = \frac{a+c}{b+d}$, as exemplified on level $i = 2$. The general rules are:

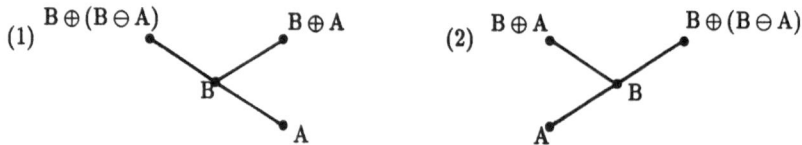

We now state two important properties of numbers assigned to nodes along certain paths in the n-RT. The first shows where solutions of $ps = qr \pm n$ occur. The second shows how the numbers in 1-RT relate to simple continued fractions, a property to be generalized.

Solution sets for the diophantine equations $ps = qr \pm n$

A branch to the left in the n-RT as shown in Figure 4, we call a **left hockey-stick**. It forms a pictorial representation of those positive integer solutions of $ps = qr + n$ which occur in the n-RT as p/q and r/s.

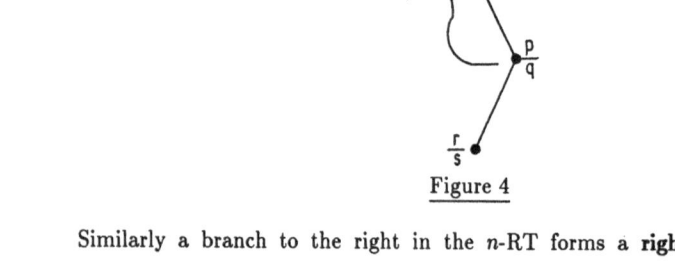

Figure 4

Similarly a branch to the right in the n-RT forms a **right hockey-stick**, which is a pictorial representation of the positive integer solutions in the n-RT, as p/q and r/s, of the equation $ps = qr - n$.

Simple continued fractions

In the 1-RT there is a direct relationship between the path from the trunk node 1/1 to the node bearing p/q, and the simple continued fraction (SCF) representing p/q.

Suppose this path is denoted by the symbol string $R^{u_0} L^{u_1} R^{u_2} \ldots R^{u_k}$ (or L^{u_k}, depending on the direction of the last branch). Then the SCF of p/q is $[u_0; u_1, u_2, \ldots, u_k, 1]$; see [1] for proof.

Similarly, the path from the trunk node 1/1 to the node p/q in the n-RT can be associated with a new kind of SCF, uniquely representing the rational number p_n/q_n. This new SCF leads us to a generalization of Euclid's algorithm (in Section 4), and later to a new type of coprimeness for the integers.

4. GENERALIZED EUCLID'S ALGORITHM AND CONTINUED FRACTIONS

In what follows, we shall refer to the well-known Euclid's algorithm by EA_1; we shall see how a similar process related to paths in the n-RT gives an analogous algorithm which we shall refer to as EA_n.

In the n-RT, with $n > 1$, we have to consider the left subtree from the trunk node $1/1$ separately from the right subtree from the trunk node.

The algorithm EA_n in the L-subtree $(p_n < q_n)$

Consider a node, with weight p_n/q_n, in the L-subtree of the n-RT. Let the path from the trunk to this node be $R^{u_0}L^{u_1}R^{u_2}...R^{u_k}$ (L^{u_k}), where $u_0 = 0$ since the first branch is a left-step.

We define the simple continued fraction CF_n associated with this node to be:

$$CF_n \equiv [0;u_1,u_2,...,u_k,1]_n \equiv \left(\frac{p_n}{q_n}\right)_n$$
$$\equiv 1/(u_1 + 1/(u_2 + \cdots + 1/(u_{k-1} + 1/(u_k + 1)))).$$

It can be shown (see [3]) that this continued fraction is equal to the rational number $\frac{p^*}{q^*} = \dfrac{p_n}{\left[\dfrac{q_n + (n-1)p_n}{n}\right]}$, where $p^* = p_n$ and $q^* = [q_n + (n-1)p_n]/n$.

Note that CF_n may be obtained by applying EA_1 to p^*/q^*; so the algorithm EA_n for the L-subtree is as set out in Table 1.

The algorithm EA_n in the R-subtree $(p_n > q_n)$

Consider a node, with weight p_n/q_n, in the R-subtree of the n-RT. Let the path from the trunk to this node be $R^{u_0}L^{u_1}R^{u_2}...R^{u_k}$ (L^{u_k}), where $u_0 > 0$ since the first branch is a right-step.

We define the simple continued fraction CF_n associated with this node to be:

$$CF_n \equiv [u_0;u_1,u_2,...,u_k,1]_n \equiv \left(\frac{p_n}{q_n}\right)_n$$
$$\equiv u_0 + 1/(u_1 + 1/(u_2 + \cdots + 1/(u_{k-1} + 1/(u_k + 1)))).$$

It can be shown (see [3]) that this continued fraction is equal to the rational number

$$\frac{p^*}{q^*} = \frac{\left[\dfrac{p_n + (n-1)q_n}{n}\right]}{q_n}, \text{ where } p^* = [p_n + (n-1)q_n]/n \text{ and } q^* = q_n.$$

Note again that CF_n may be obtained by applying EA_1 to p^*/q^*; so the algorithm EA_n for the R-subtree is as set out in Table 1.

L-subtree $(p_a/q_a)_n = [0; u_1, u_2, \ldots, u_k, 1]_n$				R-subtree $(p_a/q_a)_n = [4; u_1, u_2, \ldots, u_k, 1]_n$				
$u_0 = 0$	$p_a + \frac{q_a - p_a}{n}$	p_a	Let $p_a + \frac{q_a - p_a}{n} = v_1$					Let $q_a + \frac{p_a - q_a}{n} = v$
u_1	p_a	v_1	$v_1 - u_1 p_a = v_2$	u_0	q_a	$q_a + \frac{p_a - q_a}{n}$	$v - u_0 q_a = w_1$
u_2	v_2	p_a	$p_a - u_2 v_2 = v_3$	u_1	w_1	q_a	$q_a - u_1 w_1 = w_2$
u_3	v_3	v_2	$v_2 - u_3 v_3 = v_4$	u_2	w_2	w_1	$w_1 - u_2 w_2 = w_3$
$\vdots$	$\vdots$	v_3	etc	u_3	w_3	w_2	$w_2 - u_3 w_3 = w_4$
		$\vdots$			w_4	w_3	etc
				$\vdots$	$\vdots$	$\vdots$	
u_k	v_k	v_{k-1}		u_k	w_k	w_{k-1}	
1	v_k	v_k		1	w_k	w_k	
$(g.c.d.)_n$				$(g.c.d.)_n$			

Table 1. *Tabular forms for EA_n*

Greatest common divisor; coprimeness

Referring to Table 1 above, we can now explain the generalization of the concept of greatest common divisor.

If (from the L-subtree) $v_k = (g.c.d.)_n(p_n, q_n)$, we can define v_k to be the g.c.d. of p_n and q_n in the n-RT arithmetic.

Similarly for $w_k = (g.c.d.)_n(p_n, q_n)$ in the R-subtree.

We may now define *coprimeness* in the arithmetic of the n-RT (i.e. in *order-n arithmetic*).

Definition: for any rational number $\frac{p_n}{q_n}$ on a node of n-RT, if

$$g.c.d._n(p_n, q_n) = 1,$$

then p_n and q_n are coprime in order-n arithmetic.

Computing convergents in the n-RT

We may compute convergents in the n-RT, which appear on the nodes r_n/s_n occurring at the base-points of successive hockey-sticks along the path from the trunk to the node p_n/q_n.

The formulae for these convergents are given next.

Convergents from continued fractions of the L-subtree

The *continued fraction* in n-RT:

$$[0; u_1, u_2, \ldots, u_k, 1]_n = \left(\frac{p_n}{q_n}\right)_n \quad \text{(the path)}$$

The *formulae* for the sequence $\{C_i\}$ of *convergents* are:

$$C_0 = \frac{0}{n}$$

$$C_1 = \frac{1}{(u_1 - 1)n + 1}$$

$$C_2 = \frac{u_2}{u_2[(u_1 - 1)n + 1] + n}$$

$$C_3 = \frac{u_3 u_2 + 1}{u_3\{u_2[(u_1 - 1)n + 1] + n\} + [(u_1 - 1)n + 1]}$$

and so on, by the usual procedure, until

$$C_{k+1} = p_n/q_n \text{ is reached.}$$

We can then write this rational number as follows:

$$\frac{p_n}{q_n} = 1/([(u_1 - 1)n + 1] + n/(u_2 + 1/(u_3 + \cdots + 1/(u_k + 1))))$$

Convergents from continued fractions of the R-subtree

The *continued fraction* in n-RT

$$[u_0; u_1, u_2, \ldots, u_k, 1]_n = \left(\frac{p_n}{q_n}\right)_n \quad \text{(the path)}$$

The *formulae* for the sequence $\{C_i\}$ of *convergents* are:

$$C^* = \frac{n}{0}$$

$$C_0 = \frac{(u_0 - 1)n + 1}{1}$$

$$C_1 = \frac{u_1[(u_0 - 1)n + 1] + n}{u_1}$$

$$C_2 = \frac{u_2\{u_1[(u_0 - 1)n + 1] + n\} + [(u_0 - 1)n + 1]}{u_2 u_1 + 1}$$

and so on, until $C_{k+1} = p_n/q_n$ is reached.

We can write the final convergent in the form:

$$\frac{p_n}{q_n} = [(u_0 - 1)n + 1] + n/(u_1 + 1/(u_2 + \cdots + 1/(u_k + 1)))$$

The proofs of all the above formulae and algorithms are very simply established by induction.

5. SOME NUMBER FUNCTIONS

Having generalized the notions of Euclid's algorithm and simple continued fractions, and arrived at the concept of order-n arithmetics, it is natural that we should consider the consequences with regard to coprimeness and primeness, when using the various arithmetics. The following brief notes indicate directions which we believe have potential for fruitful further study. Further details may be found in [3].

The two number functions S and T

Consider a rational number $\alpha p/\alpha q \in Q^+$, $\alpha \in N$, with $g.c.d._1(p,q) = 1$; and its equivalence class in Q^+ which we shall denote by $[p/q]$.

Definitions

(i) Let S be a functions, with argument $[p/q]$, mapping the rational number equivalence classes to N, the set of natural numbers, in the following way:

$$S([p/q]) = c,$$

where c is the maximum number of n-RTs in which any representative of the equivalence class occurs.

[N.B. This is the maximum, over α, of the function T to be defined in (ii)]

(ii) Let T be a function, with argument $\alpha p/\alpha q$, mapping the positive rational numbers to N in the following way:

$$T(\alpha p/\alpha q) = t, \quad \text{where } \alpha \in N, \text{ and } g.c.d._1(p,q) = 1,$$

such that t is the number of n-RTs to which $\alpha p/\alpha q$ belongs.

Computing S([p/q])

We show in [3] that the value of $S([p/q])$ is equal to the number of divisors of $|p-q|$.

Computing T(αp/αq)

It is immediately evident that

$$T(\alpha p/\alpha q) \leq S([p/q]) \text{ for all } \alpha \in N$$

The value of $T(\alpha p/\alpha q)$ is obtained by finding the number of values of n which satisfy

(i)   for $\alpha p < \alpha q$,

$$g.c.d._1\left(\alpha p, \frac{\alpha(q-p)}{n}\right) = 1$$

(ii)   for $\alpha p > \alpha q$,

$$g.c.d._1\left(\alpha q, \frac{\alpha(p-q)}{n}\right) = 1.$$

Example

Let $p/q = 25/37$.   Then $|p - q| = 12$, which has divisors 1,2,3,4,6,12.   Hence $S([25/37]) = 6$. Note also that the rational number 25/37 occurs as a weight in just the trees 1-RT, 2-RT, 3-RT, 4-RT, 6-RT and 12-RT.

Consider now $\alpha p / \alpha q$ with $\alpha = 3$.

Then $T(3 \times 25/3 \times 37)$ $= T(75/111)$

$= 3$, since

$g.c.d._1\left(75, \frac{36}{n}\right) = 1$ when $n = 9,\ 18,\ 36$. Note, then, that 75/111 occurs only in the trees 9-RT, 18-RT and 36-RT.

$g.c.d._n$, coprimeness, generalized Euler functions

We first indicate reasons for the term $g.c.d._n$ and for calling its results a "greatest common divisor" in order-n arithmetic.

The n-RTs are all generalizations from the 1-RT (regular knot tree, RKT) and the 2-RT (regular Moebius knot tree, RMKT), which supply the arithmetics for modeling the processes of tying single-string, regular, cylindrical braids and single-string Moebius braids respectively.

In the 1-RT, for all rational numbers p/q on its nodes, we have $g.c.d._1(p,q) = 1$. In braiding theory, that indicates that the regular cylindrical braid can be made with one string. All the p/q occur in "lowest form", and so EA_1 is the same as the classical Euclid's algorithm. If p/q were not in "lowest form" ($g.c.d._1(p,q) = d > 1$), that would indicate that to make the corresponding regular cylindrical braid would require d strings.

Similar properties hold for the rational numbers in the 2-RT, with regard to $g.c.d._2(p,q) = 1$ and the numbers of strings required to tie regular Moebius braids.

We are led to say that, if p/q is in the 2-RT, then $g.c.d._2(p,q) = 1$, by the EA_2 algorithm. If we find by EA_2 that for some given p/q the $g.c.d._2(p,q) = d > 1$ then d strings will be required to tie the corresponding regular Moebius braid.

It is natural, then, for us to say that whenever $g.c.d._n(p,q) = 1$, for a rational number p/q occurring in the n-RT, then p and q are **coprime in order-n arithmetic**.

Since primeness of a number (in a set N^*) is evidently a consequence of the condition of coprimeness amongst the numbers in this set N^*, we can now discuss the concept of primeness of integers in N^* (with regard to order-n arithmetics). First we need to note that the n-RT, by considering the pairs (p,q) of integers (which occur as p/q on its nodes), automatically partitions the set N of natural numbers into n classes. Each class is associated with one of the integers in the set $\{1,2,\cdots,n\}$; we define the classes thus:

$$N_{n,1} = \{1 + in \mid i = 0,1,2,\cdots\}$$
$$N_{n,2} = \{2 + in \mid i = 0,1,2,\cdots\}$$
$$\vdots \qquad \vdots$$
$$N_{n,n} = \{n + in \mid i = 0,1,2,\cdots\}$$

Any pair (p,q) occurs "within" just one of these classes; none can occur in such a way that $p \in N_{n,i}$ and $q \in N_{n,j}$ with $i \neq j$. Thus only for $n = 1$ (and hence only in 1-RT) can we speak of comparing p and q in all $(p,q) \in N \times N$.

Note that $N_{n,i} \cap N_{n,j} = \emptyset$, and $\bigcup\limits_{i=1}^{n} N_{n,i} = N$.

We shall call these classes **Integer Classes** within the order-n arithmetic. We can now define **primeness** generally.

Definition: Let $m \in N_{n,j}$. Then m is a prime within the integer class $N_{n,j}$ iff $g.c.d._n(m,w) = 1$ for any $w \in N_{n,j}$ and $w < m$.

Example:

Consider the integer 53 in the order-6 arithmetic. We find which Integer Class it belongs to by computing $|53|_6 = 5$ (using ordinary modular arithmetic). Then the Integer Class of 53 is:

$$N_{6,5} = \{5,11,17,23,29,35,41,47,53,\cdots\}.$$

Now 53 is prime *iff* $g.c.d._6 (53,w) = 1$ for each of $w = 5,11,\cdots,47$. We must test each case, using the $g.c.d._6$ formula, derived from EA_6 in Section 7.

Thus:

$$
\begin{aligned}
g.c.d._6(53,5) &= g.c.d._1(5,13) &= 1\\
g.c.d._6(53,11) &= g.c.d._1(11,18) &= 1\\
g.c.d._6(53,17) &= g.c.d._1(17,23) &= 1\\
g.c.d._6(53,23) &= g.c.d._1(23,28) &= 1\\
g.c.d._6(53,29) &= g.c.d._1(29,33) &= 1\\
g.c.d._6(53,35) &= g.c.d._1(35,38) &= 1\\
g.c.d._6(53,41) &= g.c.d._1(41,43) &= 1\\
g.c.d._6(53,47) &= g.c.d._1(47,48) &= 1
\end{aligned}
$$

Since all the $g.c.d._6$ values are 1, the integer 53 is prime in order-6 arithmetic.

Note that 53 is also prime in order-1 arithmetic; this does not provide a rule, however. For counter-examples, in order-6 arithmetic the integers 8,9,10,12 and 15 are all primes within their respective Integer Classes. Thus, to take 15 as a last example, we have

$$15 \in N_{6,3} \text{ which is } \{3,9,15,\ldots\}, \text{ and}$$
$$\left. \begin{aligned} g.c.d._6(15,3) = g.c.d._1\ (3,5)\ &= 1\\ g.c.d._6(15,9) = g.c.d._1(9,10) &= 1 \end{aligned} \right\} \text{ hence 15 is prime.}$$

Note that with this generalization of primeness, it is only meaningful to talk about a given integer being prime **within its own Integer Class.**

All manner of questions may now be asked about the existence and frequencies of occurrence of these new primes within Integer Classes.

We end these notes by defining generalized Euler totient functions. It will be evident that such a function can be defined on each Integer Class.

Definition:

Let $N_{n,j}$ be a given Integer Class. Then

$$\phi_{n,j}(m) \text{ is the count of numbers}$$
$$w \in N_{n,j}, \text{ with } w < m, \text{ such that}$$
$$g.c.d._n[m,w] = 1$$

An evident result is that m is prime in $N_{n,j}$ *iff* $\phi_{n,j}(m)$ is the greatest integer less than m/n.

Example:

The generalized Euler functions in order–6 arithmetic (some values)					
$N_{6,1}$	$\phi_{6,1}(7)=1$	$\phi_{6,1}(13)=2$	$\phi_{6,1}(19)=3$	$\phi_{6,1}(25)=4$	$\phi_{6,1}(31)=5$ $\cdots$
$N_{6,2}$	$\phi_{6,2}(8)=1$	$\phi_{6,2}(14)=1$	$\phi_{6,2}(20)=2$	$\phi_{6,2}(26)=2$	$\phi_{6,2}(32)=3$ $\cdots$
$N_{6,3}$	$\phi_{6,3}(9)=1$	$\phi_{6,3}(15)=2$	$\phi_{6,3}(21)=2$	$\phi_{6,3}(27)=3$	$\phi_{6,3}(33)=4$ $\cdots$
$N_{6,4}$	$\phi_{6,4}(10)=1$	$\phi_{6,4}(16)=1$	$\phi_{6,4}(22)=2$	$\phi_{6,4}(28)=2$	$\phi_{6,4}(34)=3$ $\cdots$
$N_{6,5}$	$\phi_{6,5}(11)=1$	$\phi_{6,5}(17)=2$	$\phi_{6,5}(23)=3$	$\phi_{6,5}(29)=4$	$\phi_{6,5}(35)=4$ $\cdots$
$N_{6,6}$	$\phi_{6,6}(12)=1$	$\phi_{6,6}(18)=1$	$\phi_{6,6}(24)=1$	$\phi_{6,6}(30)=1$	$\phi_{6,6}(36)=2$ $\cdots$

Note: In the number theory texts, Euler's function $\phi(x)$ is normally defined as having value 1 at $x=1$. In our generalized Euler functions, we leave undefined the values of $\phi_{n,j}(j)$ (note that j is the least integer in the class $N_{n,j}$), until more is known on the properties of these functions.

REFERENCES

[1]   Schaake, A. G. and Turner, J. C. *A New Theory of Braiding (RR1/1)*, Research Report No. 165. University of Waikato, Hamilton, N.Z. (1988): 43 pp.

[2]   Schaake, A. G. and Turner, J. C. *A New Theory of Braiding (RR1/2)-Algorithms for Regular Knots*, Research Report No. 168. University of Waikato, Hamilton, N.Z. (1988): 41 pp.

[3]   Schaake, A. G. and Turner, J. C. *Generalizing Euclid's Algorithm, via the Regular and Moebius Knot Trees. Order n-Arithmetics* Research Report No. 196. University of Waikato, Hamilton, N.Z. (1990): 61 pp.

[4]   Schaake, A. G. and Turner, J. C. *New Methods for Solving Quadratic Diophantine Equations (Part 1 and Part II)*, Research Report No. 192. University of Waikato, Hamilton, N.Z. (1989): 99 pp.

[5]   Schaake, A. G. and Turner, J. C. *A New Chapter for Pythagorean Triples.* University of Waikato, Hamilton, N.Z. (1989): 155 pp.

GENERALIZED FIBONACCI AND LUCAS FACTORIZATIONS

A. G. Shannon, R. P. Loh and A. F. Horadam

1. INTRODUCTION

Brillhart, Montgomery and Silverman [2] have developed tables of Fibonacci and Lucas factorizations. In doing so they have used primitive parts of these numbers. These are analogous to the proper divisors defined by Horadam, Loh and Shannon [5]. The study of the multiplicative structure of the Fibonacci and Lucas numbers F_n and L_n by Brillhart et. al. was the starting point for our investigation of the solutions of the functional equation

$$f(2 - x^2) = f(x)f(-x),\qquad(1.1)$$

We show in this paper how these are related to the generalized Fibonacci and Lucas numbers $A_n(x)$ (defined in (2.1)) which we have previously related to the study of proper divisors [6]. Furthermore, we show that (1.1) is a generalization of the familiar result

$$F_{2n} = F_n L_n \qquad(1.2)$$

which was fundamental to the multiplicative properties in Brillhart et. al. While Brillhart et. al. were concerned with the computational aspects we are concerned with the recurrence relations and their associated divisibility properties.

2. GENERALIZED FIBONACCI AND LUCAS NUMBERS

Consider the sequence $\{A_n(x)\}$ where $x \neq 2$ is an integer:

$$A_n(x) = xA_{n-2}(x) - A_{n-4}(x) \qquad (n \geq 4) \qquad(2.1)$$

271

G. E. Bergum et al. (eds.), Applications of Fibonacci Numbers Volume 4, 271–276.

with initial conditions $A_0(x) = 0, A_1(x) = A_2(x) = 1, A_3(x) = x - 1$. The first few polynomials of this sequence are given in Table 1.

n	$A_n(x)$
0	0
1	1
2	1
3	$x - 1$
4	x
5	$x^2 - x - 1$
6	$x^2 - 1$
7	$x^3 - x^2 - 2x + 1$
8	$x^3 - 2x$
9	$x^4 - x^3 - 3x^2 + 2x + 1$
10	$x^4 - 3x^2 + 1$
11	$x^5 - x^4 - 4x^3 + 3x^2 + 3x - 1$
12	$x^5 - 4x^3 + 3x$
13	$x^6 - x^5 - 5x^4 + 4x^3 + 6x^2 - 3x - 1$

Table 1. $A_n(x)$

We observe that $\{A_{2n-3}(\pm 3)\}$ and $\{A_{2n-3}(-7)\}$ are related to the Fibonacci and Lucas numbers as can be seen in Table 2.

n	$A_{2n-3}(3)$	$A_{2n-3}(-3)$	$A_{2n-3}(-7)$
1	1	1	1
2	1	1	1
3	2	-4	-8
4	5	11	55
5	13	-29	-377
6	34	76	2584

Table 2. $A_{2n-3}(x)$ for $x = \pm 3, -7$

Table 2 illustrates how $f(x) = A_{2n-3}(x), x = \pm 3, -7$ satisfies (1.1) and (1.2); for example, if $n = 5$, then $A_7(3)A_7(-3) = 13. \; -29 = -377 = A_7(-7)$.

For notational convenience, we utilize the generalized Fibonacci-type sequence $\{u_n(x)\} \equiv \{u_n\}$ defined by $u_0 = 0, u_1 = 1$, and $u_{n+2} = xu_{n+1} - u_n$ for $n > 1$. Then it can be readily proved using mathematical induction that

$$A_{2n}(x) = u_n \tag{2.2}$$

$$A_{2n-3}(x) = u_{n-1} - u_{n-2}, \; n \geq 2 \tag{2.3}$$

$$A_{2n-3}(-x) = (-1)^n(u_{n-1} + u_{n-2}), \; n \geq 2 \tag{2.4}$$

$$A_{2n-3}(2 - x^2) = \frac{(-1)^n}{x} (u_{2n-4} + u_{2n-2}), \; n \geq 2. \tag{2.5}$$

For example, when $n = 3$, $\quad A_3(2 - x^2) = 1 - x^2$

$$\frac{(-1)^3}{x}(u_2 + u_4) = \frac{-1}{x}(x + x^3 - 2x) = 1 - x^2.$$

Following Horadam [3] we have

$$u_{n+1} = xu_n - u_{n-1}$$
$$= u_2 u_n - u_1 u_{n-1}$$
$$u_{n+2} = xu_{n+1} - u_n$$
$$= (x^2 - 1)u_n - xu_{n-1}$$
$$= u_3 u_n - u_2 u_{n-1}$$

whence it can be established by induction that

$$u_{n+r-1} = u_r u_n - u_{r-1} u_{n-1} \tag{2.6}$$

which agrees with Horadam's equation (3.14) [3] after a suitable shift in subscripts.

We are now in a position to prove:

Theorem: If $n > 2$, $f(x) = A_{2n-3}(x)$ satisfies (1.1).

Proof: $\qquad A_{2n-3}(x) \, A_{2n-3}(-x) = (-1)^n \, (u_{n-1}^2 - u_{n-2}^2), \; n \geq 2$ (from (2.3), (2.4))

$$A_{2n-3}(2 - x^2) = (-1)^n \, u_{n-3}, \; n \geq 3, \; \text{(from (2.6))}$$

and $\qquad\qquad\qquad u_{2n-3} = u_{n-1}^2 - u_{n-2}^2, \; n \geq 2,$

on replacing n by $n - 1$ and r by $n - 1$ in (2.6).

There are other ways of proving this but we have chosen this approach because it includes generalizations of results familiar to Fibonacci researchers. The theorem establishes that (1.1) is a generalization of (1.2) in so far as the $A_n(x)$ are generalizations of the Fibonacci numbers.

3. PROPER DIVISORS

The first few values of $A_{2n-3}(x)$ are displayed in Table 3. Further information about the sequence $\{A_n(x)\}$ may be found in [6]. By analogy with the primitive divisors of [2] we quote some results on proper divisors proved in [5]. We extend Vorob'ev's notion of a proper divisor [7] in the following definition. For any sequence $\{u_n\}$, $n \geq 1$, where $u_n \in Z - \{0\}$ or $u_n(x) \in Z(x)$, the proper divisor w_n is the quantity implicitly defined for $n \geq 1$ by $w_1 = u_1$ and $w_n = \text{l.c.m.} \ \{d : d \mid u_n \text{ and g.c.d. } (d, w_m) = 1 \text{ for every } m < n\}$.

n x	1	2	3	4	5	6	7	8
1	1	1	0	− 1	− 1	0	1	1
2	1	1	1	1	1	1	1	1
3	1	1	2	5	13	34	89	233
4	1	1	3	11	41	153	571	2131
5	1	1	4	19	91	436	2089	10009
6	1	1	5	29	169	985	5741	33461
7	1	1	6	41	281	1926	13201	90481
8	1	1	7	55	433	3409	26839	211303
9	1	1	8	71	631	5608	49841	442961
10	1	1	9	89	881	8721	86329	854569
11	1	1	10	109	1189	12970	141481	1543321
12	1	1	11	131	1561	18601	221651	2641211
13	1	1	12	155	2003	25884	334489	4322473
14	1	1	13	181	2521	35113	489061	6811741

Table 3. Values of $A_{2n-3}(x)$

Proper divisors for the sequence of polynomials $\{A_n(x)\}$ are shown in Table 4.

n	$A_n(x)$	$w_n(x)$
3	$x - 1$	$x - 1$
4	x	x
5	$x^2 - x - 1$	$x^2 - x - 1$
6	$x^2 - 1$	$x + 1$
7	$x^3 - x^2 - 2x + 1$	$x^3 - x^2 - 2x + 1$
8	$x^3 - 2x$	$x^2 - 2$
9	$x^4 - x^3 - 3x^2 + 2x + 1$	$x^3 - 3x - 1$
10	$x^4 - 3x^2 + 1$	$x^2 + x - 1$
11	$x^5 - x^4 - 4x^3 + 3x^2 + 3x - 1$	$x^5 - x^4 - 4x^3 + 3x^2 + 3x - 1$
12	$x^5 - 4x^3 + 3x$	$x^2 - 3$
13	$x^6 - x^5 - 5x^4 + 4x^3 + 6x^2 - 3x - 1$	$x^6 - x^5 - 5x^4 + 4x^3 + 6x^2 - 3x - 1$

Table 4. Proper divisors for $\{A_n(x)\}$

From the definition of proper divisors we have that for $A_n(x)$ defined in (2.1)

$$A_n(x) = \prod_{d \mid n} w_d(x).$$

For example,
$$A_{21}(x) = w_1(x)w_3(x)w_7(x)w_{21}(x)$$

where
$$w_{21}(x) = x^6 + x^5 - 6x^4 - 6x^3 + 8x^2 + 8x + 1$$

By analogy with the results of Brillhart et. al. for the multiplicative structure of the ordinary Fibonacci and Lucas numbers, we have

$$w_n(x) = \prod_{d \mid n} A_d(x)^{\mu(n/d)} \tag{3.1}$$

where μ is the Möbius function.

This is proved in [5] where associated properties are also developed. Thus, for instance,

$$w_{60}(x) = \frac{A_4(x)A_6(x)A_{10}(x)A_{60}(x)}{A_2(x)A_{12}(x)A_{20}(x)A_{30}(x)} \tag{3.2}$$

4. CONCLUDING COMMENTS

What we have done here is to relate the well-known identity (1.2) to the functional equation (1.1) and this in turn to the fourth order linear recurring sequence $\{A_n(x)\}$. In this way we have tried to view the factorization processes of Brillhart et. al. from the perspective of generalized Fibonacci sequences. We have not considered all solutions of (1.1). The work also suggests the consideration of the more general functional equation

$$f(2k - x^2) = f(x)f(-x). \tag{4.1}$$

We have chosen some aspects of the case $k = 1$. Some features of the case $k = 0$ have been investigated by Wall [8]. A more complete study of the pattern for general k is in progress.

Thanks are due to an anonymous referee for meticulous comments on earlier versions of this paper.

REFERENCES

[1]   Abramowitz, M. and Stegun, I. A. (eds.) Handbook of Mathematical Functions, Dover, New York, 1965.

[2]   Brillhart, J., Montgomery, P. L. and Silverman, R. D. "Tables of Fibonacci and Lucas Factorizations." *Mathematics of Computation, 50*, No. 181, (1988): pp. 251-260.

[3]   Horadam, A. F. "Basic Properties of a Certain Generalized Sequence of Numbers." *The Fibonacci Quarterly, 3*, No. 3 (1965): pp. 161-176.

[4]   Horadam, A. F. "Tschebyscheff and Other Functions Associated with the Sequence $\{w_n(a,b;p,q)\}$." *The Fibonacci Quarterly, 7*, No. 1 (1969): pp. 14-22.

[5]   Horadam, A. F., Loh, R. P. and Shannon, A. G. "Divisibility Properties of Some Fibonacci-type Sequences" in A. F. Horadam and W. D. Wallis (eds.), Combinatorial Mathematics VI, Springer-Verlag, Heidelberg, 1979, pp. 55-64.

[6]   Shannon, A. G., Horadam, A. F. and Loh, R. P. "Proper Divisors of Lenakel Numbers." *Mathematics Student, 45*, No. 2 (1977): pp. 63-73.

[7]   Vorob'ev, N. N. Fibonacci Numbers. Pergamon, Oxford, 1961.

[8]   Wall, G. E. "On the Conjugacy Classes in the Unitary, Symplectic and Orthogonal Groups." *Journal of the Australian Mathematical Society, 3*, No. 1 (1963): pp. 1-62.

ON EVEN FIBONACCI PSEUDOPRIMES

Lawrence Somer

1. INTRODUCTION

Throughout this paper, p will denote an odd prime and $\{F_n\}$ and $\{L_n\}$ will denote the Fibonacci and Lucas numbers respectively. It is well-known (see [4, page 44]) that

$$L_p \equiv 1 \ (mod \ p). \tag{1}$$

In [10], the stronger result that

$$L_{p^n} \equiv 1 \ (mod \ p) \tag{2}$$

for $n \geq 0$ was proven. Following Di Porto and Filipponi in [6], we will call the composite positive integer M a Fibonacci pseudoprime (FPSP) if congruence (1) holds for M, that is

$$L_M \equiv 1 \ (mod \ M). \tag{3}$$

In [6], Di Porto and Filipponi listed all the Fibonacci pseudoprimes up to 1,000,000. They noted that all these FPSP's are odd and conjectured that there are no even FPSP's, providing additional constraints on their existence. In this paper, we will extend these constraints on the existence of even FPSP's. For the remainder of this paper, N will denote an even FPSP.

2. PRELIMINARIES

The periods of $\{L_n\}$ and $\{F_n\}$ modulo p will be denoted by $\mu(p)$ and $\mu'(p)$ respectively. It is well-known (see [5, pages 344-345]) that both $\{L_n\}$ and $\{F_n\}$ are purely periodic modulo p. The restricted period of $\{L_n\}$ modulo p, denoted by $\alpha(p)$, is the least positive integer t such that

277

G. E. Bergum et al. (eds.), Applications of Fibonacci Numbers Volume 4, 277–288.

$$L_{n+t} \equiv sL_n \ (mod \ p) \tag{4}$$

for all n and some non-zero residue s. The residue s is called the multiplier of $\{L_n\}$ modulo p. The restricted period $\alpha'(p)$ and the multiplier s' of $\{F_n\}$ modulo p are defined similarly. If $k = \alpha(p)$, it is clear that

$$L_{ki+n} \equiv s^i L_n \ (mod \ p) \tag{5}$$

for all n.

It is easy to see that

$$\beta(p) = \mu(p)/\alpha(p)$$

and

$$\beta'(p) = \mu'(p)/\alpha'(p)$$

are the exponents of the multiplier modulo p for $\{L_n\}$ and $\{F_n\}$ respectively. It is known (see [8, pages 17-18]) that if $p \neq 5$, then

$$\mu(p) = \mu'(p), \ \alpha(p) = \alpha'(p), \ s \equiv s' \ (mod \ p), \ \text{and} \ \beta(p) = \beta'(p), \tag{6}$$

while

$$\mu(5) = 4, \ \alpha(5) = 1, \ s \equiv 3 \ (mod \ 5), \ \beta(5) = 4, \tag{7}$$

and

$$\mu'(5) = 20, \ \alpha'(5) = 5, \ s' \equiv 3 \ (mod \ 5), \ \beta'(5) = 4. \tag{8}$$

The following two theorems give the possibilities for $\mu(p)$, $\alpha(p)$, $\beta(p)$, and $s(mod \ p)$:

Theorem 1: (i) $\alpha(p) \mid p - (5/p)$, where $(5/p)$ denotes the Legendre symbol.

(ii) If $p \equiv 1 \ (mod \ 4)$ and $p \neq 5$, then

$$\alpha(p) \mid (p - (5/p))/2.$$

(iii) If $p \equiv 3 \ (mod \ 4)$, then

$$\alpha(p) \nmid (p - (5/p))/2.$$

(iv) If $(5/p) = 1$, then

$$\mu(p) \mid p - 1.$$

Proof: (i) and (iv) These follow from (6), (7), (8), and [4, pages 44-45].
(ii) and (iii) These follow from (6), (7), (8), and [7, page 441].

Theorem 2: (i) $\beta(p) = 1$, 2, or 4; $s \equiv 1$, -1, $\sqrt{-1}$, or $-\sqrt{-1}$ (mod p), where $\sqrt{-1}$ denotes the residue r such that $r^2 \equiv -1$ (mod p) and $1 \leq r \leq (p-1)/2$.

(ii) $\beta(p) = 1$ iff $\alpha(p) \equiv 2$ (mod 4).

(iii) $\beta(p) = 2$ iff $\alpha(p) \equiv 0$ (mod 4).

(iv) $\beta(p) = 4$ iff $\alpha(p) \equiv 1$ (mod 2).

(v) If $p \equiv 11$ or 19 (mod 20), then $\beta(p) = 1$.

(vi) If $p \equiv 3$ or 7 (mod 20), then $\beta(p) = 2$.

(vii) If $p = 5$ or $p \equiv 13$ or 17 (mod 20), then $\beta(p) = 4$.

(viii) If $p \equiv 21$ or 29 (mod 40), then $\beta(p) = 1$ or 4 and both possibilities occur.

(ix) If $p \equiv 1$ or 9 (mod 40), then $\beta(p)$ may be 1, 2, or 4 and all these possibilities occur.

Proof: Parts (i)-(ix) follow from (6), (7), (8), and [12, Theorems 2 and 4].

The following results are well-known and will be needed for our further work.

$$F_{2n} = F_n L_n. \tag{9}$$

$$F_{2n+1} = F_n^2 + F_{n+1}^2. \tag{10}$$

$$L_{2n} = L_n^2 + 2(-1)^{n-1}. \tag{11}$$

$$L_n^2 = 5F_n^2 + 4(-1)^n. \tag{12}$$

$$L_n = F_{n-1} + F_{n+1}. \tag{13}$$

$$L_{m+n} = F_n L_{m-1} + F_{n+1} L_m. \tag{14}$$

$$F_n \mid F_{dn}. \tag{15}$$

$$L_{2n+4} = 3L_{2n+2} - L_{2n}. \tag{16}$$

Identities (9) and (11) follow from results given in [4, page 32]. Identities (10), (13), and (14) can be proved by use of mathematical induction. Identity (12) is proved in [4, page 36]. The result given in (15) is proved in [4, page 33]. Identity (16) follows as a special case from [9].

3. THE MAIN THEOREMS

Theorems 3-12 provide constraints on the existence of an even Fibonacci pseudoprime N.

Theorem 3: N is not a multiple of 4.

Theorem 4: N has at least two distinct odd prime factors.

Theorem 5: Every odd prime factor of N must be $\equiv 1$ (*mod* 4).

The following constraint was proved by Di Porto and Filipponi in Theorem 5 of [6].

Theorem 6: (Di Porto and Filipponi): No even FPSP can be twice an odd FPSP.

Theorem 7: Suppose that $p > 5$ and $\alpha(p)$ is odd. Suppose that there does not exist an integer n such that $2 \le n \le (\alpha(p) - 1)/2$ and $L_n \equiv \pm 1$ or $\pm\sqrt{-1}$ (*mod* p). Then p does not divide any even FPSP.

Theorem 8: Suppose that $p \ge 7$ and $\alpha(p)$ is even. Suppose that there does not exist an even integer $2n$ such that $2 \le 2n \le (\alpha(p) - 2)/2$ and $L_{2n} \equiv \pm 1$ (*mod* p). Then p does not divide any even FPSP.

Theorem 9: No odd prime Lucas or Fibonacci number can divide any even FPSP.

Theorem 10: Let $k = \alpha(p)$. Suppose that there exists an integer r such that $2 \le r < k/2$ and $L_r \equiv \pm 1$ or $\pm\sqrt{-1}$ (*mod* p). Suppose further that either $4 \mid (r, k)$ or that there exists an odd prime p_1 not dividing any even FPSP such that $p_1 \mid (r, k)$. Then p does not divide any even FPSP.

Theorem 11: Let $k = \alpha(p)$.

    (i) Suppose $k = 3p_1 d$, where p_1 is an odd prime not dividing any even FPSP. Then p does not divide any even FPSP.

    (ii) Suppose $12 \mid k$. Then p does not divide any even FPSP.

Utilizing the constraints given in the above theorems, David Banks of the University of North Carolina at Chapel Hill performed computer calculations to find a lower bound for the set of primes which divide some even FPSP. Upon noting by Theorem 4 that any even FPSP has at least two distinct odd prime divisors, we have the following theorem:

Theorem 12: Let N be an even FPSP. If $p \mid N$ and p is odd, then $p > 3,797,117$. Moreover, we must have that

$$N > 2(3,797,117)^2 = 28,836,195,023,378.$$

In searching for odd primes less than or equal to 3,797,117 which could possibly divide some even FPSP, Banks needed to examine only those primes which are greater than 13 and which are congruent to 1 (*mod* 4) by Theorems 9 and 5. In examining these primes p, Banks' computer program first determined $\alpha'(p) = \alpha(p)$ by finding the least n such that $F_n \equiv 0 \ (mod \ p)$. The computer program then examined only those primes less than or equal to 3,797,117 for which $3 \nmid \alpha(p)$. For these primes p, the program calculated L_n^2 modulo p for $2 \leq n < \alpha(p)/2$. For those primes $p \leq 3,797,117$ such that $2 \nmid \alpha(p)$ and $3 \nmid \alpha(p)$, the program found no terms $L_n^2 \equiv \pm 1$ modulo p. For those primes $p \leq 3,797,117$ such that $2 \mid \alpha(p)$ and $3 \nmid \alpha(p)$, the program found no terms $L_n^2 \equiv 1 \ (mod \ p)$. It then followed from Theorems 7 and 8 that if $p \leq 3,797,117$ and $3 \nmid \alpha(p)$, then p does not divide any even FPSP. Thus, the only possible primes $p \leq 3,797,117$ which can divide some even FPSP are those for which $p \equiv 1$ $(mod \ 4)$ and $3 \mid \alpha(p)$. By inspection, one sees that $\alpha(p) \geq 9$ in this case. However, if $3 \mid \alpha(p)$ and $p \leq 3,797,117$, then every prime divisor q of $\alpha(p)/3$ must also satisfy $q \equiv 1 \ (mod \ 4)$ and $3 \mid \alpha(q)$ by Theorems 5 and 11. However, then $q \leq (p+1)/3$ by Theorem 1. By the method of infinite descent, one then sees that the primes $p \leq 3,797,117$ such that $p \equiv 1 \ (mod \ 4)$ and $3 \mid \alpha(p)$ also do not divide any even FPSP.

Remark: In Theorem 7 of [6], Di Porto and Filipponi showed that if N is an even FPSP and $p \mid N$, then $p \geq 23$.

4. NECESSARY LEMMAS

The following lemmas will be needed for the proofs of Theorems 3-12, excluding the proof of Theorem 6.

Lemma 1: (i) If $\alpha(p)$ is even, then p divides L_n for $0 \leq n \leq \alpha(p)$ if and only if $n = \alpha(p)/2$.

(ii) If $\alpha(p)$ is odd, then p does not divide L_n for $0 \leq n \leq \alpha(p)$.

Proof: Parts (i) and (ii) follow from the definition of $\alpha(p)$ and from (9).

Lemma 2: Let c be a fixed integer such that $1 \leq c \leq \alpha(p) - 1$. Then the ratios L_{n+c}/L_n are all distinct modulo p for $0 \leq n \leq \alpha(p) - 1$ if the ratio $L_{n+c}/0$ is denoted by ∞.

Proof: It follows from the definition of $\alpha(p)$ and the linearity of the second-order recursion relation defining $\{L_n\}$ that if

$$L_i/L_{i-1} \equiv L_j/L_{j-1} \ (mod \ p),$$

then $\alpha(p) \mid (j - i)$. Now assume that

$$L_{i+c}/L_i \equiv L_{j+c}/L_j \ (mod \ p) \tag{17}$$

for $0 \leq i \leq j \leq \alpha(p) - 1$. We will show that $i = j$. It follows from Lemma 1 that if $L_i \equiv 0 \ (mod \ p)$, then $i = j$. Thus, we can assume that neither L_i nor L_j is congruent to

$0 \pmod{p}$. By (14) we have that

$$L_{i+c} \equiv F_c L_{i-1} + F_{c+1} L_i \pmod{p}, \tag{18}$$

$$L_{j+c} \equiv F_c L_{j-1} + F_{c+1} L_j \pmod{p}. \tag{19}$$

From (17), (18), and (19), we have that

$$(F_c L_{i-1} + F_{c+1} L_i) L_j \equiv (F_c L_{j-1} + F_{c+1} L_j) L_i \pmod{p} \tag{20}$$

or

$$F_c L_{i-1} L_j \equiv F_c L_{j-1} L_i \pmod{p}. \tag{21}$$

Since $\alpha'(p) = \alpha(p)$ if $p \neq 5$ and $\alpha'(5) > \alpha(5)$, it follows that $F_c \not\equiv 0 \pmod{p}$. Thus, we obtain from (21) that

$$L_i / L_{i-1} \equiv L_j / L_{j-1} \pmod{p}. \tag{22}$$

Since

$$0 \leq j - i \leq \alpha(p) - 1,$$

we must have that $i = j$.

Remark: The statement and proof of Lemma 2 generalizes results obtained by Bruckner [3, pages 217-219].

Lemma 3: Let $k = \alpha(p)$. Let s be the multiplier of $\{L_n\}$ *modulo* p. Then

$$L_{k-n} \equiv (-1)^n s L_n \pmod{p}$$

for $0 \leq n \leq k$.

Proof: This follows by induction through the use of the recursion relation defining L_n upon noting that $L_k \equiv 2s$, $L_{k+1} \equiv s$, and $L_{k-1} \equiv s - 2s \equiv -s \pmod{p}$.

Lemma 4: Let $k = \alpha(p)$. Let c be a fixed integer such that $1 \leq c < k/2$. Let n be a non-negative integer such that $n + c < k/2$. Then

$$(L_{n+c} / L_n)(L_{k-n} / L_{k-n-c}) \equiv (-1)^c \pmod{p}.$$

Proof: By Lemma 1, neither L_{n+c} nor L_n is congruent to 0 *modulo* p. The result now follows from Lemma 3.

Lemma 5: Let $k = \alpha(p)$. Let c be a fixed integer such that $1 \leq c < k/2$. Let n be a non-negative integer such that $n + c < k/2$.

(i) If c is even, then

$$L_{n+c} \not\equiv \pm L_n \ (mod \ p).$$

(ii) If c is odd, then

$$L_{n+c} \not\equiv \pm \sqrt{-1} \, L_n \ (mod \ p).$$

Proof: (i) Suppose that c is even and $L_{n+c} \equiv \pm L_n \ (mod \ p)$. By Lemma 4,

$$(L_{n+c}/L_n)(L_{k-n}/L_{k-n-c}) \equiv (-1)^c \equiv 1 \ (mod \ p).$$

This implies that

$$L_{n+c}/L_n \equiv L_{k-n}/L_{k-n-c} \ (mod \ p),$$

contradicting Lemma 2. The result now follows.

(ii) This proof is similar to that of (i).

Lemma 6: (i) If $3 \le 2n-1 \le \alpha(p)/2$, then $L_{2n-1} \not\equiv \pm 1 \ (mod \ p)$.

(ii) If $2 \le 2n < \alpha(p)/2$, then $L_{2n} \not\equiv \pm \sqrt{-1} \ (mod \ p)$.

Proof: (i) This follows from Lemma 5 (i) using the fact that $L_1 \equiv 1 \ (mod \ p)$.

(ii) This follows from Lemma 5 (ii) using the fact that $L_1 \equiv 1 \ (mod \ p)$.

Lemma 7: There exists at most one integer n such that $2 \le n \le \alpha(p)/2$ and $L_n \equiv \pm 1$ or $\pm \sqrt{-1} \ (mod \ p)$.

Proof: Let $k = \alpha(p)$. First note that by Lemma 1 (i), if $2 \mid k$, then $L_{k/2} \not\equiv \pm 1$ or $\pm \sqrt{-1}$ $(mod \ p)$. Now suppose that there exist integers g and h such that $2 \le g < h < k/2$, $L_g \equiv \pm 1$ or $\pm \sqrt{-1} \ (mod \ p)$, and $L_h \equiv \pm 1$ or $\pm \sqrt{-1} \ (mod \ p)$. First assume that $L_g \equiv \pm L_h \ (mod \ p)$. By Lemma 5 (i), we must have that g and h have different parity. Since $L_g \equiv \pm 1$ or $\pm \sqrt{-1}$ $(mod \ p)$, it follows from Lemma 6 that g and h have the same parity. This is a contradiction.

Thus, we must have that $L_g \equiv \pm \sqrt{-1} \, L_h \ (mod \ p)$. By Lemma 5, g and h have the same parity. However, since $L_g \equiv \pm 1$ or $\pm \sqrt{-1} \ (mod \ p)$, it follows from Lemma 6 that g and h have different parity, which again is a contradiction. The lemma now follows.

Lemma 8: Let N be an even FPSP. Suppose $p \mid N$, where $p \ge 7$. Let $k = \alpha(p)$. Suppose $N \equiv \pm r \ (mod \ p)$, where $0 \le r \le k/2$. Then $r \ne 0$ or 1. Moreover, if $2 \mid k$, then $r \ne k/2$.

Proof: First suppose that $r = 0$. Then $N = ki$ for some non-negative integer i. Note that $L_0 \equiv 2 \ (mod \ p)$. Since $p \ge 7$,

$$2 \not\equiv \pm 1 \text{ or } \pm\sqrt{-1} \ (mod \ p). \tag{23}$$

By (5),

$$L_N \equiv 2s^i \ (mod \ p). \tag{24}$$

Since $s \equiv \pm 1$ or $\pm\sqrt{-1}$ (mod p) by Theorem 2 (i), it follows from (23) and (24) that $N \not\equiv 1$ (mod p). This is a contradiction and r cannot equal 0.

Now suppose that $r = 1$. Then $N = kj \pm 1$ for some non-negative integer j. We cannot have that k is even, for then N would be odd. Thus, we must have that k is odd. By Theorem 2 (iv), $s \equiv \pm\sqrt{-1}$ (mod p). It then follows from (5) and Lemma 3 that j must be even. Thus, N is odd, which is a contradiction. Hence $r \neq 1$.

Finally, suppose that $2 \mid k$ and $r = k/2$. Then by (5) and Lemma 1 (i), $L_N \equiv 0$ (mod p), which is a contradiction. The lemma now follows.

Lemma 9: Let N be an even FPSP. Suppose $p \mid N$, where $p \geq 7$. Let $k = \alpha(p)$. Suppose $N = ki \pm r$, where i is a non-negative integer and $0 \leq r \leq k/2$. Let s be the multiplier of $\{L_n\}$ modulo p. Then

$$L_r \equiv \pm s^j \ (mod \ p)$$

for some non-negative integer j.

Proof: Note that $L_N \equiv 1$ (mod p). The result now follows from (5) and Lemma 3.

Lemma 10: Let $k = \alpha(p)$.

(i)   If $k \equiv 3$ (mod 6), then $L_{k/3} \equiv \pm\sqrt{-1}$ (mod p).

(ii)   If $k \equiv 6$ (mod 12), then $L_{k/3} \equiv -1$ (mod p).

(iii)   If $k \equiv 0$ (mod 12), then $L_{k/3} \equiv 1$ (mod p).

Proof: Parts (i)-(iii) follow from Theorem 6 in [11].

5. PROOFS OF THE MAIN THEOREMS

We are finally ready for the proofs of Theorems 3-12.

Proof of Theorem 3: Suppose that $4 \mid N$. Then

$$L_N \equiv 1 \ (mod \ 4) \tag{25}$$

because N is a FPSP. By inspection, one sees that the period of $\{L_n\}$ modulo 4 is equal to 6 and that $L_n \equiv 1$ (mod 4) if and only if $n \equiv 1$ (mod 6). Thus, by (25), we must have that $N \equiv 1$ (mod 6), which is a contradiction. Thus $4 \nmid N$ and the theorem follows.

Remark: In Proposition 2 of [6], Di Porto and Filipponi provide an alternate proof of Theorem 3.

Proof of Theorem 4: By Theorem 3, it suffices to prove that N is not of the for $2p^n$. By (2),

$$L_{p^n} \equiv 1 \ (mod \ p).$$

Thus, by (11),

$$L_{2p^n} \equiv L_{p^n}^2 + 2 \equiv 3 \not\equiv 1 \ (mod \ p).$$

A fortiori,

$$L_{2p^n} \not\equiv 1 \ (mod \ 2p^n).$$

Thus, $2p^n$ is not a FPSP and the theorem follows.

Remark: Di Porto and Filipponi in Theorem 3 of [6] have proved the special case of Theorem 4 that N is not of the form $2p$.

Proof of Theorem 5: Let $N = 2m$. Suppose that $p \mid N$. By Theorem 3 and (11),

$$L_N = L_m^2 + 2 \equiv 1 \ (mod \ N). \tag{26}$$

This implies that

$$L_m^2 \equiv -1 \ (mod \ p). \tag{27}$$

This can hold only if $p \equiv 1 \ (mod \ 4)$ since p is an odd prime.

Proof of Theorem 7: This follows from Theorem 2 (i) and Lemmas 8 and 9.

Proof of Theorem 8: By parts (ii) and (iii) of Theorem 2, $s \equiv \pm 1 \ (mod \ p)$. By Lemma 6 (i), if $2 \leq r \leq \alpha(p)/2$ and $L_r \equiv \pm 1 \ (mod \ p)$, then r is even. The theorem now follows from Lemmas 8 and 9.

Remark: Note that the computations used in the test of Theorem 8 can be performed by use of identity (16).

Proof of Theorem 9: The first few odd primes p such that $p = L_n$ for some n are $3 = L_2$, $11 = L_4$, and $29 = L_7$. It follows from Theorem 5 that 3 does not divide any even FPSP. By (6), (9), and Lemma 1, it follows that $\alpha(p) = 2n$. Since $L_{n-2} \geq L_2 = 3$, it follows that $L_{n-1} < L_n - 1 = p - 1$. Since $L_1 = 1$ and $\{L_n\}$ is strictly increasing for $n \geq 1$, it follows from Theorem 8 that p does not divide any even FPSP.

The first few odd primes p such that $p = F_n$ for some n are $3 = F_4$, $5 = F_5$, $13 = F_7$, and $89 = F_{11}$. By Theorem 5, 3 does not divide any even FPSP. By inspection, one sees that

$L_n \equiv 1 \ (mod \ 5)$ if and only if $n \equiv 1 \ (mod \ 4)$. Thus, 5 cannot divide any even FPSP. Using Theorem 7, one sees by inspection that 13 cannot divide any even FPSP. Since $F_2 = 1$, it follows from (15) and the fact that $\{F_n\}$ is strictly increasing for $n \geq 2$ that F_n can be a prime only if $n = 4$ or n is an odd prime.

Now suppose that $n \geq 11$ and $F_n = p \geq 89$. Let $n = 2t + 1$. Then $t \geq 5$. By (10),

$$F_t^2 + F_{t+1}^2 = F_{2t+1} = F_n = p. \tag{28}$$

Let $\tau = (1 + \sqrt{5})/2$. It is well-known that if $r \geq 1$, then

$$\lim_{r \to \infty} F_{r+1}/F_r = \tau = 1.618033989..., \tag{29}$$

$$F_{2r}/F_{2r-1} < F_{2r+2}/F_{2r+1} < \tau, \tag{30}$$

and

$$F_{2r+1}/F_{2r} < F_{2r+3}/F_{2r+2} > \tau. \tag{31}$$

By (30) and (31),

$$F_7/F_6 = 13/8 = 1.625 \geq F_{t+1}/F_t \geq F_6/F_5 = 8/5 = 1.6. \tag{32}$$

Thus, by (28) and (32),

$$F_n = p = F_t^2 + F_{t+1}^2 \geq F_t^2 + (1.6)^2 F_t^2 = 3.56 \ F_t^2 \tag{33}$$

and

$$F_n = p = F_t^2 + F_{t+1}^2 \leq F_t^2 + (1.625)^2 F_t^2 < 3.65 \ F_t^2. \tag{34}$$

By (30) and (31),

$$F_{t-2}/F_t = (F_{t-2}/F_{t-1})(F_{t-1}/F_t) < (F_3/F_4)(1/\tau) = (2/3)(1/\tau) < .42. \tag{35}$$

Thus, by (13), (33), and (35),

$$1 < L_{t-1}^2 = (F_{t-2} + F_t)^2 < (1.42 F_t)^2 < 2.1 F_t^2 \leq 2.1 p/3.56 < p - 1. \tag{36}$$

By (12), (33), and (34),

$$L_t^2 \leq 5 F_t^2 + 4 \leq (5p/3.56) + 4 < 1.41p + 4 < 2p - 1 \tag{37}$$

and

$$L_t^2 \geq 5 F_t^2 - 4 > (5p/3.65) - 4 > 1.36p - 4 > p + 1. \tag{38}$$

Since $L_1 = 1$ and $\{L_n\}$ is strictly increasing for $n \geq 1$, it follows from (36), (37), (38), and Theorem 7 that p does not divide any even FPSP.

Remark: It is well-known that L_n can be a prime only if n is a prime or n is a power of 2.

Proof of Theorem 10: Suppose that p divides some even FPSP N. By Lemma 1 (i) and Lemma 7, r is the only integer n such that $2 \leq n \leq k/2$ and $L_n \equiv \pm 1$ or $\pm \sqrt{-1}$ (*mod p*). It then follows from Theorem 2 (i) and Lemmas (8) and (9) that

$$N \equiv \pm r \ (mod \ k).$$

Thus, either $4 \mid N$, which contradicts Theorem 3, or $p_1 \mid N$, which contradicts the hypothesis. Hence, p cannot divide any even FPSP.

Proof of Theorem 11: (i) By Lemma 10,

$$L_{k/3} = L_{p_1 d} \equiv \pm 1 \text{ or } \pm \sqrt{-1} \ (mod \ p).$$

It now follows from Theorem 11 that p does not divide any even FPSP.

  (ii) By Lemma 10 (iii),

$$L_{k/3} \equiv 1 \ (mod \ p).$$

Since $4 \mid (k/3)$, it follows from Theorem 11 that p does not divide any even FPSP.

Remark: The tables in [1] and [2] list the values $\alpha(p) = \alpha'(p)$ for all primes p such that $p \neq 5$ and $p < 100,000$.

6. ACKNOWLEDGMENT

I wish to thank the referee for his careful reading of this paper and his helpful suggestions which improved the presentation of this paper.

REFERENCES

[1]   Brousseau, A. Tables of Fibonacci Entry Points, Part One. Santa Clara, California: Fibonacci Association, 1965.

[2]   Brousseau, A. Tables of Fibonacci Entry Points, Part Two. Santa Clara, California: Fibonacci Association, 1965.

[3]   Bruckner, G. "Fibonacci Sequence Modulo a Prime $p \equiv 3$ (*mod* 4)." *The Fibonacci Quarterly 8,* No. 2 (1970): pp. 217-220.

[4]   Carmichael, R. D. "On the Numerical Factors of the Arithmetic Forms $\alpha^n \pm \beta^n$." *Ann. Math.,* Second Series 15 (1913): pp. 30-70.

[5]   Carmichael, R. D. "On Sequences of Integers Defined by Recurrence Relations." *Quart. J. Pure Appl. Math. 48* (1920): pp. 343-372.

[6]   Di Porto, A. and Filipponi, P. "More on the Fibonacci Pseudoprimes." *The Fibonacci Quarterly 27*, No. 3 (1989): pp. 232-242.

[7]   Lehmer, D. H. "An Extended Theory of Lucas' Functions." *Ann. Math.*, Second Series 31 (1930): pp. 419-448.

[8]   Somer, L. "The Divisibility and Modular Properties of kth-Order Linear Recurrences Over the Ring of Integers of an Algebraic Number Field with Respect to Prime Ideals." Ph.D. Thesis. The University of Illinois at Urbana-Champaign, 1985.

[9]   Somer, L. Solution to Problem H-377. *The Fibonacci Quarterly 24*, No. 3 (1986): pp. 284-285.

[10]   Somer, L. "Congruence Relations for kth-Order Linear Recurrences." *The Fibonacci Quarterly 27*, No. 1 (1989): pp. 25-31.

[11]   Somer, L. "Possible Restricted Periods of Certain Lucas Sequences Modulo p. To appear.

[12]   Vinson, J. "The Relation of the Period Modulo m to the Rank of Apparition of m in the Fibonacci Sequence." *The Fibonacci Quarterly 1*, No. 1 (1963): pp. 37-45.

POSSIBLE RESTRICTED PERIODS OF CERTAIN
LUCAS SEQUENCES MODULO p

Lawrence Somer

1. INTRODUCTION

Let $(u) = u(a, b)$, called the Lucas sequence of the first kind (LSFK), be the second-order linear recurrence satisfying

$$u_{n+2} = au_{n+1} + bu_n,$$ (1)

where $u_0 = 0$, $u_1 = 1$, and the parameters a and b are integers. Let $(v) = v(a, b)$, called the Lucas sequence of the second kind (LSSK), be the recurrence satisfying (1) with the initial terms $v_0 = 2$, $v_1 = a$. Let

$$x^2 - ax - b$$ (2)

be the characteristic polynomial of $u(a, b)$ and $v(a, b)$ with characteristic roots r_1 and r_2. Let $D = a^2 + 4b = (r_1 - r_2)^2$ be the discriminant of $u(a, b)$ and $v(a, b)$. By the Binet formulas,

$$u_n = \frac{r_1^n - r_2^n}{r_1 - r_2} = \frac{r_1^n - r_2^n}{\pm\sqrt{D}}$$ (3)

if $D \neq 0$ and

$$v_n = r_1^n + r_2^n.$$ (4)

Throughout this paper, p will denote an odd prime and P will denote a prime ideal of $Q(\sqrt{D})$ dividing p.

G. E. Bergum et al. (eds.), Applications of Fibonacci Numbers Volume 4, 289–298.

The period of $u(a, b)$ modulo p will be denoted by $\mu(p)$. It is known (see [5, pages 344-345]) that if $p \nmid b$, then $u(a, b)$ is purely periodic modulo p. The *restricted period* of $u(a, b)$ modulo p, denoted by $\alpha(p)$ is the least positive integer t such that $u_{n+t} \equiv s u_n \pmod{p}$ for all non-negative integers n and some non-zero residue s. Then s is called the *multiplier* of (u) modulo p. It is easy to see that $\beta(p) = \mu(p)\alpha(p)$ is the exponent of the multiplier of (u) modulo p. Note that $\alpha(p)$ is also the least positive integer n such that $u_n \equiv 0 \pmod{p}$ and is the *rank of apparition* of p in (u).

For the Fibonacci sequence $(F_n) = u(1, 1)$, Vinson [9] found congruence conditions for p which give constraints for $\alpha(p)$ modulo 4. In particular, he proved the following theorem:

Theorem 1: Consider the Fibonacci sequence $u(1, 1)$.

(i)   If $p = 5$ or $p \equiv 13$ or $17 \pmod{20}$, then $\alpha(p) \equiv 1 \pmod{2}$.

(ii) If $p \equiv 11$ or $19 \pmod{20}$, then $\alpha(p) \equiv 2 \pmod{4}$.

(iii) If $p \equiv 3$ or $7 \pmod{20}$, then $\alpha(p) \equiv 0 \pmod{4}$.

(iv) If $p \equiv 21$ or $29 \pmod{40}$, then $\alpha(p) \equiv 1 \pmod{2}$ or $\alpha(p) \equiv 2 \pmod{4}$, and both these possibilities occur.

(v)   If $p \equiv 1$ or $9 \pmod{40}$, then all the possibilities $\alpha(p) \equiv 1 \pmod{2}$ or $\alpha(p) \equiv 0$ or $2 \pmod{4}$ occur.

In [7], Theorem 1 was generalized to the LSFK's $u(a, 1)$ and $u(a, -1)$ by obtaining congruence conditions for $\alpha(p)$ modulo 4 in terms of quadratic characters of p, which by virtue of the law of quadratic reciprocity translate into congruence conditions for p.

In this paper, we will consider the recurrences $u(a, 1)$, $v(a, 1)$, $u(a, -1)$, and $v(a, -1)$. We will extend the results cited above for the LSFK's $u(a, 1)$ and $u(a, -1)$ by obtaining constraints for $\alpha(p)$ modulo 24 in terms of quadratic characters of p. These results will be specialized to the Fibonacci sequence (F_n). Moreover, if k is the restricted period of $u(a, \pm 1)$ $(mod\ p)$, we will find $u_{k/d}$ and $v_{k/d}$ modulo p, where $d = 2, 3, 4$ or 6 and $d|k$.

2. PRELIMINARIES

Before going on, we will need the following known results.

Theorem 2: Let $u(a, b)$ be a LSFK. Suppose $p \nmid b$. Then

$$\alpha(p) \mid p - \left(\frac{D}{p}\right),\tag{5}$$

where (D/p) denotes the Legendre symbol. Further, if $p \nmid D$, then

$$\alpha(p) \mid \frac{p - \left(\frac{D}{p}\right)}{2}\tag{6}$$

if and only if $(-b/p) = 1$. Moreover, if $p \nmid b$ and $(D/p) = 1$, then

$$\mu(p) \mid p - 1 .\tag{7}$$

Proof: Proofs of (5) are given in [4, pages 44-45] and [1, pages 315-317]. Proofs of (6) are given in [6, page 441] and [1, pages 318-319]. A proof of (7) is given in [8, pages 312-313].

Theorem 3: Let $u(a, b)$ be a LSFK. Suppose $p \nmid bD$.

(i) $\mu(p)$ is the least common multiple of the exponents of r_1 and r_2 *modulo* P.

(ii) $\alpha(p)$ is the exponent of r_1/r_2 *modulo* P.

Proof: This is proved in [7, page 319].

Theorem 4: Consider the LSFK $u(a, 1)$.

(i) $\beta(p) = 1, 2,$ or 4.

(ii) $\beta(p) = 1$ iff $\alpha(p) \equiv 2 \ (mod\ 4)$.

(iii) $\beta(p) = 2$ iff $\alpha(p) \equiv 0 \ (mod\ 4)$.

(iv) $\beta(p) = 4$ iff $\alpha(p) \equiv 1 \ (mod\ 2)$.

(v)   If $(-1/p) = -1$ and $(D/p) = 1$, then $\alpha(p) \equiv 2 \ (mod\ 4)$.

(vi)   If $(-1/p) = (D/p) = -1$, then $\alpha(p) \equiv 0 \ (mod\ 4)$.

(vii)   If $(-1/p) = 1$ and $(D/p) = -1$, then $\alpha(p) \equiv 1 \ (mod\ 2)$.

(viii) If $(D/p) = 0$, then $\alpha(p) \equiv p$.

(ix)   If $a \equiv 0 \ (mod\ 2)$, then $\alpha(2) = 2$ and $\beta(2) = 1$. If $a \equiv 1 \ (mod\ 2)$, then $\alpha(2) = 3$ and $\beta(2) = 1$.

(x)   If $(-1/p) = (D/p) = 1$ and $(2/p) = -1$, then $\alpha(p) \equiv 1 \ (mod\ 2)$ or $\alpha(p) \equiv 2 \ (mod\ 4)$.

(xi)   If $(-1/p) = (D/p) = (2/p) = 1$, then $\alpha(p)$ may be congruent to $1 \ (mod\ 2)$, $0 \ (mod\ 4)$, or $2 \ (mod\ 4)$.

Proof: This is proved in [7, Theorem 13].

Theorem 5: Consider the LSFK $u(a, -1)$. Then $(-D/p) = ((2 - a)/p)((2 + a)/p)$.

(i) $\beta(p) = 1$ or 2.

(ii)   If $\alpha(p) \equiv 0 \ (mod \ 2)$, then $\beta(p) = 2$.

(iii)   If $\alpha(p) \equiv 1 \ (mod \ 2)$, then $\beta(p)$ may be 1 or 2.

(iv)   If $((2 - a)/p) = ((2 + a)/p) = -1$, then $\alpha(p) \equiv 0 \ (mod \ 2)$.

(v)   If $((2 - a)/p) = 1$ and $((2 + a)/p) = -1$, then $\alpha(p) \equiv 1 \ (mod \ 2)$ and $\beta(p) = 2$.

(vi)   If $((2 - a)/p) = -1$ and $((2 + a)/p) = 1$, then $\alpha(p) \equiv 1 \ (mod \ 2)$ and $\beta(p) = 1$.

(vii)   If $(D/p) = 0$, then $((2 - a)/p) = 1$ or $((2 + a)/p) = 1$. If $((2 - a/p) = 1$ and $((2 + a)/p) = 0$, then $\alpha(p) = p$ and $\beta(p) = 2$. If $((2 + a)/p) = 1$ and $((2 - a/p) = 0$, then $\alpha(p) = p$ and $\beta(p) = 1$.

(viii) If $a \equiv 0 \ (mod \ 2)$, then $\alpha(2) = 2$ and $\beta(2) = 1$. If $a \equiv 1 \ (mod \ 2)$, then $\alpha(2) = 3$ and $\beta(2) = 1$.

(ix)   If $((2 - a)/p) = ((2 + a)/p) = 1$, then $\alpha(p)$ may be congruent to 0 or $1 \ (mod \ 2)$ and $\beta(p)$ may be 1 or 2.

Proof: Part (vii) follows from the proof of [7, Lemma 4]. The rest of the theorem follows from [7, Theorem 16].

3. THE MAIN THEOREMS

Let $u(a, \pm 1)$ be a LSFK. Let ζ_n denote a primitive nth root of unity. We will proceed by utilizing those primitive nth roots of unity ζ_n which are simple quadratic expressions of rational numbers and such that $\zeta_n \pm 1/\zeta_n$ is the square root of an integer. These are the primitive roots

$$\zeta_1 = 1, \ \zeta_2 = -1, \ \zeta_3 = (-1 \pm \sqrt{-3})/2, \ \zeta_4 = \pm\sqrt{-1}, \ \zeta_6 = (1 \pm \sqrt{-3})/2,$$
$$\zeta_8 = (\pm \sqrt{2} \pm \sqrt{-2})/2, \ \text{and} \ \zeta_{12} = (\pm \sqrt{3} \pm \sqrt{-1})/2.$$

We will then show that if $\alpha(p) \equiv 0 \ (mod \ d)$, where d is an appropriate divisor of n, then $\zeta_n \pm 1/\zeta_n$ must be a residue *modulo* p. This will enable us to find congruence conditions for $\alpha(p)$ *modulo* 24 for $u(a, 1)$ and for $\alpha(p)$ *modulo* 12 for $u(a, -1)$ given certain quadratic characters for p.

Let $k = \alpha(p)$ for the LSFK $u(a, \pm 1)$. In theorems 6 and 7, we will determine $u_{k/d}$ and $v_{k/d}$, where $d \mid k$ and $d = 2, 3, 4$, or 6. If r is a quadratic residue *modulo* p, then $\sqrt{r} \pmod{p}$ will denote the unique residue m such that $m^2 \equiv r \pmod{p}$ and $0 \leq m \leq (p - 1)/2$.

Theorem 6: Consider the LSFK $u(a, 1)$ and the LSSK $v(a, 1)$. Let k be the restricted period *modulo* p of $u(a, 1)$. Suppose $p \nmid D$.

    (i)   If $k \equiv 2 \pmod 4$, then $v_{k/2} \equiv 0$ and $u_{k/2} \equiv \pm 2/\sqrt{D} \pmod{p}$.

    (ii)   If $k \equiv 0 \pmod 4$, then $v_{k/2} \equiv 0$ and $u_{k/2} \equiv \pm 2\sqrt{-1/D} \pmod{p}$.

    (iii)   If $k \equiv 3 \pmod 6$, then $v_{k/3} \equiv \pm\sqrt{-1}$ and $u_{k/3} \equiv \pm\sqrt{3/D} \pmod{p}$.

    (iv)   If $k \equiv 6 \pmod{12}$, then $v_{k/3} \equiv -1$, $u_{k/3} \equiv \pm\sqrt{-3/D}$, $v_{k/6} \equiv \pm\sqrt{-3}$, and $u_{k/6} \equiv \pm 1/\sqrt{D} \pmod{p}$.

    (v)   If $k \equiv 0 \pmod{12}$, then $v_{k/3} \equiv 1$, $u_{k/3} \equiv \pm\sqrt{-3/D}$, $v_{k/6} \equiv \pm\sqrt{3}$, and $u_{k/6} \equiv \pm\sqrt{-1/D} \pmod{p}$.

    (vi)   If $k \equiv 4 \pmod 8$, then $v_{k/4} \equiv \pm\sqrt{-2}$ and $u_{k/4} \equiv \pm\sqrt{2/D} \pmod{p}$.

    (vii) If $k \equiv 0 \pmod 8$, then $v_{k/4} \equiv \pm\sqrt{2}$ and $u_{k/4} \equiv \pm\sqrt{-2/D} \pmod{p}$.

Proof: We first note by Theorem 3(i) that $r_1^{k/d}$ and $r_2^{k/d}$ are both roots of unity *modulo* P of order n_1 and n_2 respectively, where

$$[n_1, n_2] = d \cdot \beta(p).$$

It follows from Theorem 3 (ii) that if $p \nmid D$, then $r_1^{k/d} \not\equiv r_2^{k/d} \pmod{P}$ for $d \geq 2$. We further note that $r_1 r_2 = -1$.

    (i)   By Theorem 4 (ii), $\beta(p) = 1$. Thus, $r_1^{k/2} \equiv \pm 1$ and $r_2^{k/2} \equiv \mp 1 \pmod{P}$. Hence, by (3) and (4), $v_{k/2} \equiv 0 \pmod{P} \equiv 0 \pmod{p}$ and $u_{k/2} \equiv \pm 2/\sqrt{D} \pmod{P} \equiv \pm 2/\sqrt{D} \pmod{p}$, since both $v_{k/2}$ and $u_{k/2}$ are rational integers.

    (ii)   By Theorem (iii), $\beta(p) = 2$. Also, $(r_1 r_2)^{k/2} = (-1)^{k/2} = 1$. Hence, $r_1^{k/2} \equiv \pm\sqrt{-1}$ and $r_2^{k/2} \equiv \mp\sqrt{-1} \pmod{P}$, and we continue as in (i).

    (iii)   By Theorem (iv), $\beta(p) = 4$. Note that $(r_1 r_2)^{k/3} = (-1)^{k/3} = -1$. Thus, $r_1^{k/3} \equiv (\pm\sqrt{3} \pm\sqrt{-1})/2$ and $r_2^{k/3} \equiv (\mp\sqrt{3} \pm\sqrt{-1})/2 \pmod{P}$.

(iv)   By Theorem 4 (ii), $\beta(p) = 1$. Also, $(r_1 r_2)^{k/3} = (-1)^{k/3} = 1$ and $(r_1 r_2)^{k/6} = (-1)^{k/6} = -1$. We now see that we can take $r_1^{k/3} \equiv (-1 \pm \sqrt{-3})/2$, $r_2^{k/3} \equiv (-1 \mp \sqrt{-3})/2$, $r_1^{k/6} \equiv (1 \pm \sqrt{-3})/2$, and $r_2^{k/6} \equiv (-1 \pm \sqrt{-3})/2 \pmod{P}$.

(v)   By Theorem 4 (iii), $\beta(p) = 2$. Moreover, $(r_1 r_2)^{k/3} = (r_1 r_2)^{k/6} = 1$. Thus, $r_1^{k/3} \equiv (1 \pm \sqrt{-3})/2$, $r_2^{k/3} \equiv (1 \mp \sqrt{-3})/2$, $r_1^{k/6} \equiv (\pm \sqrt{3} \pm \sqrt{-1})/2$, and $r_2^{k/6} \equiv (\pm \sqrt{3} \mp \sqrt{-1})/2 \pmod{P}$.

(vi)   By Theorem 4 (iii), $\beta(p) = 2$. Also $(r_1 r_2)^{k/4} = (-1)^{k/4} = -1$. Thus, $r_1^{k/4} \equiv (\pm \sqrt{2} \pm \sqrt{-2})/2$ and $r_2^{k/4} \equiv (\mp \sqrt{2} \pm \sqrt{-2})/2 \pmod{P}$.

(vii)   By Theorem 4 (iii), $\beta(p) = 2$. Further $(r_1 r_2)^{k/4} = (-1)^{k/4} = 1$. Hence, $r_1^{k/4} \equiv (\pm \sqrt{2} \pm \sqrt{-2})/2$ and $r_2^{k/4} \equiv (\pm \sqrt{2} \mp \sqrt{-2})/2 \pmod{P}$.

Theorem 7: Consider the LSFK $u(a, -1)$ and the LSSK $v(a, -1)$. Let k be the restricted period *modulo* p of $u(a, -1)$. Suppose $p \nmid D$.

(i)   If $k \equiv 0 \pmod 2$, then $v_{k/2} \equiv 0$ and $u_{k/2} \equiv \pm 2\sqrt{-1/D} \pmod p$.

(ii)   If $k \equiv 0 \pmod 6$, then $v_{k/3} \equiv 1$, $u_{k/3} \equiv \pm\sqrt{-3/D}$, $v_{k/6} \equiv \pm\sqrt{3}$, and $u_{k/6} \equiv \pm\sqrt{-1/D} \pmod p$.

(iii)   If $k \equiv 3 \pmod 6$ and $\beta(p) = 1$, then $v_{k/3} \equiv -1$ and $u_{k/3} \equiv \pm\sqrt{-3/D} \pmod p$.

(iv)   If $k \equiv 3 \pmod 6$ and $\beta(p) = 2$, then $v_{k/3} \equiv 1$ and $u_{k/3} \equiv \pm\sqrt{-3/D} \pmod p$.

(v)   If $k \equiv 0 \pmod 4$, then $v_{k/4} \equiv \pm\sqrt{2}$ and $u_{k/4} \equiv \pm\sqrt{-2/D} \pmod p$.

Proof: Note that $(r_1 r_2)^m = 1^m = 1$ for all integers m.

(i)   By Theorem 5 (ii), $\beta(p) = 2$. Thus, $r_1^{k/2} \equiv \pm\sqrt{-1}$ and $r_2^{k/2} \equiv \mp\sqrt{-1} \pmod{P}$.

(ii)   By Theorem 5 (ii), $\beta(p) = 2$. Therefore, $r_1^{k/3} \equiv (1 \pm \sqrt{-3})/2$, $r_2^{k/3} \equiv (1 \mp \sqrt{-3})/2$, $r_1^{k/6} \equiv (\pm \sqrt{3} \pm \sqrt{-1})/2$ and $r_2^{k/6} \equiv (\pm \sqrt{3} \mp \sqrt{-1})/2 \pmod{P}$.

(iii)   Since $\beta(p) = 1$, we must have that $r_1^{k/3} \equiv (-1 \pm \sqrt{-3})/2$ and $r_2^{k/3} \equiv (-1 \mp \sqrt{-3})/2 \pmod{P}$.

(iv)   By Theorem 5 (ii), $\beta(p) = 2$. Thus, $r_1^{k/4} \equiv (\pm \sqrt{2} \pm \sqrt{-2})/2$ and $r_2^{k/4} \equiv (\pm \sqrt{2} \mp \sqrt{-2})/2 \pmod{P}$.

Theorems 8 and 9 will constrain those primes which can lead to certain congruence conditions for $\alpha(p)$ for the LSFK $u(a, \pm 1)$.

Theorem 8: Consider the LSFK $u(a, 1)$. Suppose $p \nmid D$.

(i)   If $\alpha(p) \equiv 1 \ (mod \ 2)$, then $(-1/p) = 1$.

(ii)   If $\alpha(p) \equiv 3 \ (mod \ 6)$, then $(-1/p) = 1$ and $(D/p) = (3/p)$.

(iii) If $\alpha(p) \equiv 2 \ (mod \ 4)$, then $(D/p) = 1$.

(iv)   If $\alpha(p) \equiv 4 \ (mod \ 8)$, then $(-2/p) = (-D/p) = 1$.

(v)   If $\alpha(p) \equiv 0 \ (mod \ 8)$, then $(2/p) = (-D/p) = 1$.

(vi)   If $\alpha(p) \equiv 6 \ (mod \ 12)$, then $(-3/p) = (D/p) = 1$.

(vii) If $\alpha(p) \equiv 0 \ (mod \ 12)$, then $(3/p) = (-D/p) = 1$.

Proof: (i) This follows by Theorem 4 (iv) since we must have that $\beta(p) = 4$ and thus that $s \equiv \pm\sqrt{-1} \ (mod \ p)$, where s is the multiplier of $u(a, 1) \ (mod \ p)$.

Parts (ii)-(vii) follow from Theorem 6.

Theorem 9: Consider the LSFK $u(a, -1)$. Suppose $p \nmid D$.

(i)   If $\alpha(p) \equiv 1 \ (mod \ 2)$ and $\beta(p) = 1$, then $((2 + a)/p) = 1$.

(ii)   If $\alpha(p) \equiv 1 \ (mod \ 2)$ and $\beta(p) = 2$, then $((2 - a)/p) = 1$.

(iii) If $\alpha(p) \equiv 3 \ (mod \ 6)$, then $(3/p) = (-D/p)$.

(iv)   If $\alpha(p) \equiv 2 \ (mod \ 4)$, then $(-D/p) = 1$.

(v)   If $\alpha(p) \equiv 0 \ (mod \ 4)$, then $(2/p) = (-D/p) = 1$.

(vi)   If $\alpha(p) \equiv 0 \ (mod \ 6)$, then $(3/p) = (-D/p) = 1$.

Proof: Assertions (i)-(vi) follow from Theorems 5 and 7.

Theorem 10 and its corollary and Theorem 11 will be our principal results. Theorems 10 and 11 will provide constraints for $\alpha(p)$ for the LSFK $u(a, \pm 1)$ given certain quadratic characters that p must satisfy. The corollary to Theorem 10 will be specialized to the Fibonacci sequence. For Theorem 11 which treats the LSFK $u(a, -1)$, we will determine $\beta(p)$ in addition to $\alpha(p)$, since $\beta(p)$ may be 1 or 2 if $\alpha(p) \equiv 1 \ (mod \ 2)$. For the LSFK $u(a, 1)$, $\beta(p)$ is completely determined if $\alpha(p)$ is known modulo 4. Theorems 10 and 11 will be derived from Theorem 8 and 9 upon noting that the congruence conditions given for $\alpha(p)$ in these two theorems provide covering sets for the integers.

In Theorems 10 and 11, we will consider residue classes for $\alpha(p)$ *modulo* 24 and *modulo* 12 respectively. For conciseness, we will use the smallest defining moduli for these residue classes. For example, if we have that $r \equiv 4$, 12, or 20 (*modulo* 24), we will denote this by saying that $r \equiv 4$ (*mod* 8).

Theorem 10: Consider the LSFK $u(a, 1)$. The following table provides constraints for $\alpha(p)$ depending on p.

p	$\alpha(p)$
$(-1/p) = 1, (3/p) = (D/p) = -1$	1 (*mod* 2)
$(D/p) = 0$ or $(3/p) = (-1/p) = 1, (D/p) = -1$	1, 5 (*mod* 6)
$(-3/p) = (D/p) = 1, (-1/p) = -1$	2 (*mod* 4)
$(D/p) = 1, (-3/p) = (-1/p) = -1$	2, 10 (*mod* 12)
$(-2/p) = (3/p) = (-D/p) = 1, (-1/p) = -1$	4 (*mod* 8)
$(-2/p) = (-3/p) = (-D/p) = 1, (-1/p) = -1$	4, 20 (*mod* 24)
$(2/p) = (3/p) = (-D/p) = 1, (-1/p) = -1$	0 (*mod* 8)
$(2/p) = (-3/p) = (-D/p) = 1, (-1/p) = -1$	8, 16 (*mod* 24)
$(-1/p) = (D/p) = 1, (2/p) = (3/p) = -1$	1, 2, 5, 7, 10, 11 (*mod* 12)
$(-1/p) = (D/p) = (3/p) = 1, (2/p) = -1$	1, 2, 3 (*mod* 4)
$(-1/p) = (D/p) = (2/p) = 1, (3/p) = -1$	1, 2 (*mod* 3)
$(-1/p) = (D/p) = (2/p) = (3/p) = 1$	0, 1 (*mod* 2)

Table 1.

Possible Values *Modulo* 24 for the Restricted Period $\alpha(p)$ of $u(a, 1)$

Given Certain Quadratic Characters of p

Proof: This follows from Theorem 4 (viii) and Theorem 8.

The following corollary sharpens Theorem 10 for the case of the Fibonacci sequence and generalizes Theorem 1.

Corollary: The following table provides constraints for the restricted period $\alpha(p)$ for the Fibonacci sequence $(F_n) = u(1, 1)$ depending on p, where p is a prime.

p	$\alpha(p)$
17, 53 (*mod* 60)	1 (*mod* 2)
13, 37 (*mod* 60)	1, 5 (*mod* 6)
19, 31 (*mod* 60)	2 (*mod* 4)
11, 59 (*mod* 60)	2, 10 (*mod* 12)
83, 107 (*mod* 120)	4 (*mod* 8)
43, 67 (*mod* 120)	4, 20 (*mod* 24)
23, 47 (*mod* 120)	0 (*mod* 8)
7, 103 (*mod* 120)	8, 16 (*mod* 24)
29, 101 (*mod* 120)	1, 2, 5, 7, 10, 11 (*mod* 12)
61, 109 (*mod* 120)	1, 2, 3 (*mod* 4)
41, 89 (*mod* 120)	1, 2 (*mod* 3)
1, 49 (*mod* 120)	0, 1 (*mod* 2)

Table 2.
Possible Values *Modulo* 24 for the Restricted Period $\alpha(p)$ of (F_n)
Given Congruence Conditions for p *Modulo* 120

Proof: Note that $D = 5$ for the Fibonacci sequence. The Corollary now follows from Theorem 10 upon using the law of quadratic reciprocity.

Remark: The tables in [2] and [3] show that the Corollary is complete in the sense that all the given values for $\alpha(p)$ are actually realized.

Theorem 11: Consider the LSFK $u(a, -1)$. Let $g = 2 + a$ and $h = 2 - a$. The following table provides constraints for $a(p)$ and $\beta(p)$ depending on p.

p	$\alpha(p)$	$\beta(p)$
$(g/p) = 1, (3/p) = (h/p) = -1$	1 (*mod* 2)	1
$(h/p) = 1, (3/p) = (g/p) = -1$	1 (*mod* 2)	2
$(3/p) = (g/p) = 1, (h/p) = -1$	1, 5 (*mod* 6)	1
$(3/p) = (h/p) = 1, (g/p) = -1$	1, 5 (*mod* 6)	2
$(3/p) = 1, (2/p) = (g/p) = (h/p) = -1$	2 (*mod* 4)	2
$(2/p) = (3/p) = (g/p) = (h/p) = -1$	2, 10 (*mod* 12)	2
$(2/p) = (3/p) = 1, (g/p) = (h/p) = -1$	0 (*mod* 2)	2
$(2/p) = 1, (3/p) = (g/p) = (h/p) = -1$	2, 4 (*mod* 6)	2
$(g/p) = (h/p) = 1, (2/p) = (3/p) = -1$	1, 2, 5, 7, 10, 11 (*mod* 12)	1 or 2
$(2/p) = (g/p) = (h/p) = 1, (3/p) = -1$	1, 2 (*mod* 3)	1 or 2
$(3/p) = (g/p) = (h/p) = 1, (2/p) = -1$	1, 2, 3 (*mod* 4)	1 or 2
$(2/p) = (3/p) = (g/p) = (h/p) = 1$	0, 1 (*mod* 2)	1 or 2

Table 3.
Possible Values for $\alpha(p)$ *Modulo* 12 and for $\beta(p)$ for $u(a, -1)$
Given Certain Quadratic Characters of p

Proof: This follows from Theorem 9.

4. ACKNOWLEDGMENT

I wish to thank the referee for his careful reading of this paper and his helpful suggestions which improved the presentation of this paper.

REFERENCES

[1]   Backstrom, R. "On the Determination of the Zeros of the Fibonacci Sequence." *The Fibonacci Quarterly 4*, No. 4 (1966): pp. 313-322.

[2]   Brousseau, A. Tables of Fibonacci Entry Points, Part One. Santa Clara, California: Fibonacci Association, 1965.

[3]   Brousseau, A. Tables of Fibonacci Entry Points, Part Two. Santa Clara, California: Fibonacci Association, 1965.

[4]   Carmichael, R. D. "On the Numerical Factors of the Arithmetic Forms $\alpha^n \pm \beta^n$." *Ann. Math.*, Second Series 15 (1913): pp. 30-70.

[5]   Carmichael, R. D. "On Sequences of Integers Defined by Recurrence Relations." *Quart. J. Pure Appl. Math. 48* (1920): pp. 343-372.

[6]   Lehmer, D. H. "An Extended Theory of Lucas' Functions." *Ann. Math.*, Second Series 31 (1930): pp. 419-448.

[7]   Somer, L. "Divisibility Properties of Primary Second-Order Recurrences with Respect to Primes." *The Fibonacci Quarterly 18*, No. 4 (1980): pp. 316-334.

[8]   Somer, L. "Possible Periods of Primary Fibonacci-Like Sequences with Respect to a Fixed Odd Prime." *The Fibonacci Quarterly 20*, No. 4 (1982): pp. 311-333.

[9]   Vinson, J. "The Relation of the Period *Modulo m* to the Rank of Apparition of *m* in the Fibonacci Sequence." *The Fibonacci Quarterly 1*, No. 2 (1963): pp. 37-45.

USING MATRIX TECHNIQUES TO ESTABLISH PROPERTIES OF A GENERALIZED TRIBONACCI SEQUENCE

Marcellus E. Waddill

1. INTRODUCTION

We consider the generalized Tribonacci Sequence, $\{V_n\}$, defined as follows:

$$V_n = rV_{n-1} + sV_{n-2} + tV_{n-3} \quad (n \geq 3), \tag{1}$$

where V_0, V_1, V_2 are arbitrary integers and r, s, t, are non-zero integers.

This sequence has been studied by Jarden [4], Shannon and Horadam [6], Yalavigi [7], Pethe [5] and Gerdes [3]. If we set $r = s = t = 1$ and $V_0 = 0$, $V_1 = V_2 = 1$, then $\{V_n\}$ is the well-known Tribonacci Sequence which has been considered extensively. See, for example, [1] and [2] for the earliest work.

By means of the S-matrix defined below, we are able to find and prove properties of $\{V_n\}$ and identities involving terms of $\{V_n\}$ which are heretofore undiscovered.

2. BASIC PROPERTIES AND IDENTITIES

For our purposes, the most useful technique for generating $\{V_n\}$ is by means of what we call the S-matrix which has been defined and used in [6] and is a generalization of the R-matrix defined in [7]. The S-matrix is defined as

$$S = \begin{bmatrix} r & s & t \\ 1 & 0 & 0 \\ 0 & 1 & 0 \end{bmatrix} \tag{2}$$

G. E. Bergum et al. (eds.), Applications of Fibonacci Numbers Volume 4, 299–308.
© 1991 Kluwer Academic Publishers.

We list for reference two fundamental matrix equations which are found in [6]:

$$\begin{bmatrix} V_n \\ V_{n-1} \\ V_{n-2} \end{bmatrix} = \begin{bmatrix} r & s & t \\ 1 & 0 & 0 \\ 0 & 1 & 0 \end{bmatrix}^{n-2} \begin{bmatrix} V_2 \\ V_1 \\ V_0 \end{bmatrix} \tag{3}$$

and

$$\begin{bmatrix} r & s & t \\ 1 & 0 & 0 \\ 0 & 1 & 0 \end{bmatrix}^n = \begin{bmatrix} U_{n+1} & sU_n + tU_{n-1} & tU_n \\ U_n & sU_{n-1} + tU_{n-2} & tU_{n-1} \\ U_{n-1} & sU_{n-2} + tU_{n-3} & tU_{n-2} \end{bmatrix} = \begin{bmatrix} U_{n+1} & W_n & tU_n \\ U_n & W_{n-1} & tU_{n-1} \\ U_{n-1} & W_{n-2} & tU_{n-2} \end{bmatrix} \tag{4}$$

where we define

$$W_i = sU_i + tU_{i-1}. \tag{5}$$

The sequence $\{U_n\}$ is the special case of $\{V_n\}$ where $V_0 = 0$, $V_1 = 1$, $V_2 = r$, and (4) indicates how $\{U_n\}$ arises or is generated "naturally" as we consider successive powers of S.

Jarden [4] examined another special case of $\{V_n\}$ which we call $\{J_n\}$ where

$$V_0 = J_0 = 3, \ V_1 = J_1 = r, \ V_2 = J_2 = r^2 + 2s.$$

If we let $r = s = t = 1$, the sequence $\{J_n\}$ becomes the third order sequence, $\{J'_n\}$, which bears the same relation to the Tribonacci Sequence as does the Lucas Sequence to the Fibonacci Sequence. That is, if r_1, r_2, r_3 are roots of

$$x^3 - x^2 - x - 1 = 0,$$

then

$$J'_n = r_1^n + r_2^n + r_3^n ,$$

and it also follows that if s_1, s_2, s_3 are roots of

$$x^3 - rx^2 - sx - t = 0,$$

then

$$J_n = s_1^n + s_2^n + s_3^n .$$

The wealth of literature on the Lucas Sequence suggests the importance of the sequences $\{J_n\}$ and $\{J'_n\}$ for the theory of third order recurring sequences.

By (1), the terms of $\{V_n\}$ are defined only for $n \geq 0$. That restriction can be removed with the result that V_n is defined for all n. We omit the details.

As an application of (3), we have

$$
\begin{bmatrix} V_{n+p} \\ V_{n+p-1} \\ V_{n+p-2} \end{bmatrix} = \begin{bmatrix} r & s & t \\ 1 & 0 & 0 \\ 0 & 1 & 0 \end{bmatrix}^{n-1} \begin{bmatrix} r & s & t \\ 1 & 0 & 0 \\ 0 & 1 & 0 \end{bmatrix}^{p-1} \begin{bmatrix} V_2 \\ V_1 \\ V_0 \end{bmatrix} = \begin{bmatrix} U_n & W_{n-1} & tU_{n-1} \\ U_{n-1} & W_{n-2} & tU_{n-2} \\ U_{n-2} & W_{n-3} & tU_{n-3} \end{bmatrix} \begin{bmatrix} V_{p+1} \\ V_p \\ V_{p-1} \end{bmatrix} \tag{6}
$$

from which we conclude that

$$
V_{n+p} = U_n V_{p+1} + W_{n-1} V_p + tU_{n-1} V_{p-1}, \tag{7}
$$

$$
V_{n+p} = U_p V_{n+1} + W_{p-1} V_n + tU_{p-1} V_{n-1}. \tag{8}
$$

Equation (7) in slightly different form was derived by Jarden [4, p. 86]. More generally we can show as above that for any k,

$$
V_{n+p} = U_{n-k} V_{p+k+1} + W_{n-k-1} V_{p+k} + tU_{n-k-1} V_{p+k-1}, \tag{9}
$$

$$
V_{n+p} = U_{p-k} V_{n+k+1} + W_{p-k-1} V_{n+k} + tU_{p-k-1} V_{n+k-1}. \tag{10}
$$

Using (10) with k = 0, we have what might be called a generalized version of (6),

$$
\begin{bmatrix} V_{n+p+q} \\ V_{n+p} \\ V_{n+1} \end{bmatrix} = \begin{bmatrix} U_{p+q} & W_{p+q-1} & tU_{p+q-1} \\ U_p & W_{p-1} & tU_{p-1} \\ 1 & 0 & 0 \end{bmatrix} \begin{bmatrix} V_{n+1} \\ V_n \\ V_{n-1} \end{bmatrix}. \tag{11}
$$

The transpose of the S-matrix demonstrates an interesting and useful relation among the terms of the sequence $\{V_n\}$. We have the matrix equation

$$
\begin{bmatrix} V_n \\ Y_{n-1} \\ tV_{n-1} \end{bmatrix} = \begin{bmatrix} r & 1 & 0 \\ s & 0 & 1 \\ t & 0 & 0 \end{bmatrix}^{n-2} \begin{bmatrix} V_2 \\ Y_1 \\ tV_1 \end{bmatrix} \tag{12}
$$

where $Y_i = sV_i + tV_{i-1}$. Hence we may view S^T as generating the sequence

$$
V_0, \, tV_0, \, Y_0, \, V_1, \, tV_1, \, Y_1, \, V_2, \, tV_2, \, Y_2, \, ... \tag{13}
$$

which, in effect, is a meshing of three sequences, $\{V_i\}$, $\{tV_i\}$ and $\{Y_i\}$.

3. QUADRATIC, CUBIC AND HIGHER ORDER IDENTITIES

The first quadratic identity we establish is

$$V_n^2 + sV_{n-1}^2 + 2tV_{n-1}V_{n-2} = V_2V_{2n-2} + Y_1V_{2n-3} + tV_1V_{2n-4}. \tag{14}$$

Proof of 14: After recognizing the left side of (14) as the dot product of two appropriate vectors, we use (3) and (12) to obtain

$$V_n^2 + sV_{n-1}^2 + 2tV_{n-1}V_{n-2}$$

$$= [V_n, Y_{n-1}, tV_{n-1}] \cdot [V_n, V_{n-1}, V_{n-2}]$$

$$= [V_2, Y_1, tV_1] \begin{bmatrix} r & s & t \\ 1 & 0 & 0 \\ 0 & 1 & 0 \end{bmatrix}^{2n-4} \begin{bmatrix} V_2 \\ V_1 \\ V_0 \end{bmatrix}$$

$$= [V_2, Y_1, tV_1] \cdot [V_{2n-2}, V_{2n-3}, V_{2n-4}]$$

$$= V_2V_{2n-2} + Y_1V_{2n-3} + tV_1V_{2n-4} .$$

For the sequence $\{U_n\}$, (14) becomes

$$U_n^2 + sU_{n-1}^2 + 2tU_{n-1}U_{n-2} = U_{2n-1}, \tag{15}$$

which also was derived by Jarden [4, p. 86].

By using the characteristic equation of the S-matrix and the Cayley-Hamilton Theorem, we can establish other more general quadratic identities based on (14). For example, we prove the following:

$$V_{n+1}^2 - rV_{n+1}V_n + (t-rs)V_nV_{n-1} + sV_n^2 - rt(V_nV_{n-2} + V_{n-1}^2) - stV_{n-1}V_{n-2}$$

$$- t^2(V_{n-1}V_{n-3} + V_{n-2}^2) = s[V_2V_{2n-2} + (sV_1+tV_0)V_{2n-3} + tV_1V_{2n-4}] . \tag{16}$$

Proof of (16): The characteristic equation of the S-matrix is

$$x^3 - rx^2 - sx - t = 0 .$$

Hence by the Cayley-Hamilton Theorem, we have

$$S^3 - rS^2 - sS - tI = 0$$

or

$$sS = S^3 - rS^2 - tI. \tag{17}$$

We adopt the notation

$$M^T = [V_2, Y_1, tV_1], \quad N^T = [V_2, V_1, V_0]$$

and consider the expression

$$M^T [S^{2n-5}(S^3 - rS^2 - tI)]N$$

$$= M^T S^{2n-2} N - r M^T S^{2n-3} N - t M^T S^{2n-5} N$$

$$= (M^T S^{n-1})(S^{n-1} N) - r(M^T S^{n-2})(S^{n-1} N) - t(M^T S^{n-3})(S^{n-2} N)$$

$$= [V_{n+1}, Y_n, tV_n] \cdot [V_{n+1}, V_n, V_{n-1}]$$

$$- r[V_n, Y_{n-1}, tV_{n-1}] \cdot [V_{n+1}, V_n, V_{n-1}]$$

$$- t[V_{n-1}, Y_{n-2}, tV_{n-2}] \cdot [V_n, V_{n-1}, V_{n-2}],$$

which is the left side of (16). Also by (17), we have

$$M^T[S^{2n-5}(S^3 - rS^2 - tI)]N = s M^T S^{2n-4} N = s M^T \cdot [V_{2n-2}, V_{2n-3}, V_{2n-4}].$$

which is the right side of (16), and the proof is complete.

For $\{U_n\}$, the identity (16) becomes

$$U_{n+1}^2 - rU_{n+1}U_n + (t - rs)U_n U_{n-1} + sU_n^2 - rt(U_n U_{n-2} + U_{n-1}^2) - stU_{n-1}U_{n-2}$$

$$- t^2(U_{n-1}U_{n-3} + U_{n-2}^2) = sU_{2n-1} \qquad (18)$$

As might be expected from observing the right side of (18), the left side may be simplified. By a proper grouping of terms and by using (1) repeatedly in appropriate ways, we can reduce (18) to

$$U_{n+1}U_{n-1} + tU_{n-1}U_{n-2} + U_n^2 - rU_n U_{n-1} = U_{2n-1}$$

which may be reduced further to (15).

The technique above may be used to establish any number of additional quadratic identities.

We may use matrix techniques also to write cubic, quartic and higher order identities which are analogous to (15). To obtain a cubic, for example, we use

$$S^{3n} = (S^n)^3.$$

By (4), the element in the third row, first column of S^{3n} is U_{3n-1}. The element in the third row, first column of $(S^n)^3$ is

$$U_{n-1}(R_1 \cdot C_1) + W_{n-2}(R_2 \cdot C_1) + tU_{n-2}(R_3 \cdot C_1) ,$$

where R_i is the ith row of S^n and C_1 is the first column of S^n. Hence we have the identity

$$U_{3n-1} = U_{n-1}(U_{n+1}^2 + W_n U_n + tU_{n-1}U_n) + W_{n-2}(U_n U_{n+1} + W_{n-1}U_n + tU_{n-1}^2)$$

$$+ tU_{n-2}(U_{n-1}U_{n+1} + W_{n-2}U_n + tU_{n-2}U_{n-1}). \qquad (19)$$

In order to generalize (19) for $\{V_n\}$, we replace U_i and W_i on the right side by V_i and Y_i respectively and then use (7) repeatedly with an appropriate value for p followed by repeated use of (19) to obtain the result. The process is somewhat tedious!

We next consider generalizations of two equivalent cubic identities which appear in [6]:

$$\begin{vmatrix} U_{n+1} & W_n & tU_n \\ U_n & W_{n-1} & tU_{n-1} \\ U_{n-1} & W_{n-2} & tU_{n-2} \end{vmatrix} = t^n \qquad (20)$$

and

$$\begin{vmatrix} U_{n+1} & U_{n-1} & U_n \\ U_n & U_{n-2} & U_{n-1} \\ U_{n-1} & U_{n-3} & U_{n-2} \end{vmatrix} = t^{n-2} , \qquad (21)$$

each of which is clearly independent of r and s and dependent on n only as the power of t.

Discovery and proof of these generalizations require a lemma found in [7] which, for completeness, we state here.

Lemma: Let A be any 3x3 matrix and let $\mathbf{x}$ and $\mathbf{y}$ be vectors in R^3; then the following equation holds:

$$(A\mathbf{x}) \times (A\mathbf{y}) = \text{cof} (A) (\mathbf{x} \times \mathbf{y}),$$

where cof (A) is the cofactor matrix of A and $\times$ designates cross product in the vector space R^3.

The identity we prove, written as a determinant equation, is

$$\begin{vmatrix} V_{n+m+h} & V_{n+p+h} & V_{n+h} \\ V_{n+m+q} & V_{n+p+q} & V_{n+q} \\ V_{n+m} & V_{n+p} & V_n \end{vmatrix} = t^n \begin{vmatrix} U_{h-1} & U_h \\ U_{q-1} & U_q \end{vmatrix} \begin{vmatrix} V_{m+2} & V_{m+1} & V_m \\ V_{p+2} & V_{p+1} & V_p \\ V_2 & V_1 & V_0 \end{vmatrix} \qquad (22)$$

Clearly the right side of (22), except for the power of t, is independent of n.

Proof of (22): We adopt the notation

$$B = \begin{bmatrix} U_{h+1} & W_h & tU_h \\ U_{q+1} & W_q & tU_q \\ 1 & 0 & 0 \end{bmatrix}$$

and recognize the left side of (22) as the triple scalar product

$$\begin{bmatrix} V_{n+m+h} \\ V_{n+m+q} \\ V_{n+m} \end{bmatrix} \times \begin{bmatrix} V_{n+p+h} \\ V_{n+p+q} \\ V_{n+p} \end{bmatrix} \cdot \begin{bmatrix} V_{n+h} \\ V_{n+q} \\ V_n \end{bmatrix} = BS^{n-2} \begin{bmatrix} V_{m+2} \\ V_{m+1} \\ V_m \end{bmatrix} \times BS^{n-2} \begin{bmatrix} V_{p+2} \\ V_{p+1} \\ V_p \end{bmatrix} \cdot BS^{n-2} \begin{bmatrix} V_2 \\ V_1 \\ V_0 \end{bmatrix}$$

$$= [\mathrm{cof}\,(BS^{n-2})] \begin{bmatrix} V_{m+2} \\ V_{m+1} \\ V_m \end{bmatrix} \times \begin{bmatrix} V_{p+2} \\ V_{p+1} \\ V_p \end{bmatrix} \cdot BS^{n-2} \begin{bmatrix} V_2 \\ V_1 \\ V_0 \end{bmatrix}$$

$$= \mid BS^{n-2} \mid [(BS^{n-2})^{-1}]^T \begin{bmatrix} V_{m+2} \\ V_{m+1} \\ V_m \end{bmatrix} \times \begin{bmatrix} V_{p+2} \\ V_{p+1} \\ V_p \end{bmatrix} \cdot BS^{n-2} \begin{bmatrix} V_2 \\ V_1 \\ V_0 \end{bmatrix}$$

$$= t^{n-2} \mid B \mid \left[\begin{bmatrix} V_{m+2} \\ V_{m+1} \\ V_m \end{bmatrix} \times \begin{bmatrix} V_{p+2} \\ V_{p+1} \\ V_p \end{bmatrix} \right]^T (BS^{n-2})^{-1}(BS^{n-2}) \begin{bmatrix} V_2 \\ V_1 \\ V_0 \end{bmatrix}$$

$$= t^n \begin{vmatrix} U_{h-1} & U_h \\ U_{q-1} & U_q \end{vmatrix} \begin{vmatrix} V_{m+2} & V_{m+1} & V_m \\ V_{p+2} & V_{p+1} & V_p \\ V_2 & V_1 & V_0 \end{vmatrix} ,$$

where we use (11) and the Lemma at the appropriate places; and this completes the proof.

If we let $p = q = 1$, $m = h = 2$ in (22), we have

$$\begin{vmatrix} V_{n+4} & V_{n+3} & V_{n+2} \\ V_{n+3} & V_{n+2} & V_{n+1} \\ V_{n+2} & V_{n+1} & V_n \end{vmatrix} = t^n \begin{vmatrix} V_4 & V_3 & V_2 \\ V_3 & V_2 & V_1 \\ V_2 & V_1 & V_0 \end{vmatrix} \qquad (23)$$

which appears in [6] also and is a generalized version of (21).

For another special case of (22), if we let $p = q = n$ and $h = m = 2n$, we have

$$\begin{vmatrix} V_{5n} & V_{4n} & V_{3n} \\ V_{4n} & V_{3n} & V_{2n} \\ V_{3n} & V_{2n} & V_n \end{vmatrix} = t^n \begin{vmatrix} U_{2n-1} & U_{2n} \\ U_{n-1} & U_n \end{vmatrix} \begin{vmatrix} V_{2n+2} & V_{2n+1} & V_{2n} \\ V_{n+2} & V_{n+1} & V_n \\ V_2 & V_1 & V_0 \end{vmatrix} . \qquad (24)$$

Finally, if we let $p = q = 2n$, $h = m = 4n$, we have

$$\begin{vmatrix} V_{9n} & V_{7n} & V_{5n} \\ V_{7n} & V_{5n} & V_{3n} \\ V_{5n} & V_{3n} & V_n \end{vmatrix} = t^n \begin{vmatrix} U_{4n-1} & U_{4n} \\ U_{2n-1} & U_{2n} \end{vmatrix} \begin{vmatrix} V_{4n+2} & V_{4n+1} & V_{4n} \\ V_{2n+2} & V_{2n+1} & V_{2n} \\ V_2 & V_1 & V_0 \end{vmatrix} . \qquad (25)$$

4. IDENTITIES INVOLVING LINEAR SUMS

We next derive some identities involving linear sums of terms of $\{U_n\}$ and $\{V_n\}$ where matrix techniques along with the Cayley-Hamilton Theorem are essential for their discovery and useful for their proof. The procedure is first to establish the identities for $\{U_n\}$ and then generalize for $\{V_n\}$.

The following have been proved:

$$(rs+t) \sum_{i=1}^{n} r^{n-i} U_i = rU_{n+2} + tU_n - r^{n+2} \qquad (26)$$

$$(rs+t) \sum_{i=1}^{n} s^{n-i} U_{2i-1} = sU_{2n} + tU_{2n-1} \qquad (27)$$

$$(rs+t) \sum_{i=1}^{n-1} s^{n-1-i} U_{2i} = sU_{2n-1} + tU_{2n-2} - s^n \qquad (28)$$

$$(rs+t) \sum_{i=1}^{n} r^{n-i} V_i = rV_{n+2} + tV_n - r^n(rV_2 + tV_0) \qquad (29)$$

$$(rs+t) \sum_{i=1}^{n} s^{n-i} V_{2i-1} = sV_{2n} + tV_{2n-1} + s^n(rV_1 - V_2) \tag{30}$$

$$(rs+t) \sum_{i=1}^{n-1} s^{n-1-i} V_{2i} = sV_{2n-1} + tV_{2n-2} + s^n(rV_0 - V_1) \tag{31}$$

Proof of (26): We use (17) to conclude that

$$S^3 - rS^2 - sS + rsI - tI - rsI = 0$$

or

$$(S - rI)(S^2 - sI) = (rs+t)I.$$

Hence

$$(S - rI)^{-1} = (rs+t)^{-1}(S^2 - sI) . \tag{32}$$

Since

$$S^n - r^n I = (S - rI)(r^{n-1}I + r^{n-2}S + \cdots + rS^{n-2} + S^{n-1}),$$

by using (32), we conclude that

$$\sum_{i=1}^{n} r^{n-i}S^{i-1} = (rs+t)^{-1}(S^2 - sI)(S^n - r^n I) = (rs+t)^{-1}[S^{n+2} - sS^n - r^n(S^2 - sI)]. \tag{33}$$

We obtain (26) by equating corresponding terms in the first row, first column on both sides of this matrix equation.

The validity of (32) requires that $t \neq -rs$. However, it can be shown directly that if $t = -rs$, then (26) holds with both sides being 0.

The proof of (27) and (28) is similar but requires the equation

$$S^{2n} - s^n I = (S^2 - sI)(s^{n-1}I + s^{n-2}S^2 + \cdots + S^{2n-2}) .$$

Proof of (29): By (33) we have

$$(rs+t) \sum_{i=1}^{n} r^{n-i}V_i = (rs+t) [0, 1, 0] \left(\sum_{i=1}^{n} r^{n-i}S^{i-1} \right) \begin{bmatrix} V_2 \\ V_1 \\ V_0 \end{bmatrix}$$

$$= [0, 1, 0] (rs+t) \frac{\left(S^{n+2} - sS^n - r^n(S^2 - sI)\right)}{(rs+t)} \begin{bmatrix} V_2 \\ V_1 \\ V_0 \end{bmatrix}$$

$$= [0, 1, 0] \left\{ \begin{bmatrix} V_{n+4} \\ V_{n+3} \\ V_{n+2} \end{bmatrix} - s \begin{bmatrix} V_{n+2} \\ V_{n+1} \\ V_n \end{bmatrix} - r^n \begin{bmatrix} V_4 \\ V_3 \\ V_2 \end{bmatrix} + r^n s \begin{bmatrix} V_2 \\ V_1 \\ V_0 \end{bmatrix} \right\}$$

$$= V_{n+3} - sV_{n+1} - r^n V_3 + r^n s V_1$$

$$= rV_{n+2} + tV_n - r^n (rV_2 + tV_0).$$

The proofs of (30) and (31) are similar.

ACKNOWLEDGEMENT

The author gratefully acknowledges the excellent comments and suggestions offered by the referee, particularly his suggesting the alternate proof of (29) which appears above.

REFERENCES

[1] Agronomoff, M. "Sur Une Suite Recurrente." *Mathesis 34* (1914): pp. 125-126.

[2] Feinberg, Mark. "Fibonacci-Tribonacci." *The Fibonacci Quarterly 1*, 3 (1963): pp. 71-74.

[3] Gerdes, Walter. "Generalized Tribonacci Numbers and Their Convergent Sequences." *The Fibonacci Quarterly 16*, 3 (1978): pp. 269-275.

[4] Jarden, Dov. "Third Order Recurring Sequences." *Recurring Sequences.* Jerusalem: Riveon Lematematika, 1966.

[5] Pethe, S. "Some Identities for Tribonacci Sequences." *The Fibonacci Quarterly 26*, 2 (1988): pp. 144-151.

[6] Shannon, A. G. and Horadam, A. F. "Some Properties of Third-Order Recurrence Relations." *The Fibonacci Quarterly 10*, 2 (1972): pp. 135-146.

[7] Waddill, Marcellus E. and Sacks, Louis. "Another Generalized Fibonacci Sequence." *The Fibonacci Quarterly 5*, 3 (1967): pp. 209-222.

[8] Yalavigi, C. C. "Properties of Tribonacci Numbers." *The Fibonacci Quarterly 10*, 3 (1972): pp. 231-246.

A Fast Algorithm of the Chinese Remainder Theorem 242
Admissible Set 141
Algebraic Equations 189, 192, 200
Amplitude 194, 201
Average Number of Comparisons 75
b-Adic Tree 127, 128, 131
Balanced Tree 70
Basic Property of SDS 10
Binary Tree 248
Binet's Formula 209, 217
Binomial Coefficient, Generalized 111
Biomathematics 167
Braiding-New Theory of 248
Cagley-Hamilton Theorem 302, 306
Carry Sequence 110
Characteristic Equation 70, 72
Characteristic Polynomial 38, 178
Characteristic Roots 38
Chinese Remainder Theorem 241
Classification-Tree of Primitive Triples 254, 255
Co-Related Sequence 122
Companion Matrix 38
Complementary Argument 151
Complete Graph 133
Completeness of the Fibonacci Sequence 168
Complex-Subscripted Fibonacci Numbers 51
Computer Graphics-Rapid Rendering 2
Congruence 34
Congruent Line-Sequence 203, 208
Conjugate Line-Sequence 211, 212
Conjugate Number 209
Constant Term 193, 201
Continued Fraction 172
Continued Fractions Generalized 263
Contraction Rule for a PSCF 252
Convergents Generalized 264, 265
Cyclic Group 207
d-Ocagne, Theorem of 192
Damped Sinusoids 200, 201, 202
Decision Tree 70, 253
DeMoivre-Type Identities 215
Derangements 83
Descending Diagonal Functions 162
Difference Equations 89
Difference Triangles 88

Digital Filters   194
Dilation   204
Diophantine Equation   65, 232, 240
Divisibility Pattern   119
Dominant Root   70, 72
Doubly Magic Cube   20-22
Duples   253
Electronic Signals   198
Electronics   190, 194
Encodings of Integers   70
Euclid's Algorithm   248
Euclid's Algorithm Generalized   263
Euclidean Algorithm   173, 241
Euclidean Space   133
Euler's Totient Function Generalized   267, 269
Exponent Matrix   114
Exponent of the Multiplier   290
Exponential Functions   189
Factorizations   271
Ferrari's Method   215
Fibonacci   121
Fibonacci Distance   136, 137
Fibonacci Edge   137
Fibonacci Group   64
Fibonacci Identities   102
Fibonacci Length   136, 137
Fibonacci Line-Sequence   203
Fibonacci Mosaic Graph   127
Fibonacci Numbers   1, 45-46, 133
Fibonacci Numbers of Degree r   70
Fibonacci Numbers of Degree r and Order k   69
Fibonacci Polynomials   99
Fibonacci Pseudoprime   277
Fibonacci Ray-Sequence   203
Fibonacci Representation   133, 135-138
Fibonacci Search   69, 75
Fibonacci Search of Degree r   69, 70
Fibonacci Sequence   65, 78, 177, 231
Fibonacci Tree   70, 127, 128, 131, 132
Fibonacci Tree of Degree r   70
Fibonacci Tree of Degree r and Order k   70
Fibonacci Triangle   133, 136
Flat Parity Chart   206, 210
Frequency   194, 201
Functions of Matrices   52
Functionsl Equation   271
General Rational Number Tree (n-RT)   261
Generalized Bernoulli Polynomials   145
Generalized Binomial Coefficient   9
Generalized Euler Polynomials   145
Generalized Fibonacci Identities   93
Generalized Fibonacci Numbers   69, 91, 99

Generalized Fibonacci Polynomials | 48
Generalized Genocchi Polynomials | 146
Generalized Lucas Numbers | 91, 99
Generalized Modified Pascal Triangle | 13
Generalized Multinomial Coefficient | 9
Generalized Pascal Pyramid | 11
Generalized Pascal Triangle | 10
Generating Functions | 87
Generating Number | 204
Generator | 204
Golden Mean | 2
Golden Rectangle | 171
Golden Triangle | 45, 171
Great Common Divisor | 239
Greatest Common Divisor (GCD) | 219
Greatest Common Divisor Generalized | 267
Group Presentation | 63
Higman-Sims Group | 64
Hyperbolic Identities | 94
Hyperbolic Sine and Cosine | 189, 191
Inner Product | 205
Integer Solution | 240
Integral Representation of a Graph | 133
Integral Triangles | 41-50
Jacobi Symbol | 237
Kummer Carry Theorem | 109
Lame's Theorem | 241
Latin Cube | 17-20
Least Common Multiple | 239
Length of Line-Sequence | 205
Line, Curve Drawing | 4
Line-Sequence | 203
Line-Sequential Vector | 205
Linear Congruential Pseudo Random Numbers | 6
Linear Recurrence | 177
Linear Recurrence, Second Order | 37
Lucas | 121
Lucas Line-Sequence | 205, 210
Lucas Number | 232
Lucas Polynomials | 99
Lucas Sequence | 37, 63, 231
Lucas Sequence of the First Kind | 289
Lucas Sequence of the Second Kind | 289
Magic Cube | 17
Matrix Exponential | 53
Matrix Fibonacci Identities | 56
Matrix-Subscripted Fibonacci Numbers | 53
Maximal Admissible Set | 141
Mobius Function | 275
Modulo | 233, 234, 237, 238
Moebius Braids | 258, 260
Monoid | 139

Monos   253, 254
Mosaic Graph   127, 128, 131, 132
Multinomial Triangles   77, 79, 80, 84, 85
Multiples   253
Multiplicative Structure   271
Multiplier   278, 290
n-RT   261
Nearly Isosceles Triangles   45-46, 49-50
Octahedron Graph   136
Odd Parity   205
One-Relator Product   63
Order-n Arithmetics   258, 268
Order-1 Arithmetic   258
Order-2 Arithmetic   258
Orthogonal Cube   18
Orthogonal Line-Sequence   205
p-Adic Expansion   110
p-Adic Integer   110
p-Integral Rational Number   110
Parity Chart   206
Parity Relation   209
Partitions of Numbers   81, 82
Pascal's Triangle   77, 79-81, 84-87, 197, 219
Pell Line-Sequence   210, 212
Pell Numbers   45-46
Perfect Cube   17, 19
Perfect Square   232, 234, 237, 239
Period   233, 235
Period-Preserving Map   37, 38
Periodic   37
Permutations   83
Phase Angle   201
Plant Biology   167
Power Series   33
Primeness in Integer Classes   268
Primitive Triples   247, 248
Principal Subtree   71, 73, 74
Proper Divisors   274
Pseudo-Random Numbers   1
Pythagorean Simple Continued Fraction (PSCF)   249, 250, 253
Pythagorean Triples   45-46, 247
Quadratic Diophantine Equations   248
Quadratic Nonresidue   39, 235, 236
Quadratic Residue   37, 39
Quasi-Symmetric Simple Continued Fraction   250
r-Fibonacci Numbers   69, 73
r-Fibonacci Numbers of Order k   69
r-Fibonacci Search   69, 72, 73, 75
r-Fibonacci Search Code   75
r-Fibonacci Tree   70, 71, 72
r-Fibonacci Tree of Order k   70
Rank of Apparition   290

Real-Subscripted Fibonacci Numbers 51
Realization 128, 130, 131, 132
Recreational Amusement 167-170
Recurrence 236
Recurrence Relation 177, 189, 203, 208
Regular Knot Tree (RKT) 258, 259
Regular Moebius Knot Tree (RMKT) 258, 259
Residue Class 235, 236
Residue Sequence 233, 234
Restricted Period 278, 290
Rising Diagonal Functions 163
S-Matrix 299, 301, 302
Sampling Period 201
Search Tree 70
Shannon's Theorem 75
Similar Line-Sequence 204
Simple Continued Fractions 247
Sinusoids 189, 191
Skeleton Tree 71, 74
Star of David Theorem 219
Stirling Number of the First Kind 119
Strong Divisibility Sequence (SDS) 9
Submonoid 139
Summand Pair 140
Superposition 191
Symmetry 205
System of Linear Equations 200
Tetrabonacci Number 215
Tools for Teaching 167, 169
Transfer Function 194
Translation 207, 208
Translation Chart 207
Translation Property 206, 207
Trapezoid 135
Tree 248, 253-255
Tree Structure 208
Triangle Inequality 133, 134
Triangular Number 231, 240
Tribonacci Sequence 299, 300
Trick with the Fibonacci Sequence 167-170
Triple Scalar Product 305
Turan Number 137
Uniformity of Pseudo Random Numbers 4
Vandermonde Determinant 199
Vieta, Theorem of 198
Waring Formula 217
Z-Transform 89
z-Transformation 194, 197, 198
Zeckendorf Expansion 73
Zeckendorf's Theorem 167-170

Fibonacci Numbers and Their Applications

Volume 1

A. N. Philippou, G. E. Bergum and A. F. Horadam (eds.)

ISBN 90-277-2234-X

TABLE OF CONTENTS

EDITOR'S PREFACE — *vii*
REPORT — *ix*
CONTRIBUTORS — *xi*
FOREWORD — *xv*
THE ORGANIZING COMMITTEES — *xvii*
LIST OF CONTRIBUTORS TO THE CONFERENCE — *xix*
INTRODUCTION — *xxiii*

FIBONACCENE
Peter G. Anderson *1*

ON A CLASS OF NUMBERS RELATED TO BOTH THE FIBONACCI AND PELL NUMBERS
Nguyen-Huu Bong *9*

A PROPERTY OF UNIT DIGITS OF FIBONACCI NUMBERS
Herta T. Freitag *39*

SOME PROPERTIES OF THE DISTRIBUTIONS OF ORDER k
Katuomi Hirano *43*

CONVOLUTIONS FOR PELL POLYNOMIALS
A. F. Horadam & Br. J. M. Mahon *55*

CYCLOTOMY-GENERATED POLYNOMIALS OF FIBONACCI TYPE
A. F. Horadam & A. G. Shannon *81*

ON GENERALIZED FIBONACCI PROCESS
Daniela Jarušková *99*

FIBONACCI NUMBERS OF GRAPHS III: PLANTED PLANE TREES
Peter Kirschenhofer, Helmut Prodinger,
& Robert F. Tichy *105*

A DISTRIBUTION PROPERTY OF SECOND-ORDER LINEAR RECURRENCES
Péter Kiss *121*

ON LUCAS PSEUDOPRIMES WHICH ARE PRODUCTS OF s PRIMES
  Péter Kiss, Bui Minh Phong, & Erik Lieuwens 131

FIBONACCI AND LUCAS NUMBERS AND THE MORGAN-VOYCE POLY-
NOMIALS IN LADDER NETWORKS AND IN ELECTRIC LINE THEORY
  Joseph Lahr 141

INFINITE SERIES SUMMATION INVOLVING RECIPROCALS OF
PELL POLYNOMIALS
  Br. J. M. Mahon & A. F. Horadam 163

FIBONACCI AND LUCAS NUMBERS AND AITKEN ACCELERATION
  J. H. McCabe & G. M. Phillips 181

ON SEQUENCES HAVING THIRD-ORDER RECURRENCE RELATIONS
  S. Pethe 185

ON THE SOLUTION OF THE EQUATION $G_n = P(x)$
  Attila Pethö 193

DISTRIBUTIONS AND FIBONACCI POLYNOMIALS OF ORDER k,
LONGEST RUNS, AND RELIABILITY OF CONSECUTIVE-k-OUT-
OF-n : F SYSTEMS
  Andreas N. Philippou 203

FIBONACCI-TYPE POLYNOMIALS AND PASCAL TRIANGLES
OF ORDER k
  G. N. Philippou & C. Georghiou 229

A NOTE ON FIBONACCI AND RELATED NUMBERS IN THE THEORY
OF 2×2 MATRICES
  Gerhard Rosenberger 235

PROBLEMS ON FIBONACCI NUMBERS AND THEIR GENERALIZATIONS
  A. Rotkiewicz 241

LINEAR RECURRENCES HAVING ALMOST ALL PRIMES AS
MAXIMAL DIVISORS
  Lawrence Somer 257

ON THE ASYMPTOTIC DISTRIBUTION OF LINEAR RECURRENCE
SEQUENCES
  Robert F. Tichy 273

GOLDEN HOPS AROUND A CIRCLE
  Tony van Ravenstein, Keith Tognetti,
  & Graham Winley 293

Applications of Fibonacci Numbers

Volume 2

A. N. Philippou, A. F. Horadam and G. E. Bergum (eds.)

ISBN 90-277-2673-6

TABLE OF CONTENTS

THE SECOND INTERNATIONAL...   vii
CONTRIBUTORS   ix
FOREWORD   xiii
THE ORGANIZING COMMITTEES   xv
LIST OF CONTRIBUTORS TO THE CONFERENCE   xvii
INTRODUCTION   xix

FERMAT-LIKE BINOMIAL EQUATIONS
    Heiko Harborth   1

RECURRENCES RELATED TO THE BESSEL FUNCTION
    F. T. Howard   7

SYMMETRIC RECURSIVE SEQUENCES MOD M
    Kenji Nagasaka & Shiro Ando   17

PRIMITIVE DIVISORS OF LUCAS NUMBERS
    Peter Kiss   29

A CONGRUENCE RELATION FOR A LINEAR RECURSIVE SEQUENCE OF
ARBITRARY ORDER
    H. T. Freitag & G. M. Phillips   39

FIBONACCI NUMBERS AND GROUPS
    Colin M. Campbell, Edmund F. Robertson
    & Richard M. Thomas   45

A TRIANGULAR ARRAY WITH HEXAGON PROPERTY, DUAL TO PASCAL'S
TRIANGLE
    Shiro Ando   61

FUNCTIONS OF THE KRONECKER SQUARE OF THE MATRIX Q
    Odoardo Brugia & Piero Filipponi   69

FIBONACCI NUMBERS OF THE FORMS $PX^2 \pm 1$, $PX^3 \pm 1$, WHERE P
IS PRIME
    Neville Robbins   77

ON THE K-TH ORDER LINEAR RECURRENCE AND SOME PROBABILITY
APPLICATIONS
    George N. Philippou   89

ON THE REPRESENTATION OF INTEGRAL SEQUENCES {F_n/d} and
{L_n/D} AS SUMS OF FIBONACCI NUMBERS AND AS SUMS OF LUCAS
NUMBERS
  Herta T. Freitag & Piero Filipponi   97

PRIMES HAVING AN INCOMPLETE SYSTEM OF RESIDUES FOR A
CLASS OF SECOND-ORDER RECURRENCES
  Lawrence Somer   113

COVERING THE INTEGERS WITH LINEAR RECURRENCES
  John R. Burke & Gerald E. Bergum   143

RECURSIVE THEOREMS FOR SUCCESS RUNS AND RELIABILITY OF
CONSECUTIVE-K-OUT-OF-N: F SYSTEMS
  Andreas N. Philippou   149

ASVELD'S POLYNOMIALS $P_j(N)$
  A. F. Horadam & A. G. Shannon   163

MORE ON THE PROBLEM OF DIOPHANTUS
  Joseph Arkin & Gerald Bergum   177

ON A PROBLEM OF DIOPHANTUS
  Calvin Long & Gerald Bergum   183

THE GENERALIZED FIBONACCI NUMBERS {C_n}, $C_n = C_{n-1} + C_{n-2} + K$
  Marjorie Bicknell-Johnson & Gerald E. Bergum   193

FIRST FAILURES
  Dmitri Thoro   207

SUBJECT INDEX 211

Applications of Fibonacci Numbers

Volume 3

G. E. Bergum, A. N. Philippou and A. F. Horadam (eds.)

ISBN 0-7923-0523-X

TABLE OF CONTENTS

A REPORT ON THE THIRD INTERNATIONAL CONFERENCE... vii
LIST OF CONTRIBUTORS TO THIS PROCEEDINGS ix
FOREWORD xv
THE ORGANIZING COMMITTEES xvii
LIST OF CONTRIBUTORS TO THE CONFERENCE xix
INTRODUCTION xxiii

THE ROLE OF THE FIBONACCI SEQUENCE IN THE ISOLATION OF THE REAL ROOTS OF POLYNOMIAL EQUATIONS
    A. G. Akritas & P. G. Bradford 1
A GCD PROPERTY ON PASCAL'S PYRAMID AND THE CORRESPONDING LCM PROPERTY OF THE MODIFIED PASCAL PYRAMID
    Shiro Ando & Daihachiro Sato 7
TRANSLATABLE AND ROTATABLE CONFIGURATIONS WHICH GIVE EQUAL PRODUCT, EQUAL GCD AND EQUAL LCM PROPERTIES SIMULTANEOUSLY
    Shiro Ando & Daihachiro Sato 15
FIBONACCI LENGTH OF GENERATING PAIRS IN GROUPS
    C. M. Campbell, H. Doostie & E. F. Robertson 27
A GENERALIZATION OF FIBONACCI TREES
    Renato M. Capocelli 37
GENERALIZED FIBONACCI NUMBERS ARE ROUNDED POWERS
    Renato M. Capocelli & Paul Cull 57
ON GENERALIZED FIBONACCI NUMBERS OF GRAPHS
    Michael Drmota 63
AN INVESTIGATION OF SEQUENCES DERIVED FROM HOGGATT SUMS AND HOGGATT TRIANGLES
    Daniel C. Fielder & Cecil O. Alford 77
REPRESENTATION OF NATURAL NUMBERS AS SUMS OF FIBONACCI NUMBERS: AN APPLICATION TO MODERN CRYPTOGRAPHY
    Piero Filipponi & Emilio Montolivo 89
A NOTE ON RAMIFICATIONS CONCERNING THE CONSTRUCTION OF PYTHAGOREAN TRIPLES FROM RECURSIVE SEQUENCES
    Herta T. Freitag 101
ON THE REPRESENTATION OF $\{F_{kn}/F_n\}$, $\{F_{kn}/L_n\}$, $\{L_{kn}/L_n\}$, AND $\{L_{kn}/F_n\}$ AS ZECKENDORF SUMS
    Herta T. Freitag 107
FUNCTIONAL RECURRENCES
    Krystyna Grytczuk & Aleksander Grytczuk 115
CONCENTRIC CYCLES IN MOSAIC GRAPHS
    Heiko Harborth 123
FIBONACCI TRIANGLES
    Heiko Harborth & Arnfried Kemnitz 129
MOSAIC NUMBERS OF FIBONACCI TREES
    Heiko Harborth & Sabine Lohmann 133
FALLING FACTORIAL POLYNOMIALS OF GENERALIZED FIBONACCI TYPE
    A. F. Horadam 139

SOME NOTES ON FIBONACCI BINARY SEQUENCES
    Yasuichi Horibe 155
CONGRUENCES FOR WEIGHTED AND DEGENERATE STIRLING NUMBERS
    F. T. Howard 161
AN INVERSE THEOREM ON FIBONACCI NUMBERS
    Naotaka Imada 171
SOME RESULTS ON DIVISIBILITY SEQUENCES
    Norbert Jensen 181
ON MENTAL CALCULATION OF REPEATING DECIMALS, FINDING FIBONACCI
NUMBERS AND A CONNECTION TO PASCAL'S TRIANGLE
    Marjorie Bicknell-Johnson 191
DIOPHANTINE REPRESENTATION OF FIBONACCI NUMBERS OVER NATURAL
NUMBERS
    James P. Jones 197
ON PRIME DIVISORS OF THE TERMS OF SECOND ORDER LINEAR RECURRENCE
SEQUENCES
    Peter Kiss 203
AN ALTERNATING PRODUCT REPRESENTATION FOR REAL NUMBERS
    Arnold Knopfmacher & John Knopfmacher 209
MAXIMUM LENGTH OF THE EUCLIDEAN ALGORITHM AND CONTINUED FRACTIONS
IN F(X)
    Arnold Knopfmacher & John Knopfmacher 217
RECURRENCE RELATIONS IN SINUSOIDS AND THEIR APPLICATIONS TO SPECTRAL
ANALYSIS AND TO THE RESOLUTION OF ALGEBRAIC EQUATIONS
    Joseph Lahr 223
A RECURRENCE RELATION FOR GAUSSIAN MULTINOMIAL COEFFICIENTS
    S. L. Lee & G. M. Phillips 239
SOME BINOMIAL FIBONACCI IDENTITIES
    Calvin T. Long 241
A SURVEY OF PROPERTIES OF THIRD ORDER PELL DIAGONAL FUNCTIONS
    Br. J. M. Mahon & A. F. Horadam 255
MULTIVARIATE FIBONACCI POLYNOMIALS OF ORDER K AND THE
MULTIPARAMETER NEGATIVE BINOMIAL DISTRIBUTION OF THE SAME ORDER
    Andreas N. Philippou & Demetris L. Antzoulakos 273
LONGEST CIRCULAR RUNS WITH AN APPLICATION IN RELIABILITY VIA THE
FIBONACCI-TYPE POLYNOMIALS OF ORDER K
    Andreas N. Philippou & Frosso S. Makri 281
FIBONACCI NUMBERS AND AN ALGORITHM OF LEMOINE AND KÁTAI
    Jukka Pihko 287
GENERALIZATIONS OF SEQUENCES OF LUCAS AND BELL
    A. G. Shannon & A. F. Horadam 299
DISTRIBUTION OF RESIDUES OF CERTAIN SECOND-ORDER LINEAR RECURRENCES
MODULO P
    Lawrence Somer 311
THE FIBONACCI TREE, HOFSTADTER AND THE GOLDEN STRING
    Keith Tognetti, Graham Winley & Tony van Ravenstein 325
THREE NUMBER TREES - THEIR GROWTH RULES AND RELATED NUMBER
PROPERTIES
    J. C. Turner 335
SUBJECT INDEX 351